Disasters, Collective Behavior, and Social Organization

Disasters, Collective Behavior, and Social Organization

Edited by Russell R. Dynes and Kathleen J. Tierney

DELAWARE

Newark: University of Delaware Press
London and Toronto: Associated University Presses

Associated University Presses
440 Forsgate Drive
Cranbury, NJ 08512

Associated University Presses
25 Sicilian Avenue
London WC1A 2QH, England

Associated University Presses
P.O. Box 338, Port Credit
Mississauga, Ontario
Canada L5G 4L8

The paper used in this publication meets the requirements
of the American National Standard for Permanence of Paper
for Printed Library Materials Z39.48-1984.

Library of Congress Cataloging-in-Publication Data
Disasters, collective behavior, and social organization / edited by
Russell R. Dynes and Kathleen J. Tierney.
p. cm.
Includes bibliographical references (p.).
ISBN 0-87413-498-6 (alk. paper)
1. Disaster relief—Social aspects. 2. Disasters—Research.
3. Emergency management—Research. 4. Disaster victims—Psychology.
5. Collective behavior. I. Dynes, Russell Rowe, 1923–
II. Tierney, Kathleen J.
HV553.D572 1994
363.3′4—dc20 93-46766
CIP

THIRD PRINTING 2000

PRINTED IN THE UNITED STATES OF AMERICA

Contents

Acknowledgments 7

Foreword: The Emergence and Importance of Social Organization: The Contributions of E. L. Quarantelli 9
RUSSELL R. DYNES, KATHLEEN J. TIERNEY, and CHARLES E. FRITZ

Part One. Disaster Research

Introduction 23

Disaster in Aisle 13 Revisited 26
THOMAS E. DRABEK

Disaster Archives and Structural Analysis: Uses and Limitations 45
GARY A. KREPS

Public Risk Communication 71
COLLEEN FITZPATRICK and DENNIS S. MILETI

A Model of Evacuation Compliance Behavior 85
RONALD W. PERRY

EMS in Halifax after the 6 December 1917 Explosion: Testing Quarantelli's Theories with Historical Data 99
JOSEPH SCANLON

Postdisaster Sheltering and Housing: Social Processes in Response and Recovery 115
ROBERT BOLIN

Stress Coping and Emergency Disaster Volunteers: A Discussion of Some Relevant Factors 128
NEIL R. BRITTON, CARMEN C. MORAN, and BENJAMIN CORREY

An Ecological Approach to Disasters 145
FREDERICK L. BATES and CARLO PELANDA

Part Two. Collective Behavior and Social Movements

Introduction 163

MicroStructure and Panic: The Impact of Social Bonds on Individual Action in Collective Flight from the Beverly Hills Supper Club Fire 168
NORRIS R. JOHNSON, WILLIAM E. FEINBERG, and DRUE M. JOHNSTON

The Crowd Crush at Hillsborough: The Collective Behavior of an Entertainment Crush 190
JERRY M. LEWIS and MICHAEL L. KELSEY

Collective Behavior Theory and the Study of Mass Hysteria 207
ROBERT A. STALLINGS

The Convergence of Volunteers in a Consensus Crisis: The Case of the 1985 Mexico City Earthquake 229
DENNIS E. WENGER and THOMAS F. JAMES

Rumor as Intensified Information Seeking: Earthquake Rumors in China and the United States 244
RALPH H. TURNER

Collective Behavior and Social Movement Theory 257
BENIGNO E. AGUIRRE

Are Social Movements Irrational or Are They Collective Behavior? 273
LEWIS M. KILLIAN

An Elite-Sustained Movement: Women's Rights in the Post–World War II Decades 281
VERTA TAYLOR

Fragmentation and Cohesion in American Society 306
GARY T. MARX

Bibliography 331

List of Contributors 365

Acknowledgments

In a collective work, the various authors need to be thanked. For much of the processing of the manuscript, our thanks go to various members of the Disaster Research Center staff, including Marge Simmons, Cindi Olson, Maura Jentz, and Diane Murray. Our special appreciation goes to Elizabeth Goetz for her work on the consolidated bibliography. Most of all, our appreciation goes to our colleague, E. L. Quarantelli, whose efforts have informed all of the authors.

Foreword: The Emergence and Importance of Social Organization: The Contributions of E. L. Quarantelli

RUSSELL R. DYNES, KATHLEEN J. TIERNEY, and CHARLES E. FRITZ

The essays in this book pay tribute to the intellectual and organizational contributions of E. L. Quarantelli in the fields of disaster research, collective behavior, and social organization. The seventeen essays cover a wide variety of subjects encompassed by these fields, but all—in one way or another—acknowledge that Quarantelli has made significant theoretical and empirical contributions in all three fields. This introduction aims to sketch some mileposts in Quarantelli's professional career to provide a historical context for the contributions that follow.

As a cofounder and long-time director of the Disaster Research Center, Quarantelli has focused his career on understanding how behavior in unstructured situations becomes organized. In situations like disasters, in which behavior appears unpatterned, chaotic, and disorganized, Quarantelli has looked more closely. The reputation he acquired during that work led one reporter to characterize him as the "arch-Druid of cataclysm." Not entirely appreciated by the media commentator was the point that in "cataclysm," he was looking for continuity, not morbidity; for order, not disorder.

Inseparable from his work as a scholar has been his guidance of an enduring research organization that has been a catalyst for creating a research tradition in the sociology of disasters.

It is probably best to introduce this book with a discussion of the reference points that have guided Quarantelli's activities. Based on his over 150 publications and on his activities at the Disaster Research Center (DRC), we will try to identify propositions that we believe best reflect his orientation to sociological theory and his conclusions about the research enterprise.

1. It is important to study behavior, not attitudes, because attitudes are not good predictors of behavior.

2. It is important to explain behavior by sociological theories, rather than psychological or social-psychological principles.
3. No single sociological theory is sufficient to address the complexity of social life, but combining collective behavior theory with organizational theory may come closest to this objective.
4. The dialectical quality of social and cultural change implies that we must pay close attention to collective behavior and social movements. Conversely, collective behavior and social movements always have their roots in social organization.
5. No single research method is methodologically superior to others, but field research is best able to capture the complexity of social life.
6. In the search for explanations, it is important to focus not only on the consequences of events but also on the conditions that give rise to and characterize those events.
7. The development of theoretical understanding requires comparative research and longitudinal studies ("longitudinal" implies continuity.)
8. Because science is a collective process, it requires that the scientist be self-critical; share research ideas with others; create an international community of scholars; educate the next generation of scholars; and create resources for collective use.
9. Good theoretical research has important implications for social policy.
10. Conversely, good practical research has important implications for sociological theory.

Taking seriously the first proposition, emphasizing the importance of studying behavior rather than attitudes, we have been observing Quarantelli's career at various times since 1951, and we can illustrate many of the aforementioned propositions.

Quarantelli's identification with the field of disaster research goes back virtually to its origins in the United States. Like many of his generation, his college experience was postponed by service in World War II. After combat experience in Europe, he completed his undergraduate degree at the University of Chicago and then entered its graduate program. Then, as now, the University of Chicago was a dominant force in graduate education in sociology. Although known for many specialties, the university had a tradition of interest in what came to be called *collective behavior*. Robert Park is generally given credit for initiating the field in the United States, but much of the theoretical coherence was provided by Park's student, Herbert Blumer. Blumer's influence as a teacher was very significant for the post–World War II generations of sociologists. In fact, the continuity of the collective behavior tradition in the United States has depended on those who were educated during that period. That cohort is represented in this collection, not only by

Quarantelli but also by Ralph Turner and Lewis Killian, whose book on collective behavior, now in its third edition (1987), has remained the dominant text since 1957.

But Quarantelli's graduate school work also included more than the influence of professors. In 1951, Quarantelli was hired by Charles E. Fritz as a part-time interviewer on a disaster research project conducted by the National Opinion Research Center (NORC). Coming out of the previous decade, in which the instability of war was the defining characteristic, there was interest in looking at disaster not only as a possible peacetime prototype of social instability but also as a means for preparing for possible future wars. Two sociologists on that multi-disciplinary team, Fritz and Quarantelli, continued in their subsequent careers to maintain an interest in disasters. Fritz was the associate director and in charge of field operations on the project and Quarantelli worked with him as an interviewer supervisor and research assistant until the completion of the project in 1954.

A major part of the NORC project involved a study of the social impact of a tornado that occurred in White County, Arkansas. In that study, 342 interviews were conducted on a random sample of the preimpact population of five communities. The result of that study never received the attention it deserved: Quarantelli (1988) described it as a "fountainhead" study.

Two of Quarantelli's continuing interests—one methodological and the other theoretical—began to emerge during that project. In the White County study, the interviews were tape-recorded and then transcribed. Because the use of tape-recorded interviews was quite new at that time in the social sciences, two articles by Fritz, Quarantelli, and another research assistant, Rue Bucher (Bucher, Fritz, and Quarantelli 1956a, 1956b) discussed the research and administrative issues involved in the use of such techniques. In addition to these methodological concerns, the NORC study furthered Quarantelli's consideration of the concept of "panic" as the topic for his master's thesis. That discussion (1954) involved certain observations derived from the NORC studies as well as a search of the literature for other panic-producing situations. That treatment of panic was important for several reasons. The traditional psychological view of panic centered attention on individual behavior; panic was viewed as a psychological state produced by conditions of acute, paralyzing anxiety, that results in behavioral paralysis. In contrast, Quarantelli's treatment emphasized the social context in which panic behavior appears. He argued that the explanation for panic behavior has to be sought in the norms and relationships that structure the particular situation, and not in the individual's mental state or in some other individual-level variable. Johnson, Feinberg, and Johnston pick up those same themes in this volume in their analysis of club patrons' behavior at the Beverly Hills fire.

Equally important, Quarantelli suggests that panic behavior is extremely rare and is almost nonexistent in most disasters. Although this conclusion continues to be validated in research, many emergency managers still justify the withholding of threatening information as necessary to keep people from "panicking", others find "panic" in situations where it doesn't exist, and some journalists continue to describe rational collective flight as evidence of some "herd instinct" or regressive personality tendency. Those original themes, namely, an attention to methodological concerns and a theoretical orientation that focuses on sociological rather than psychological explanations, have persisted throughout Quarantelli's career.

The NORC experience ended in 1954. Fritz went on to join the National Academy of Science's Committee on Disaster Studies in Washington, D.C., and Quarantelli, after completing short teaching stints at Indiana University, South Bend, and Harpur College (now SUNY-Binghamton), joined the faculty at Ohio State University. There, his initial research efforts turned to understanding the professionalization process involved in the education of dentists. But in conversations with his colleagues at Ohio State, his interests turned back to the possibility of continuing the tradition of field research in disasters, modeled after the NORC experience and conforming to the recommendations of the National Academy of Science's Committee on Disaster Studies and Disaster Research Group. Quarantelli prepared a proposal to create a standby research team to study disasters. Two colleagues, Russell R. Dynes and J. Eugene Haas, also had a proposal to examine the impact of differences in organizational structure. In conversations, Quarantelli mentioned the need to understand the transformations that occur in community organizations in disaster, especially because the earlier research had focused primarily on the impact of disasters on individual victims. As a consequence, the proposals were merged and sent to a local foundation but, as an afterthought, the proposal was also sent to a federal agency. Several months later, interest developed, not on the part of the original recipient of the proposal, but rather in the Office of Civil Defense (OCD), then a part of the Department of Defense. Later, the three researchers discovered that the proposal had been brought to the attention of OCD by Charles Fritz, then a staff member of the Institute for Defense Analyses—an organization that provides analytic studies for the secretary of defense and also for the joint chiefs of staff. That proposal was modified after a conference with the OCD sponsors, and funding was subsequently assured for five years. The basic structure of what was to become the Disaster Research Center (DRC) was developed on the plane ride back to Columbus, Ohio, from Washington, D.C.

The initial focus of the research activity centered on the emergency period, with attention to the way emergency organizations respond to disaster. In addition, the focus was on urban communities as strategic research locations. The issue of what constitutes a "disaster" was left somewhat flexible.

Given those definitions, a field team could be dispatched quickly, so that observations could be made during the emergency period and officials of the organizations that were involved could be interviewed in-depth after the emergency period. Those interviews would be taped and later transcribed. Although the methods were similar to those of the NORC project, the conceptualization of the organizational impact of disaster evolved as the fieldwork was conducted. The aim of the fieldwork was to study many different disasters. By contrast, much of the earlier and even more recent research has focused on single cases. Because of the continuity of the DRC's activity, the staff was also able to study communities over a long period of time. Drabek's contribution to this volume looks retrospectively, after twenty-five years, at the impact of the Indianapolis Coliseum explosion, the first major DRC study. Kreps's study illustrates the importance of continuity and a multiple-case approach by examining the same set of variables in many different types of disasters and community settings.

As the fieldwork was begun, the DRC staff asked what theoretical perspective best fit the observations that were being obtained. One conclusion reached very early was that organizational theories rooted in classic notions of bureaucratic functioning were clearly inadequate to explain the behavior of organizations in disasters. This required the development of a set of propositions that were more dynamic and more consistent with the disaster research field observations.

One important outcome was the development of what has come to be called the *DRC typology*, which attempted to explain how much of the pre-impact structure of organizations is maintained in emergency situations and how new structures emerge in such situations. That typology was derived collectively, but it is clear, in retrospect, that the ideas relate to Quarantelli's earlier work. What are the conditions under which certain types of behavior persist? What are the conditions under which new behaviors emerge? How are these new situations socially patterned? The typology was used as a useful conceptual guide for the field teams, and when it was published in the sociological literature, it was subtitled, "A Required Convergence of Organizational and Collective Behavior Perspectives" (Dynes and Quarantelli 1968).

The Alaskan earthquake in 1964 offered a unique opportunity for research and demanded a considerable investment of staff time and effort. That research eventually became the basic social science data included in the National Academy of Sciences report on the earthquake. (See, for example, Anderson 1970.) During that same year, an earthquake in Niigata, Japan, provided the opportunity for a cross-cultural study.

Because the initial research effort focused on the response of emergency organizations, the emergence of widespread civil disturbances in American cities provided the opportunity to observe similarities and differences between consensus disasters—"natural" disasters—and "conflict" disasters—civil disturbances. The results of those studies revealed quite different outcomes in crises that others tend to consider equivalent (*American Behavioral Scientist* 1970, 1973).

The studies comparing consensus disasters and civil disturbances also offered the opportunity to extend the understanding of emergent behavior. The media depict both disasters and civil disturbances as instances of social disorder in which social control breaks down. In those situations, the story goes, the worst characteristics of our brutish "human nature" emerge. But Quarantelli and Dynes were able to distinguish significant differences in the patterning of looting in the two situations: (1) In civil disturbances, looting may be widespread, but in natural disasters—despite media accounts—it is quite rare. (2) Looting in civil disturbances is collective, not solitary in nature. It is not the act of "criminals" but involves many "law-abiding" citizens. (3) Looting in civil disturbances is public, not private. It is given considerable public support, even though that support may be short-lived. Rather than adopting the view that antisocial behavior emerges in the absence of social control, Quarantelli and Dynes explained the differences in terms of different emergent norms in the two situations. In disaster, norms emerge to protect property, but at the same time norms emerge supporting the notion that goods become community property if used for the "common good." In such situations, actual looting is rare, but taking property for emergency use is considered legitimate. On the other hand, in civil disturbances, looting is a prime expression of the nature of the conflict itself, and norms that define the use of property are redefined so that what had been private property can now be appropriated. Such appropriation is done openly and is supported by public opinion. Even in civil disturbances, looting is socially patterned. Some stores are looted but not others. And the fact that the looters may still obey traffic signals suggests the admixture of traditional and emergent norms. The primary point is that, even in what others see as chaotic and antisocial, there is the persistence of socially structured behavior.

Quarantelli's insistence on understanding the social roots of behavior have also caused him to be quite critical of the frequent assertions of the traumatic and long-lasting personality consequences of disasters. Frequently clinicians assume that disaster must create significant personality changes in some population category—victims, children, older people, and so forth—and that such categories will need extensive psychological help to overcome such trauma. Quarantelli has often argued that such assertions stem from the theoretical assumptions of the clinicians and not from empirical evidence. Few clinicians have extensive research experience in disasters. A major research deficiency is the lack of any baseline or preimpact data on the mental health of the "victims." For a recent version of this critique, see Quarantelli

(1987). But, more important, although no one will deny that disasters are personally unsettling, the clinical view underestimates the ability of those victims within the community to mobilize their own resources. Family units do not disintegrate nor do traditional community organizations abandon their traditional helping roles. (See Quarantelli 1960; Fritz 1961; Dynes and Quarantelli 1976.) In fact, longitudinal studies show that what others call trauma, victims see as positive learning experiences. Although disaster brings suffering, that "suffering" is more collective and public than predisaster suffering, which is isolating and private.

Such a view is not popular with many clinicians because it questions the use of their anticipated therapeutic role in disasters. There is considerable irony, however, that through the experience gained from understanding what others see as traumatic, Quarantelli continues to emphasize the protective nature of social structure and the continued appreciation of the ability of social structure to cushion stress, not just to create it.

There are, of course, other consistent themes in the publications of Quarantelli, centering on the myths of disaster behavior (Quarantelli and Dynes 1972), the social nature of disaster (1977), the adaptation of community structure in the emergency period (1971), the dimensions of emergency planning (Dynes and Quarantelli 1972), and the differences among social situations that are conventionally considered as disasters. The continuous fieldwork of the DRC staff, totaling over five hundred different studies, has shown the advantages of the continuity of research as an antidote to conclusions drawn from unique case studies and dramatic journalistic accounts. This rich empirical base has allowed Quarantelli to publish what others have called *grounded theory* about the continuation of the protective role of the family (1960), issues of organizational role expansion and simplification (Dynes and Quarantelli 1986), emergent leadership in riots (Anderson, Dynes, and Quarantelli 1974), and problems of first responders (Gray and Quarantelli 1984), all of which are seen from a sociological perspective. It has allowed him to discuss warning systems, emergency medical systems (1983), the handling of the dead in mass casualty situations (Blanshan and Quarantelli 1981), the role of the media in disaster (1980, 1988), issues of evacuation and relocation (1984), and the attribution of blame in disasters (Drabek and Quarantelli 1969). It has allowed him to examine a wide variety of collective behavior phenomena, such as student disturbances (Dynes and Quarantelli 1981), civil disturbances (Quarantelli, Ponting, and Fitzpatrick 1974), cult behavior (Wenger and Quarantelli 1973), crowd formation and behavior (Quarantelli and Hundly 1969), the spread of such fads as streaking (Aguirre, Quarantelli, and Mendosa 1988), chemical emergencies (Stallings and Quarantelli 1988), hazardous waste controversies (1988), and emergent citizen groups (1985b).

In recent years, with his administrative responsibilities reduced, Quarantelli has devoted increasing time and effort to encouraging scholars in other countries to study disasters, and to working with international agencies that wish to incorporate disaster mitigation elements into their development programs. He is also giving attention to chronicling the intellectual history of disaster research and, in particular, developing oral histories from other pioneering disaster researchers and policymakers. All of these efforts indicate his commitment to strengthening the research tradition for the future.

One of Quarantelli's prime accomplishments has been in the organization and the continuity of the DRC. Although others have been involved at various times, the major explanation of the DRC's success is Quarantelli's central role. Several early decisions shaped the DRC: first, to keep the DRC integrated with an academic department in a university location; second, to use graduate students as the basis for field-team operations. By contrast, it could have been structured in an independent multidisciplinary DRC staffed with "professionals."

Additionally, the DRC faced the choice of ending after the initial funding was over or becoming a more enduring institution. Certain actions taken early tended to institutionalize the DRC. First, the DRC managed to acquire, with the help of Charles E. Fritz, the library resources of the earlier research operations of the Committee on Disaster Studies and the Disaster Research Group at the National Academy of Sciences. That provided the nucleus for a central location for fugitive research materials. Such materials were made available for the use of other researchers. Second, the research efforts of the DRC staff, in particular with the Alaskan earthquake, provided considerable national and international visibility, especially with the use of that research by the National Academy of Sciences. Third, the DRC was able to recruit and interest some capable graduate students, many of whom continued their research interests after they left the DRC. The work of several of these individuals is included here. While the DRC was at Ohio State University, thirty graduate students supported by DRC received Ph.D.'s, and many other graduates and undergraduates were also trained there. Many graduates have coauthored articles with Quarantelli, largely as a result of his encouragement and collaboration.

In continuing the DRC's research program, Quarantelli made concerted efforts to extend and expand the research community. In addition to his efforts to encourage and help other scholars around the world, he organized bilateral conferences first with Japanese social scientists in 1972 (Quarantelli ed. 1972) and then with Italian social scientists in 1986 (Quarantelli and Pelanda 1989). He provided the major impetus for the creation of International Sociological Association's Research Committee on Disasters, and was its first president. From 1975–79 he coedited *Mass Emergencies.* He initiated the *International Journal of Mass Emergencies and Disasters,* which he coedited for its first five years. Quarantelli also served as chair of the section on

Collective Behavior and Social Movements of the American Sociological Association in 1986–87.

Another constant theme in Quarantelli's work was his insistence that policy research should be used to improve sociological theory. Thus, for example, he would applaud the efforts of Bates and Pelanda, in this volume, to place disaster in the context of an overall ecological theory, even though he might not agree with their overall thesis. He would also appreciate Gary Marx's chapter on the fragmentation and cohesion in American society, because of the dialectical qualities of these concepts.

One other strong commitment that Quarantelli has passed to his students is the need to be self-critical. Articles written jointly with Weller (1973) and with Aguirre (1983) give particular attention to intellectual debates concerning the nature of collective behavior. The chapters by Aguirre and Killian continue in this tradition.

In addition to his contributions to collective behavior and social movement theory, Quarantelli's impact on public policy has also been significant, although more indirect. His work and the work of the DRC are well known nationally and internationally. Quarantelli's work has been translated into French, Spanish, Italian, German, Arabic, Farsi, Armenian, Russian, Japanese, and Chinese. As a result, the importance of social science research for disaster policy is recognized not only in the United States but also in many other countries. Through his contact with colleagues and policymakers, both directly and indirectly, through agencies like the United Nations and the World Bank, his research has also influenced disaster policy in Canada, France, Great Britain, Sweden, Italy, Mexico, Australia, Japan, and perhaps even the Peoples' Republic of China and the former USSR. Because he is a member of the National Academy's Board of Natural Disaster, which has overall responsibility for the U.S. effort during the International Decade of Natural Disaster Reduction, we expect this influence to continue.

Disasters, Collective Behavior, and Social Organization

Part One
Disaster Research

Introduction

The term *disaster* carries with it considerable ambiguity. Most attention is given to its novelty, its uniqueness, its dramatic quality. Many are impressed with the physical aspects of disaster. A recent book about earthquakes and volcanos describe them as "agents of chaos." The interest of the social sciences is, of course, on the social dimensions of disaster. That research tradition has made less of "chaos" than of social stability and continuity and of adaptability and response. Although disasters evoke extensive public discussion, the systematic study of disasters is recent. Much of the past understanding was based on isolated and episodic accounts, usually by untrained observers and resulted in a popular literature that emphasized both tragedy and heroism. Knowledge about disaster behavior no longer has to depend on such accounts because a literature has developed that is more systematic and comparative. The first two essays reflect the cumulative data sources that are now available to enhance our understanding. *Disaster in Aisle 13* was the first case study published by the Disaster Research Center in 1968. The author, Thomas Drabek, revisits that study to examine how changes in theory, methodology, and application have evolved since the study was published. Drabek, who has continued his research on various aspects of disaster since his initial study, indicates how knowledge has gradually been incorporated into the increased professionalism now found among emergency managers.

The development of more systematic data bases has allowed other ways to analyze the materials that had been gathered over time by the staff of the DRC. Kreps describes his efforts to use those archives to develop theoretical ideas about the origins and restructuring of organization as well as the place social networks have in the organizing process. Through the development of structural code, he was able to use the archives, selecting fifteen different disaster events, to develop certain theoretical notions about the evolution of social structure. Much of the research derived from these efforts is reported in his book *Social Structure and Disaster* (1989).

In addition to the contributions that disaster research can provide for the development of theory, there are important implications for public policy. In recent years, ways of communicating information about risk have developed into a sub-field, often designated as risk communication. One part of risk communication centers around providing public information intended to create awareness of the consequences of certain hazards. Much of the motivation for such effort has been in the context of the public's "right to know." A second aspect of risk communication centers around what might be called the "warning" process—How do you communicate risk in impending emergencies? These processes have often been characterized in terms of simple stimulus and response. Mileti has been one of the most active scholars in researching this area and, with Colleen Fitzpatrick, he provides an important summary of research of this area of concern. Closely tied to this issue of risk communication is research on evacuation. One of the most effective methods of reducing the effects of hazard incidents is the evacuation of populations from areas with high risks. Much of the conventional wisdom about the reactions to evacuation has implied that people will be frightened by warnings and will act irrationally if provided information about impending risk. Instead most of the research shows that rather than people acting in a helter-skelter fashion, the problem is to get the public to take some appropriate action. Ron Perry has, over the years, devoted considerable attention to this problem, and in his contribution here, he developed a model of evacuation compliance behavior and tested it in two different settings.

In addition to the response to warning and evacuation as actions that will reduce disaster impact, there is a growing knowledge of behavior during the postdisaster emergency response period. One critical area within the emergency response is the medical system. Much of the traditional thinking about emergency medical care suggests that a system based on techniques used to handle daily routine emergencies will be adequate if expanded in emergencies. In the DRC studies of EMS, Quarantelli (1983) has pointed out the inadequacy of that assumption. Joseph Scanlon has been able to examine the model that Quarantelli has suggested in another context. Most scholars identify Samuel Henry Prince's study of the Halifax explosion as the first systematic social science effort to understand disaster. In addition to his own work on various aspects of disaster, Scanlon has developed an interest in the origins of the Price study and in the Halifax explosion itself. Scanlon's knowledge of the sources of information about that explosion allows him to reconstruct the EMS system and to compare those developments to the theoretical observations Quarantelli developed many years later. Scanlon's contribution suggests the value of historical materials in testing more contemporary theoretical positions.

In addition to the concerns and activities of the emergency period, disasters can also be studied in the short- and long-term recovery period. Robert Bolin has done extensive analysis of the sheltering and housing of disaster victims, both in the United States and elsewhere. He underscores the importance of the social processes for recovery and indicates the important effect of various governmental programs relating to housing in the overall recovery.

The last two chapters in this section deal with other important themes. The chapter by Britton, Moran, and Correy examines the variables that are thought to influence individual capacity to cope with stress. In this context, disaster is anticipated to be a significant stressor not only for victims but also for those personnel who have responsibility as members of emergency organizations. Too, such stress might vary depending on the degree of involvement in different roles. For example, Dynes (1970) examined the consequences for personnel in four different organizations. In the next section of the book, Wenger and James look at the role of volunteers in the Mexico City earthquake. Britton et al. look at what they call the *permanent volunteer*. Every disaster response involves many different types of personnel victims themselves, casual volunteers, full time personnel in emergency organizations, but many organizations that are relevant in emergencies cannot afford to maintain permanent staffing to deal with such infrequent events as disasters. In many organizations, then, *permanent* volunteers are used and, in some countries, such as Australia, such personnel comprise critical human resources in mobilizing an emergency response. Britton et al. examine the potential stressors disasters pose for such personnel.

The concluding chapter on an ecological approach to disaster raises an important and very complex issue. Disasters can be approached by many disciplines, and those different disciplinary views often provide varied and perhaps even inconsistent views of reality. Bates and Pelanda outline their approach, using systems theory, to develop what they consider an integrated ecological approach. They place the notion of disaster in the context of sociocultural evolution that illustrates the way in which both biological and social systems evolve. The essay presents an important conceptual approach for the future.

Disaster in Aisle 13 Revisited

THOMAS E. DRABEK

It was a Halloween night. It was raining throughout the Indianapolis area. Inside the Coliseum, 4,327 people were enjoying the final moments of the "Holiday on Ice" show. Without warning—at 11:06 P.M. to be exact—a massive explosion rocked the building. This Coliseum is situated inside the Indiana State Fairgrounds, about two hundred yards from the main entrance. Some spectators were blown from their seats onto the ice rink. Others dropped down into a newly opened cavern that seconds earlier had been a small commissary area where popcorn was being prepared for the next day's performance. The explosion killed fifty-four outright; twenty-seven more died shortly thereafter; nearly four hundred others were injured. The Indianapolis emergency response network mobilized quickly. Within twenty-four hours local and state officials had things under control, more or less. By then, they realized they were in the process of recovering from the worst disaster in the state's history (see Drabek 1968).

During the afternoon following the explosion, relatives of victims continued the process of identifying their loved ones' bodies and belongings. It was then that a four-person field team from the Ohio State University Disaster Research Center (DRC) arrived on the scene and began conducting interviews with representatives from the twelve organizations most heavily involved in the response. DRC had been established about six weeks prior to the explosion, and initially the fieldwork had been viewed as a training exercise for the research team. Because of the richness of the materials obtained and the high level of cooperation provided by local officials who supplemented in-depth interviews with numerous organizational documents and

I wish to thank Ruth Ann Drabek for her work on this paper and her suggestion that I return to the Indianapolis study as a focus for this essay. I also wish to acknowledge the assistance received from staff members of the Indianapolis/Marion County Emergency Management Planning Division. Partial support was provided by NSF Grant Number CES-8511603. Any opinions, findings, conclusions, or recommendations expressed in this publication are those of the author and do not necessarily reflect the views of the National Science Foundation.

records, the field experience was transformed into a detailed case study. One year later, a two-person team from the DRC returned to Indianapolis to confirm observations made initially and to collect information regarding subsequent policy changes. Publication of results was delayed for several years, in part because of resistance by one of the agencies that funded the DRC, and also because of administrative difficulties in establishing a monograph series. In the fall of 1968, nearly five years after the Coliseum explosion, *Disaster in Aisle 13* was released.

In this essay I use this work—the first monograph published by the DRC, which was relocated to the University of Delaware in 1985—as a vehicle to review selected aspects of the evolution of social science research on human responses to disaster. Following a brief discussion of the study conclusions, three contrasts are made. First, subsequent developments in three aspects of organizational theory are discussed, that is, interorganizational coordination, organizational environment, and blame assignation. Second, several issues in research methodology are summarized, especially the concept of disaster taxonomy. Third, developments in the application of research results, including changes in the state of readiness within the Indianapolis area, are reviewed. Collectively, these three discussions illustrate how disaster research and our understanding of community preparedness have progressed during the past three decades.

Study Conclusions

Twenty-six conclusions are discussed in the final chapter of *Disaster in Aisle 13*. They are organized around five general themes: (1) organizational environment, (2) mobilization, (3) communication, (4) coordination, and (5) preplanning. Many of these themes reflect aspects of the central core of knowledge that defines the empirical base for the emerging profession of emergency management.

Organizational Environment

Personnel from law enforcement, fire, medical, and other emergency organizations confronted an environment that was characterized by several important factors. Each constrained their responses in different ways. For example, despite a popular myth of victim disorganization, emergency responders reported neither panic flight nor hysterical breakdown. Instead, Coliseum attendees reacted. They did not wait for directions from authorities. Rather, using emergent definitions of risk and appropriate adaptive responses, they proceeded to initiate rescue, recovery, and transportation activities. Given the available information, these actions were rational, desirable, and reflective of an emergent norm of altruism.

Two other factors defined the response environment. First, much of the public response reflected convergence behavior, including material, personal, and informational. Second, various spatial and temporal qualities tended to minimize organizational problems. I observed, for example, that "the presence of the enclosure around the Fairgrounds with a limited number of gates made security in the area of the Coliseum infinitely less difficult than otherwise would have been the case" (Drabek 1968, 150–51). Because it was Halloween, many additional police were on duty to curtail vandalism. Also, because the explosion occurred at 11:06 P.M., hospitals, fire, and law enforcement organizations had double shifts of personnel available.

Mobilization

Three conclusions were highlighted regarding mobilization: two were descriptive, and one was analytic. Critical to understanding the community response and especially certain problem areas, the report emphasized that some organizations had not been advised of the explosion. Personnel in three of the five hospitals, for example, learned of it only when victims arrived in the emergency room. This complicated subsequent mobilization efforts. In contrast, those organizations that were notified quickly experienced few difficulties.

Shifting from these two descriptive conclusions, field interviews supported a proposition advanced by Form and Nosow (1958, 136–216), who had assessed the community response to a tornado in Michigan. Their data indicated that certain characteristics of organizational structure, for example, degree of formalization and size, were associated with emergency response. Although this line of reasoning was not pursued in detail, one analytic proposition was highlighted. "The greater the proportion of paid personnel (as contrasted to volunteers) in an organization, the greater the speed of mobilization" (Drabek 1968, 154).

Communication

Communication inadequacies were the major difficulty encountered with the Coliseum response. Five observations were offered that were interrelated. Two of these pertained to the hospitals. The lack of communication among them contributed to both intra- and interorganizational problems, especially an overload of their separate communication facilities. These two observations were specific aspects of the more general issues. No central communications center was established to facilitate cross-agency interaction. Also, informational convergence on telephones impaired efforts with interorganizational communication.

The follow-up field trip made on the one-year anniversary of the explosion provided documentation of a positive adaptation. Largely through the leadership provided by local Red Cross officials, an interhospital communication system was established. Thus, the event precipitated a positive change in the community's state of disaster readiness.

Coordination

Given the communication inadequacies just noted, coordination difficulties followed logically. The first of the five conclusions offered, however, emphasized a key idea. Although the field interviews uncovered a few instances to the contrary, nearly all emergency personnel cooperated with one another. Thus, this case illustrated what others have since documented elsewhere (Drabek 1979). Cooperation is a necessary, but not a sufficient condition, for coordination. Much more is required.

Following this more abstract conclusion, four others of a more specific nature were discussed. Victim dispersal from the explosion site was a key task that was not coordinated. Some key tasks were duplicated needlessly, while others were neglected. Personnel tended to rely heavily on internal resources rather than seeking assistance from other organizations. Once the high-priority task of helping the injured had been met, however, the level of interagency coordination increased. These observations underscored the importance of preparedness planning rooted within a communitywide perspective. They illustrated a fundamental axiom: the emergency response effectiveness of any single agency is directly dependent on the level of coordination within the emergent multiagency network.

Control

As is typical following large-scale disasters, responding agencies represented varied sponsorship bases. Many, of course, were divisions of local government, for example, Indianapolis Police and Fire departments. Others were specialized private sector organizations such as ambulance firms or voluntary agencies like the Red Cross and Salvation Army. Because the Coliseum was within the state fairgrounds, the state police eventually were recognized as the responsible authority. During the immediate aftermath, however, no overall authority was recognized. Hence, the interagency control processes were weak or nearly nonexistent.

In part, this situation reflected inconsistent, ambiguous, and overlapping organizational policy statements and local and state laws. These normative prescriptions had little relevance in guiding the actions taken by emergency officials during the response. Rather, reflecting an emergent norm of altruism, officials made reasoned judgments. Their efforts were hindered by communication inadequacies rather than incompatible policy prescriptions.

Preplanning

Field data provided support and clarification of a fundamental axiom that had not been recognized previously except among a few research specialists. "There is a marked difference in organization planning for day-to-day emergencies and planning for post-disaster community responses" (Drabek 1968, 170). Like many of their peers throughout the nation at that point in time, all but a few organizational executives in Indianapolis reflected an "add on" planning perspective. Because they successfully handled miniature emergencies daily, they believed they were prepared for large-scale events. If more personnel were needed, they could be obtained. This view fails to recognize important qualitative distinctions and changes in the task environment associated with large-scale disasters. Disasters do not constitute a simple straight line extension of an auto accident or house fire.

Three descriptive observations also were presented. First, at the interagency level, "disaster plans were incomplete, unclear, and not widely known or rehearsed" (Drabek 1968, 171). Second, hospital and related medical specialists expressed a need for medical planning for large-scale disasters. Third, there were several important changes in intraorganizational planning that were precipitated by the explosion.

Finally, a descriptive observation was offered regarding longer term change, or rather the lack of it. "While most organizational officials expressed an interest in the development of a comprehensive disaster plan when initially interviewed after the explosion, one year later no such plan had materialized" (Drabek 1968, 176). This pattern of initial interest, followed by inaction, has been documented subsequently following other events (Drabek 1986). These windows of opportunity are relatively short. If policies are not enacted within a few months after such events, there is a strong propensity to return to older patterns of denial and inaction.

Implications for Organizational Theory

Although *Disaster in Aisle 13* was largely descriptive, aspects of organizational theory were introduced into the analysis. By juxtaposing these ideas with subsequent developments during the past three decades, we can obtain a sense of the continuity in this theoretical base and the enormous expansion that has followed. Three concepts were central to the analysis of this event: (1) interorganizational coordination, (2) blame assignation, and (3) organizational environment.

Interorganizational Coordination

As reflected in many of the conclusions previously outlined, the fundamental flaw in the multiagency response to this explosion was the relative lack of coordination. Subsequent development of this concept and the analytic conditions related to it represent an important expansion of organizational theory. Additionally, as evidenced by the range of responding organizations with differing sponsorship bases (e.g., federal, military, state police, city and county law enforcement, and private sector agencies like Red Cross and Salvation Army), this case study illustrated the need to classify disaster response organizations. Both Quarantelli (1966) and Dynes (1970), like numerous others (e.g., Stallings 1978, 1987), recognized that coordination of such multiagency responses is difficult in part because of the variety of units represented. Aspects of this variety are sorted out usefully in what came to be known as the DRC typology of organizations. By cross-tabulating two dimensions—tasks and structure—a fourfold typology is produced: (1) established (regular tasks, old structure); (2) expanding (regular tasks, new structure); (3) extending (nonregular tasks, old structure); and (4) emergent (nonregular task, new structure). Because of the absence of a shared consensus regarding multiagency responsibilities and procedures, a coordinating task force gradually emerged several hours after the explosion. Structures of this form are clearly identified within the DRC typology. Today, they are much better understood as are the conditions that give rise to their emergence (Kreps 1989; Drabek 1987a).

Interorganizational coordination was assessed in detail within a variety of nondisaster settings including both antecedents and consequences (Rogers et al. 1982). Following several disasters, emergent communication structures were mapped temporally and were related to both relative degrees of coordination and key operational problems (e.g., Drabek et al. 1981; Drabek 1985). These social-mapping techniques illustrate the significant gains that have occurred in both conceptual refinement and measurement sophistication since the analysis of the 1968 explosion.

Others have applied these techniques to the nonemergency phase so as to examine the structure of interagency networks with more precision (Gillespie et al. 1986; Caplow et al. 1984). And recent research documented a series of coping strategies used by local government officials to construct and maintain such linkage sets (Drabek 1987b; 1990). Thus, research in the United States as well as in other societies, such as Australia (Britton 1989a, 1989b; Britton and Wettenhall 1991), underscores the importance of this concept and illustrates the enormous progress made.

Blame Assignation

In the year that followed the 1963 explosion, a complex blame assignation process emerged. This was summarized in the monograph and assessed further by Quarantelli and me (1967). In this instance, not unlike the cases studied previously by Bucher (1957), the search for guilty individuals diverted attention away from fundamental flaws in the mitigative subsystems.

Although relatively few analysts have focused on this emergent process during the past three decades (Drabek 1986), several important studies have appeared. For example, Neal (1984) examined an emergent citizen group that formed because of their concerns about the possible health consequences of a pollution policy. They objected to the lack of regulations pertaining to the burning of polychlorinated biphenyls (PCBs). By linking blame assignation processes to the emergence of such citizen groups, Neal provided an important bridging mechanism. Indeed, his work, like that of Walsh (1981) and Quarantelli (1985), has pressed us forward in our understanding of the social conditions that spawn such groups. As Wenger (1987) pointed out, these types of analyses have tightened the linkages between traditional collective behavior studies and organizational analysis.

Clearly, much more needs to be known about the social process of blame assignation. But nesting future analyses within assessments of emergent citizen groups links the concept to fundamental principles of the birth or origins of such social units and, at times, a collective consensus regarding a fundamental social problem (Drabek 1989a).

Organizational Environment

As previously noted, limited aspects of the environment within which these agencies responded were highlighted. The observations made, however, mainly had to do with public responses, for example, lack of extreme personal disorganization. These observations were consistent with previous studies (e.g., Quarantelli 1954) and have been verified subsequently (Johnson 1988). Recent research suggested, however, that there may be some cultural variation in the overall pattern (Alexander 1990).

Although this component of organizational environment is highly significant for both theory and application, there has been a total redirection in organizational theory since 1968. Although a few perceptive analysts writing at that time emphasized environmental uncertainty as a key to understanding organizational dynamics (e.g., Thompson 1967), awareness accelerated rapidly. Following extensive and insightful assessments by writers like Aldrich (1979), Hannan and Freeman (1989) began to develop population ecology models wherein environmental factors became paramount. Indeed, recent texts propose these models as paradigm shattering because of their power and completeness (e.g., Carroll 1987). Whereas most regard such claims as overreactions (e.g., Perrow 1986), this early observation anticipated progress in organizational theory.

Subsumed within the discussion of agency environments, albeit vaguely and incompletely, were references to the intergovernmental system. Although here the analysis was very weak, the finding prefigures significant developments brought to disaster research and organizational theory by specialists in political science (e.g., May 1985) and public administration (Charles and Kim 1988). The web of constraints and cross-cutting sectors of strain represented by the intergovernmental system that comprised a major segment of the response network described in this monograph appear over simplified when juxtaposed against more recent analyses. Hence, the infusion of concepts and specific empirical analyses by representatives from these two disciplines especially, have significantly enriched the theory base available for future analyses of such responses.

Methodological Implications

The case study method used in the analyses of the Indianapolis explosion reflected a strong and continuing tradition in sociology and disaster studies in particular (e.g., see Kreps 1981). Critics emphasized the limitations of the excessive use of this single research strategy (e.g., Cisin and Clark 1962; Drabek 1970; Mileti 1987). Aside from a few extremists, whose attacks have bordered on the polemical at times (e.g., Rossi et al. 1981), the emergent consensus among sociologists is that case study designs remain useful for the purposes for which they are intended (Campbell 1975; Yin 1984). Yet, the fact remains that the Coliseum explosion was assessed within a context of being a one-shot case study of responses to one event, in one community, at one point in time.

Juxtaposition of this work against the literature of the past few years reveals at least three highly significant developments: (1) increased sophistication regarding sampling, (2) improved measurement techniques, and (3) greater awareness of the problematics inherent in comparative analyses.

Sophistication Regarding Sampling

Discussions of sampling during the early 1960s largely reflected issues related to victim and nonvictim pools. Although a few advocated that disaster studies could be improved through improved sampling techniques (e.g., Killian 1956), the use of even quasi-experimental designs had not yet occurred (e.g., Campbell and Stanley 1966). As controversy intensified regarding disaster impacts on victim health status following the flooding of Buffalo Creek (e.g., Erikson 1976; Tichener and Kapp 1976; Lifton and Olson 1976), critics focused on the inherent limitations of analyses limited to data bases containing only victim responses (e.g., Perry 1979). Yet, even today, the use of quasi-experimental designs that incorporate victim and nonvictim data, as well as pre- and postevent information, remains relatively rare (e.g., Drabek and Key 1984). On the other hand, there is far more widespread recognition that these more sophisticated designs will be required if the quality of the knowledge base is to be improved.

Equally important has been the growing awareness and recognition of the parallel issues associated with selecting samples of organizations and multiorganizational networks (Drabek et al. 1982; Miller et al. 1985). Actually, the very idea of conceptualizing the multiagency response to the 1963 explosion as an emergent multiorganizational network, which might be viewed as an independent unit of analysis, had to await the conceptual clarity that came nearly a decade later as theorists like Aldrich (1979) sorted out the ambiguities that prevailed. Hence, only in the early 1980s did studies appear that raised the issue of selecting samples of events and samples of multiunit response networks for comparative analysis. Within these multiagency response networks, sampling guidelines remain minimal. This juxtaposition reveals the progress made in issue awareness, but little more.

Improved Measurement Techniques

How are communication processes, coordination, preparedness planning, or control best measured? During the late 1960s and early 1970s, a large volume of studies was published assessing these and other aspects of organizational structure through a combination of survey techniques and traditional index construction procedures (e.g., see works by Blau 1970; Hall 1987; Hage and Aiken 1967). Indeed, these practices became so widely accepted that handbooks of measurement tools were published for both intra (e.g., Price 1972) and interorganizational (e.g., Morrissey et al. 1982) characteristics. These represented major steps forward. But as critics like Benson (1977) pointed out with vigor, they reflected oversimplified conceptualizations of organizational structure and continuity with a positivistic epistemological assumption base that many had begun to question.

Others have continued to press ahead and confront the measurement challenge head on (e.g., Gillespie et al. 1986). It is clear that many issues of measurement are really issues of theory. There is an enhanced sophistication today regarding the limitations inherent in complex concepts like coordination and effectiveness, especially when human service agencies or networks comprised of them, are the units of analysis. Thus, here again, the juxtaposition reveals important progress in that the issues are being taken seriously although the answers remain elusive.

Problematics of Comparative Analysis

Although no two disasters are exactly alike, it is clear that many aspects of the full life cycle of any event, including mitigation, preparedness, response and recovery activities, share some common elements. Hence, in contrast to the single event case study of the Coliseum explosion, many researchers attempted comparative analyses of disaster events through well-planned research designs (e.g., Leik et al. 1981), opportunistic reconstructions (e.g., Perry 1983) or syntheses of published results (e.g., Mileti et al., 1975; Drabek 1986). Others took a mitigative focus and examined the institutional dynamics of emergent regulatory systems that were designed to reduce risk (e.g., Perrow 1984; Kunreuther and Ley 1982).

These analyses illustrated the concerns raised by Dynes and Aguirre (1979) who had reflected on the problematics of integrating archival data amassed by the DRC staff during two decades. They also challenged Kreps and his associates (Kreps 1989b) who developed an innovative paradigm for dealing with many of the problematics. Although their "DRAT" system leaves many unanswered questions, it pressed many in the field to come face-to-face with the most fundamental issues of all; that is, what is a disaster? What classification schemes, of both disasters and social structures, should guide comparison and generalization?

Kreps's (1989b) work, like my analysis of taxonomy (Drabek 1989b), challenged all disaster researchers to examine a range of issues that simply could not have been considered at the time of the 1963 explosion. As reflected in the special issue of the *International Journal of Mass Emergencies and Disasters* that Kreps edited (1989a), many social theorists and disaster researchers are now devoting a portion of their energies toward these difficult but most fundamental issues. Depending on the rigor of their work in the near future, disaster studies may make significant leaps forward or continue to build broader and more complete data sets that will remain noncumulative. Despite the enormity of this challenge and the uncertainty of the response by the disaster research community, this juxtaposition reveals a remarkable degree of progress.

Influences on Application

The third broad axis of juxtaposition is application. Revisiting *Disaster in Aisle 13* provided opportunity for disciplined reflection on two specific aspects of application: (1) preparedness in Indianapolis and (2) emergent professionalism within emergency management.

Preparedness in Indianapolis

Since the 1963 explosion, three major social dynamics have altered the state of disaster readiness within the Indianapolis community. First, there were several additional disasters that served as reminders. Among the most severe of these are the following: tornadoes (1965, 1975, and 1989); major snow blizzards and ice storms (1978, 1988); a toxic substance release (1984); a three-alarm downtown fire (1985); and an A-7D fighter airplane crash into a motel (1987).

These events, along with numerous others of a smaller scale, required the emergency response system to be activated. These activations, and the horror of victim suffering, altered community risk perceptions and legitimated various reform proposals and increased resource allocations. Critiques helped pinpoint operational problems and required modifications in response procedures.

Second, in the early 1970s a "consolidated city" form of government, that is, "UniGov", was established to integrate governmental services within Marion County and the City of Indianapolis. Not all services were included, however. Directly relevant to the disaster response system was the decision to exclude law enforcement, fire, and emergency medical services. Also, three municipalities within the county were excluded (i.e., Lawrence, Beech Grove, and Speedway). Hence, although this administrative reform brought some services under a common administrative structure, those most relevant to disaster response remain differentiated.

Third, despite continued differentiation among law enforcement, fire, and other emergency service agencies, a series of intensive planning efforts reportedly reduced the degree of fragmentation and increased the quality of interagency coordination. Whereas several of these might be selected to illustrate this enhanced state of readiness, three are directly related to the juxtapositions made in this chapter: (1) clarification of lines of authority, (2) explicit recognition that disaster planning is a process, not a product, and (3) improved interhospital communications.

Clarification of Lines of Authority

Several clarifications have occurred at different points within the intergovernmental system. For example, prior to the opening of the 1989 Indiana State Fair, authorities signed a "Major Incident Coordination Plan" (Emergency Management Planning Division et al. 1989). Interestingly, the opening paragraph in the preface of this plan is a quotation from *Disaster in Aisle 13*. Clearly, the local institutional memory was impacted by this DRC publication.

The state fair plan is a variant of the widely used Incident Command System (ICS) that gained popularity in California in the 1980s. It defines three levels of incidents based on the number injured. This crude measure of organizational demand is juxtaposed with incident command or on-site responses and county agency response levels. For example, a Level 3 incident at the fair is defined as one in which five to ten persons are injured. This will trigger county notifications. A Level 5 incident, defined as more than twenty-five injured persons, will activate a county response of Level 1 or 2.

Lines of authority regarding activation of this plan are delineated clearly.

> An activation of the Indiana State Fair Plan at Levels 3, 4, or 5 is the responsibility of the Indiana State Police—Fairgrounds Operation (District V Commander).
>
> Because the Indiana State Fairgrounds is an island of state property inside the County of Marion, the Marion County Emergency Management Plan will be activated at standby; Level One; or Level Two. The administrator of the Emergency Management Planning Division has the authority to declare activation of standby, and Level One. Activation of the Marion County Plan above a Level One will require action by the Mayor of the Consolidated City of Indianapolis-Marion County, or the Director of Public Safety acting on his behalf. (Emergency Management Planning Division et al. 1989, 2).

A series of "response objectives" are defined within the plan. These nine functions are coordination, communications, law enforcement, fire-rescue-hazardous materials response, emergency medical, human services, public works, transportation, and utilities. Elsewhere in the plan, these functions are listed by organizational responsibility. Hence, the function of coordination is the responsibility of three member agencies: (1) Emergency Management Planning Division, Consolidated City of Indianapolis-Marion County; (2) State of Indiana Emergency Management Agency; and (3) Indiana State Fairgrounds Administration and Operations—Coordination Center Staff.

In contrast, the communications function is to be assumed by five member agencies: (1) National Weather Service—Indianapolis Forecast Office; (2) Indiana State Police, State Warning Point; (3) Indianapolis Police Department, Marion County Warning Point; (4) Emergency Broadcast System, CPCS-1, WIBC (1070 am), and CPCS02 WFBQ (94.7 fm); and (5) Marion County Radio Amateur Civil Emergency Service (RACES). One agency, the American Red Cross, Indianapolis chapter, is responsible for the human services function. The remaining six functions are to be assumed by four to seven community agencies.

This broad division of labor is augmented by the specification of three different emergency organizations that will be established. The functions and locations of each are as follows:

1. Emergency Command Center (ECC)—staff working at the on-site command post will provide direction and coordination of field response and recovery activities. This on-site post also is referred to as the "Incident Command Organization."
2. Operations Coordination Center (OCC)—personnel assigned to the operations center, located within one of the administrative buildings at the fairgrounds, will coordinate county and state support activities, incident information, and public information.
3. Emergency Operating Center (EOC)—personnel assigned to the operating center, located within a state administrative building in downtown Indianapolis, will coordinate policy and legal issues, large-scale state agency assistance, and public information (adapted from Emergency Management Planning Division et al. 1989, 4).

This organizational plan differentiates the tactical decisions that are required by on scene commanders from the strategic decisions that service agency directors must make. Both of these are differentiated from the broad policy decisions that elected officials must make with the assistance of senior agency advisors.

One function is singled out for special emphasis—media information. With regard to the Incident Command Organization (ECC), the plan states a clear prohibition: "There will be no public information released directly to the media from the Incident Command Organization" (Emergency Management Planning Division et al. 1989, 4). A similar statement is made for the OCC. Hence, all information distributed to media personnel will be centralized with the EOC. Given the tactics used by most media personnel following most disasters, this effort at centralizing the public information function may improve accuracy, precision, and consistency. Eradication and reduction of rumors may occur if the plan is implemented as written.

Disaster Planning: A Process, Not a Product

Throughout the 1990s and 1980s, numerous researchers emphasized that disaster planning must be conceptualized as a process (Drabek 1986, 46–60). Moreover, those who are going to implement the plan must be involved actively in the planning process. If they are not, the paper plan that exists in a file drawer will have minimal impact on response behavior. As Dynes, Quarantelli, and Kreps (1972) put it: "Disaster plans too often remain paper plans and are not rehearsed in whole or in part" (79).

These themes were emphasized by many, but were documented most carefully by Wenger and two associates—James and Faupel (1980). They reviewed seventy-one community disaster plans and conducted in-depth interviews with fifty planning officials. Three observations sum up the deficiencies they discovered.

> . . . there is a tendency on the part of officials to see disaster planning as a product, not a process. (Wenger et al. 1980, 156)

> Too often, disaster planning is isolated from the day-to-day planning process. It is often assigned to organizations, or units within organizations, that are divorced from traditional, institutionalized sources of social power within the community. (Wenger et al. 1980, 156)

> . . . these plans include almost no expectations for public behavior during a disaster. Furthermore, when attention is given to public response, it is generally predicated on erroneous conceptions of public behavior. (Wenger et al. 1980, 155)

The creation and rehearsal of the 1989 Indiana State Fair Plan reflected the recognition and understanding of these fundamental planning principles by the emergency management and emergency services agencies in the Indianapolis community. Informal discussions with one staff member verified the social process that the plan reflects. Furthermore, review and discussion of a preliminary sketch of the planning that took place prior to the 1990 Indianapolis 500 mile automobile race, provided further confirmation. Clearly, these planning principles have been integrated within the emergency response capacity of this community.

Improved Hospital Communications

As previously noted, one of the major changes made in the year following the 1963 Coliseum explosion was the establishment of an interhospital communication system. This system was upgraded and improved during the next decade as numerous emergency medical services planning efforts were completed. The results have paid off several times.

One illustration documents this increased capacity: the crash of the U.S. Air Force Corsair A7-D fighter jet into a motel and branch bank building, in which ten people died. The crash occurred at approximately 0915 hours (EST) on 20 October 1987. Area hospitals were alerted quickly after the incident, however.

> The Marion County Sheriffs Department, the Alternate County Warning Point, provided information over the Indianapolis Emergency Communications System (INDECS) during the activation of the Emergency Management Plan at a Level One status at 0927 hours. All emergency service communications centers, and key public safety officials have INDECS tone activated radio receivers.
>
> Hospital notification over the Indiana Hospital Emergency Radio Network (IHERN) occurred at 0925 hours, requesting immediate emergency room capabilities and bed availability data. The product of this roll call, the Patient Distribution List, was transmitted to Wishard Ambulance Supervisory by 0945 hours. (Emergency Management Planning Division 1988, 31)

This rapid response reflected that a roll call is done twice daily for all hospitals on the network (IHERN) within Marion County and the seven contiguous county hospital systems. Although all ambulances certified in Indiana are required to have radio equipment capable of receiving and broadcasting on the IHERN channel, plans have been completed to control use.

> During a major emergency—disaster in Marion County, the EMS coordination Center will control the use of the IHERN radio, reserving it primarily for inter-hospital coordination.
>
> However, since the IHERN radio is required equipment on every ambulance certified in Indiana, it may be used by ambulance personnel who have been assigned to the major incident if they do not have Med Channel 8. Like Med Channel 8, it will only be used for essential off-site communications. If the incident requires a substantial adjacent county EMS agency mutual-aid response, the EMS Coordination Center should assume the IHERN's use for inter-hospital communications will become ineffective due to ambulance requirements for it. The EMS Coordination Center should then make use of a "patch" to, and from, the hospitals via Med Channel 1, 5, or 6 for the RollCall and other essential communications. At a point early into the response phase to a mass casualty incident, the EMS Coordination Center may announce over the appropriate radio channels "a suspension of medical direction" via Med Channels 1, 5, or 6. Ambulance personnel treatment then will be based on "standing order" of their protocols. (Emergency Management Planning Division 1990, 39)

Specific procedures have been developed to insure rapid dissemination of notifications of any event in which ten or more persons are injured at a single site. Thus, the dispatchers are instructed to notify their superior and then the

Indianapolis Police Department (primary Marion County Warning Point) or the Marion County Sheriffs Department (alternate). They are to ask that the Emergency Management Duty Officer be paged. They are to then notify the Marion County Medical Exchange and ask that the Chief Medical Service Coordinator and the Public Health Group Manager be paged (Emergency Management Planning Division 1990, T-1).

These procedures are a sharp contrast to the behavior pattern that was documented in *Disaster in Aisle 13*. They are evidence of the increased emergency response capacity that has emerged within the Indianapolis community. Informal interviews with several emergency response and coordination staff members have indicated that this monograph continues to be remembered as a stimulus that many used as a reminder. Both substantively and politically, it has assisted them in their efforts to improve the disaster response capacity of this community and many others.

Emergent Professionalism With Emergency Management

Shifting to a broader scope of analysis provides a final area of juxtaposition. From the early 1960s, the entire nation has experienced a major redirection in disaster preparedness that reflects the rapid emergence of a new professional. Increasingly, local government officials have recognized the need for improved coordination within the emergency response system. And increasingly this function has been explicitly assigned to an agency directed by a professional with specialized training and job title. The new era of emergency management reflected at least four major areas of change: (1) developments in preparedness theory, (2) new training opportunities, (3) specialized innovations, and (4) increased linkages between research and practitioner communities.

Developments in preparedness theory have ranged from specialized coordination procedures like those reflected in the widely used ICS to highly sophisticated transportation planning models to assist in community evacuation. More generally, however, a series of fundamental principles have begun to emerge regarding topics like the planning process, warning responses, evacuation behavior, hazard analysis, and the like. Although certain aspects of these rapidly evolving methodologies reflect the contributions of social science research (e.g., Quarantelli 1984), others are rooted in highly bureaucratic models. Dynes (1989), Kreps (1989b) and I (Drabek 1987b, 1990) have all questioned the applicability of these bureaucratic models to emergency management. It is clear that this complex issue will require a great deal of clarification and debate in the coming decade. In the meantime, some within the emerging profession will continue to apply organizational theories that have a high probability of failing them and their communities.

Major changes have occurred in emergency management training. The institutionalization of a course sequence at FEMA's Emergency Management Institute in Emmitsburg, Maryland is the most obvious. These are complemented by other programs sponsored by the National Emergency Training Center, of which EMI is a component, like the instructional programs that thousands of practitioners throughout the world receive each month via satellite, through the Emergency Education Network (EENET).

All of these training initiatives are augmented by the growing number of university based programs that are degree based. At the North Texas State University, the University of Wisconsin, several universities in California, and elsewhere, the professionalization process is being formalized.

Many important innovations have further reinforced the professional process. Most notable are the significant changes in communications systems. Many communities have adopted voice-augmented sirens to enhance their warning capability. Similarly, satellite-based radio units permit high quality transmissions linking personnel in the field to command posts and EOCs located miles away. These developments pale, however, when contrasted to the significant impacts of the widespread adoption of microcomputers. As I have documented elsewhere (Drabek 1990, 1991), the images of local emergency management agencies have been enhanced greatly by the adoption and implementation of microcomputers. This innovation has increased their centrality within the interorganizational network and their capacity to manage resources, guide evacuations, use sophisticated storm-modeling programs, quickly ascertain toxic substances, and perform other disaster-relevant tasks. Because of these and other innovations, emergency management agencies have gained increased recognition and acceptance within most communities. The comparison to the response following the Coliseum explosion is most instructive.

A final area of juxtaposition has to do with the relationships between researchers and practitioners. The linkages here have been intensified greatly through a variety of strategies. The annual workshop sponsored by the Natural Hazards Research and Applications Information Center has been most significant. As Yin and Moore (1985) discovered, the informally based contacts that are nurtured in such settings provide the pathways that redirect research agendas and speed the pace with which findings are disseminated. Paralleling such developments have been other organized attempts to involve the users of research results, for example, research project advisory committees (Drabek 1987b, 267–68) wherein representatives from both communities can press principle investigators to sharpen their concepts and struggle with potential dissemination and application routes.

A text that was published recently within the International City Management Association's prestigious "Green Book" series established a new standard in this integration (Drabek and Hoetmer, 1991). While researchers, reflecting the disciplines of sociology, law, public administration, journalism, city and regional planning, and social work, prepared eight of the twelve chapters, practitioners from local and state emergency management agencies prepared the others. One chapter was coauthored by a long-term National Science Foundation staff member whose administrative skill has done much to facilitate the quality and quantity of research in the area. This blending of talents and perspectives reflects an emergent integration of interests and skills that today comprises an important national resource. The contrast it provides to the situation that existed when the 1963 explosion occurred in Indianapolis is dramatic, instructive, and encouraging. When placed within the context of the numerous other contrasts highlighted in this chapter, these juxtapositions provide inspiration.

Conclusion

The analyses that comprise this chapter point toward three broad conclusions. Each reflects significant changes that have occurred in disaster research and the emerging profession of emergency management during the 1970s and 1980s.

1. Juxtaposition to theory highlights the enormous progress that has been made in our understanding of interorganization coordination and organizational environments. Developments within general sociological theory have increased the quality of disaster research. Disaster studies, in turn, have added to sectors of organizational theory, especially that pertaining to interorganizational analysis. Although blame assignation processes have not received much attention, those focusing on more general issues of collective behavior, especially emergent citizens groups, have pushed our understanding forward.
2. Juxtaposition to methodological issues brings us face to face with the problematics inherent in comparative analysis. Since *Aisle 13* was a case study, these issues were not raised until others began to attempt comparative analysis using some of the data collected. The range of generalization, or external validity, was not addressed until others reported parallel or conflicting findings. Neither the social structures nor the range of events to which the conclusions of the study might apply were clear. Thus, it has only been in recent years that some disaster researchers have begun to tackle fundamental theoretical issues that underlie such methodological problems. "What is a disaster?" emerges as a perplexing problematic that awaits the energy and discipline of the disaster research community.
3. Juxtaposition to applications underscores the validity and use of the

> conclusions reached in the study of the 1963 explosion in the Coliseum at the Indiana State Fairgrounds. Nearly three decades later, the conclusions regarding such processes as notification, mobilization, authority, communication, and coordination remain useful cross-checks for local emergency managers. They also highlight the significant progress that has been made in disaster planning within the Indianapolis community. Finally, this juxtaposition affords an appraisal of the significant contribution that the disaster research community has made to the emerging profession of emergency management. With an improved knowledge base, practitioners can facilitate community planning and coordination processes with information rooted in science rather than myth.

As these basic problematics are pondered and debated, however, disaster researchers should take pride in knowing that some of their analyses have influenced policymakers and administrators. The emergency response capacity of many communities has been increased because of the insights, findings, and questions raised by disaster researchers. Although a certain degree of tension and distance must be maintained between the two sectors, that is, practitioners and researchers, a new level of cooperation has been reached. In the coming decade, emergency management professionals will benefit from this social innovation and possibly improve the quality of the contribution they can offer to those they serve. In this sense then, the tradition reflected in *Aisle 13* continues, although in a greatly expanded way. In reference to the local officials who cooperated with the study team, this tradition was noted as follows: "Their motivation, as ours, was that others might benefit from what was learned through the Indianapolis tragedy" (Drabek 1968, v). Explication of this continuity underscores a key element in the legacy of the disaster research tradition and the fundamental ethical commitment of all professional groups to those they have chosen to serve.

Disaster Archives and Structural Analysis: Uses and Limitations

GARY A. KREPS

Introduction

When the DRC was founded in 1963, its codirectors (E. L. Quarantelli, Russell Dynes, and J. Eugene Haas) had access to raw data and final reports from earlier studies at the National Research Council, the NORC, and several university-based projects. From this foundation, both a disaster research library and data archives were established at the new center. It was hoped that the library would become a major collection point for disaster publications from a host of sources, and that the data archives would expand with each new study undertaken by the DRC itself. That hope continues to be realized as the DRC begins the fourth decade of its existence.

Although the DRC's library has been a major resource for disaster scholars over the years (e.g., Mileti, Drabek, and Haas 1975; Quarantelli and Dynes 1977; Kreps 1984; Drabek 1986), the focus of this chapter is on the archives resulting from the DRC's own studies. Creating and maintaining them was no small undertaking. It meant that tape-recorded interviews with hundreds and (over time) eventually thousands of direct participants and informants had to be transcribed. Once transcribed, these interviews had to be inventoried and stored along with large numbers of documents (meetings minutes, after-action reports, communications logs, news accounts, tape-recorded direct observations) from each event studied. The transcribing of interviews continued for eleven years (1963–74), and then was largely suspended. The interview tapes and documents continued to be stored when the

---

This paper is based on work supported by the National Science Foundation under Grant Nos. CEE-9121135, CEE-8400486, and CES-8703518. Any expressed opinions, findings, conclusions, or recommendations are those of the author and do not necessarily reflect the views of the National Science Foundation. My thanks to David Aday and Susan L. Bosworth for comments and criticisms of earlier drafts of the paper. Special thanks also to Henry Quarantelli, who challenged my argument (as a graduate student) that DRC archival data were too unstructured to be useful. I was wrong.

costs of transcribing all of the interviews became too prohibitive. Recently, DRC has resumed the work of transcribing interview tapes.

The DRC archives and *possible* uses of them are nicely summarized in a recent paper by Dennis Wenger (1989), who served as a codirector of the Center from 1985–89. The following chapter complements the Wenger piece by describing and illustrating *actual* uses of the archives within the context of my own research program (for summaries and critiques of that research program, see Enoch 1988; Kreps 1989). My experiences and those of my colleagues during the past nine years are worth chronicling for at least one reason: they may shed light on the strengths and limitations of disaster archives for addressing basic conceptualization and measurement problems in the analysis of social structure.

The first section of the chapter provides a brief historical look at our research program as it evolved from my work as a research assistant at the DRC (1967–71), later through my analysis of over a thousand transcribed interviews and several hundred related documents (1981–83), and finally through collaborative studies with my research associate, Susan L. Bosworth, and ten graduate students at William and Mary. This is followed by more focused discussions of the use of the archives for unraveling social structure during the emergency period of disasters. The chapter closes with a brief reflection about the overall value of the DRC data archives and the field research methods from which they were produced.

Overview of Kreps's Archival Studies

My initial interest in the archives resulted from personal experiences as a field researcher at DRC (1968–71) and many stimulating conversations and debates among colleagues and the codirectors at the DRC about the theoretical and practical value of its research. Theoretical questions about the difference between organization and collective behavior were the most interesting to all of us at the time (Dynes 1970), and we believed that disasters were strategic research sites for addressing these issues. My own interest and belief have become much stronger after many years of archival research.

The research began with some preliminary reading of several hundred transcribed interviews, using Weller's (1969) conception of organizing as a sensitizing device for interpreting what existing or emergent organizations were doing during the immediate emergency period of disaster. The result was a preliminary statement of what I thought the archives were saying about the commonality of organization and collective behavior as basic sociological concepts (Kreps 1978). The theoretical tool to express that commonality was later to become a structural code consisting of domains (D), tasks (T), resources (R), and activities (A). Sustained funding by the National Science Foundation and The College of William and Mary during the 1980s has afforded my colleagues and me the opportunity to put that structural code to work in more detailed analyses of the DRC archives.

My initial efforts involved creating a data file on 423 instances of organizing from fifteen disaster events, using the above code to ground empirically the difference between organization (all four of the above elements evidenced) and other structural forms (one to three of the above elements evidenced). The data file was produced from 1,062 transcribed interviews and a large number of documents on the fifteen disasters. The work was labor intensive and took about fourteen months to complete. There followed a series of collaborative archival studies with Susan Bosworth and ten other graduate students. These studies involved further analysis of this data file and the creation of several new ones from the DRC archives, as well as data on social networks provided by Thomas Drabek (see Drabek et al. 1981). All of this work has been reported in a series of papers, articles, and most recently, a book (Kreps 1989).

For discussing the uses and limitations of disaster archives, I highlight four different types of studies from our research program. These include studies of (1) the origins of organization, (2) social networks as organizing, (3) the restructuring of organization, and (4) organizing and role enactment. The common thread that links these separate studies is the use of disaster archives for conceiving and measuring *structure in process.*

The Origins of Organization

We are asking three key questions in our archival studies: First, what is an organization? Second, how are organizations different from other structural forms? And third, precisely how does an organization come into being? These questions are germane to the emergency period of disaster because the existence of organization is often an issue rather than a given. Indeed, the life histories of many responding units are bounded completely by days, weeks, or months. And as these new entities emerge, many existing groups and organizations undertake nonroutine activities or temporarily stop operating.

A structural code of domains (D), tasks (T), resources (R), and activities (A) provides our answer to the first and pivotal question: What is an organization? Domains (D) are collective representations of bounded units and their reasons for being. Tasks (T) are collective representations of a division of labor for the enactment of human activities. The elements (D) and (T) are termed structural *ends* of organization. Resources (R) are individual capacities and collective technologies of human populations. Activities (A) are the conjoined actions of individuals and social units. The elements (R) and (A) are termed structural *means* of organization. Because the meaning and interpretation of these elements are elaborated and critiqued in detail elsewhere (Kreps 1983, 1985, 1989, 1991b; Saunders and Kreps 1987), that discussion will not be repeated here.

Central to answering the second and third earlier questions, we argue that the four elements are individually necessary and collectively sufficient for organization to exist. This means that each one represents a unique expression of social structure; that their mutual copresence points to the existence of organization; and that no pattern of their origins is necessarily more frequent, important, or effective. Accordingly, all logically possible combinations and permutations of the elements are expressed taxonomically as forms of association. As noted in Table 1, a derived sixty-four cell taxonomy includes four one-element forms, twelve two-element forms, twenty-four three-element forms, and twenty-four four-element forms. While only the four-element forms are sufficient for organization to exist, the remaining forty types depict stages in its origins.

Using the above code as a conceptual tool, the archives have been searched to locate instances of enacted organization (i.e., the unit of analysis) and to reconstruct precisely how that happened. The following example (adapted from Enoch 1988, 54–55) illustrates the kind of data production that has been possible (for several other examples, see Kreps 1989):

> A tornado strikes a metropolitan community on a Friday morning. It carves a three and a half mile path of destruction. Hundreds of homes and other buildings are either destroyed outright or sustain major structural damage. Recognizing the potential danger of these unsafe structures for their inhabitants, and knowing that many buildings will have to be condemned, the mayor and other city officials turn the problem over to the city building and zoning department (D). The building commissioner, knowing that he and his staff of four cannot begin to inspect effectively all of the damaged buildings, decides to enlist the assistance of a regional building association. As the only individual legally empowered to condemn buildings, and realizing that the city would soon be flooded with private contractors and insurance claim adjustors, the building commissioner prepares to take charge of what will quickly become an extensive operation. A previous agreement by members of the regional building association, one which resulted from an earlier tornado, expedites the arrival of volunteers. The volunteers are building commissioners and inspectors from surrounding communities. They begin arriving by early Saturday morning (D-R). As the person in charge, the local building commissioner then makes several immediate decisions. First, to protect "volunteers" from any legal repercussions, they are "hired" on the spot for one dollar each, with each individual receiving a temporary appointment card. These "employees" are then divided into teams, with each team led by an inspector appointed by the commissioner. Teams are assigned to different sectors of the impacted area and told to fan out in order to cover all of it. Buildings are to be categorized as belonging to one of three classes of damage (D-R-T). Three separate inspections are completed over the next few days, until such time that all those buildings which could not be salvaged are condemned and eventually destroyed (D-R-T-A). The city's fire department is assigned the problem of handling the destruction of these buildings.

TABLE 1. Taxonomy of Structural Forms

Four Element Forms: Organization	Three Element Forms	Two Element Forms	One Element Forms
D-T-R-A	D-T-R	D-T	D
D-T-A-R	D-T-A	D-R	T
D-R-A-T	D-R-A	D-A	R
D-R-T-A	D-R-T	T-R	A
D-A-T-R	D-A-T	T-A	
D-A-R-T	D-A-R	T-D	
T-R-A-D	T-R-A	R-A	
T-R-D-A	T-R-D	R-D	
T-A-D-R	T-A-D	R-T	
T-A-R-D	T-A-R	A-D	
T-D-R-A	T-D-R	A-T	
T-D-A-R	T-D-A	A-R	
R-A-D-T	R-A-D		
R-A-T-D	R-A-T		
R-D-T-A	R-D-T		
R-D-A-T	R-D-A		
R-T-D-A	R-T-D		
R-T-A-D	R-T-A		
A-D-T-R	A-D-T		
A-D-R-T	A-D-R		
A-T-D-R	A-T-D		
A-T-R-D	A-T-R		
A-R-D-T	A-R-D		
A-R-T-D	A-R-T		
24 forms	24 forms	12 forms	4 forms

Total Forms = 64

Structural Code

Structural Ends

Domains (D) are collective representations of bounded units and their reasons for being.

Tasks (T) are collective representations of a division of labor for the enactment of human activities.

Structural Means

Resources (R) are individual capacities and collective technologies of human populations.

Activities (A) are the conjoined actions of individuals and social units.

Notice that when some empirically discernible threshold for any of the four elements has been passed, a *structural form* (Wolff 1950 on Simmel) exists in time and space. As each additional element comes into play, a new form has been created. When all four elements are present, an organization has been socially constructed.

The above example depicts a D-R-T-A form of origins. In my original coding of 423 instances of enacted organization from fifteen disaster events, twenty-one of the twenty-four logically possible four-element forms (organization) were located at least once, and all but four of the remaining one-, two-, and three-element forms were located as they depict stages of origins. It must be remembered that the original DRC field studies focused on the actions of social units that had obvious responsibilities during the emergency period (e.g., police and fire departments, hospitals, departments of public works, voluntary agencies). However, often fascinating accounts of the actions of emergent organizations (no pre-event existence) were recorded in the interviews and documents. Moreover, the field teams frequently were able to record equally interesting accounts of the improvisations of units that existed prior to the events in question.

The studies done during 1964–72 form the boundaries of my archival analysis because all of the interviews from these studies had been transcribed and catalogued. Because the data were not collected with *any* structural code in mind, they are very difficult to work with for purposes of empirically grounding this particular one. The problems my colleagues and I faced are inherent in situations where data are collected with one set of objectives in mind, and then new aims are formulated later. But on the more positive side, the DRC archives are often well suited for empirically grounded theory building on organization. They reveal what active social units are doing, and how, and also the various contexts in which collective action takes place.

Even though most (all but fifty-two) of the 423 instances of organization located were enacted by established units (i.e., units that existed prior to the events), existence was *not* assumed for studying the process of organizing. Note that the DRC studied impacted communities during the aftermath of an event. For my archival analysis, I decided to use the event as a social catalyst (Dubin 1978) whose presence was necessary for examining the origins of organization. As suggested by the earlier example, however, this strategy obviously does not deny the relevance of pre-event conditions for what takes place.

The criteria I used for selecting events from the archives were as follows: (1) that they could yield interpretable data on organizational enactments; (2) that data production would continue until a sample of three hundred to four hundred cases was achieved; and (3) that a sufficient number of events would be studied so that organizational form, disaster event, and community differences could subsequently be explored. Data production was stopped with purposive samples of 15 events, 1,062 interviews, and 423 instances of enacted organization. The population parameters of organizational enactments for the events studied are not known and for some events there are data on but a few responses. Some of the events were studied with greater intensity by the DRC, and the interviews and documents vary widely in depth and quality. There is no alternative to labor intensive spadework because one never knows where or even if useful information can be found.

The key requirement in working with the archives was to document the presence and sequencing of the four structural elements in the code. For each case, my strategy initially involved locating an enacting unit. I accomplished this in the following way. First, direct participants or others involved in the emergency usually characterize nascent social units by specific names or associate collective action with already existing groups and organizations. This was of critical importance to me for establishing the identity of an organization and documenting its domain. Second, both domains and tasks are sometimes evidenced by official declaration, sometimes by written statements, and sometimes only through interpersonal exchanges. But in all cases, I documented domains and tasks with descriptions of formal or informal communications—what Mayhew (1980) termed the organization of information—as they were relayed to the interviewers by participants and informants, recorded in documents, or observed in the field. Third, the presence of resources and activities is evidenced in the archives by their numbers, location, and timing. The archives are often incomplete about these kinds of information. Moreover, organizational boundaries are difficult to determine when more than one social unit is engaged in the same domain. This means that identifying obvious discontinuities in the timing and location of organizational enactments is essential for specifying discrete entities.

Having located an instance of organization, the next problem was to determine the temporal ordering of the four elements at its origins. This required a reconstruction of what happened from the interviews and documents. Data on the fifty-two emergent units generally were adequate and the sequencing problem was not too difficult. A similar situation existed in those cases where domains or tasks were specified in formal planning. Where sequencing judgments were more uncertain, determining which element set off an instance of organization was less of a problem than determining the ordering of the remaining elements. The primary reason for this latter difficulty was that the quality of the archival data varied dramatically. This problem is inherent in doing disaster research during the immediate emergency period (Drabek 1970).

More important, my research strategy required making threshold judgments in very fluid circumstances. Developing precise threshold criteria was and remains an important goal. Although reliability in coding the sequencing of the elements is a continuing concern, one of the major advantages of the code is that it has built-in counterfactuals. That is to say, there are many alternatives for every judgment made about the sequencing of the elements. Attribute judgments, and debates about them, are part and parcel of taxonomic work (see Enoch 1988 for a discussion about relevant issues).

Social Networks as Organizing

A second aspect of our archival analysis involves using the structural code to describe dyadic social networks as they might represent stages in the origins of organization (1–3 elements present). To that end, we began by analyzing the networks of fifty-five organized responses from one disaster for which the DRC data clearly were the most complete. Each of the fifty-five cases became a focal organization, and all of its links to other groups and organizations during the emergency period were examined one at a time (dyadic pairs). Having identified a dyadic pair, we determined how many structural elements were represented by the network, as well as the sequencing of these elements. The resulting file yielded data on 465 dyadic pairs for the fifty-five cases, or an average of about eight networks per case (for a discussion of findings, see Francis and Kreps 1984).

Using the same format, we then analyzed archival data provided by Thomas Drabek from his earlier research on multiorganizational search and rescue responses following five natural disasters (Drabek et al. 1981). Drabek's data were potentially useful to us because he and his colleagues attempted to describe emergency social networks on their own terms, using block-modeling techniques (White, Boorman, and Breiger 1976). Drabek's interest was with the conventional use of block-modeling to describe the network as a whole. We were interested in exploring Drabek's case materials to compare dyads themselves as more elemental forms of association. From 1,122 possible dyads, we were able to identify 130 dyadic networks (12 percent) for which there was sufficient information to (1) describe the content of each relationship from informant accounts of what happened and (2) characterize the form of each relationship in terms of our structural code (findings reported in Kreps and Bosworth 1987).

Four structural questions are suggested by the code and can be asked about any social network: Does the social network involve conjoined action (A)? Does the social network involve a mobilization of human and material resources (R)? Does the social network have a division of labor that is agreed to by member units (T)? Are the units identified and legitimated externally as parts of a more inclusive entity (D)? Affirmative answers to one or more of these questions depict different structural forms (see Table 1), and such forms represent logically possible and empirically plausible stages in the origins of organization. The following examples, both following tornados, illustrate the process of organizing, but not organization itself. One- (T) and two-element (R-A) forms of association are depicted.

With little forewarning a tornado cuts a seven-mile path through a northwestern city. The state air national guard facilities and planes are badly damaged, as are blocks of homes and a trailer park. Emergency personnel begin responding even before the tornado leaves the city. Two volunteer agencies are distributing food, independently of each other. By the second day, their lack of coordination results in people missing meals. At this point an emergency coordinator meets with representatives of both agencies and suggests that they develop a more cooperative effort. The two agencies then work out an arrangement in which each will be responsible for food distribution on alternate days (T). This resolution is successful, and the two units continue to operate separately during the emergency.

A tornado whips through a midwestern city leaving a mile-wide path of destruction. Weather conditions had been ripe for a tornado, and a watch had been in effect four hours prior to impact. A well-planned warning system saves many lives, but the tornado destroys thousands of homes and several schools, and severely damages a shopping mall. Six or seven city police cars are damaged beyond repair and many others are temporarily out of commission. Within hours a state agency, headquartered several hundred miles away from the impacted city, offers to assist in any way it can. Within the next twelve hours it provides the impacted city's police department with transportation vehicles and drivers (R). Members of the state agency team up with members of the police department. The police know the layout of the city and can give their partners from the state agency directions. The latter relieve the police of driving responsibilities and assist with security and search and rescue as needed. Together the teams control curfew and looting and provide overall law enforcement (R-A). At no time is there an explicit division of labor evidenced in communications which links the police department and state agency during the emergency. Their conjoined actions continue for twelve days, at which time the state agency withdraws its resources.

Notice again that when some empirical threshold for an element has been passed, a structural form exists in time and space. As indicated in Table 1, a structural form can reveal one, two, three, or four elements of the code. By definition, only the four-element forms are sufficient for organization to exist. The above two examples obviously depict social structure, but not organization as we have defined the concept.

A crude measure of weak to stronger interdependence of network members is therefore possible: the greater the number of structural elements present in a social network, the greater the evidence of organization in process. The vast majority (90 percent) of the nearly six hundred dyadic networks we have examined have only one or two elements present; and the clear majority of these (85 percent) involve only resources (R) and/or activities (A). This means that the modal situations with these social networks involve transfers or exchanges of resources (R forms), conjoined actions (A forms), or mobilization of resources in association with conjoined actions (R-A or A-R forms). Only a minority of the nearly six hundred cases (less than 25 percent) include an implicit division of labor involving member units (T) or their symbolization as parts of a more inclusive entity (D). We suspect that social networks in disasters are not much different from social networks generally. That is to say, they are bounded by *means* (A, R) rather than *ends* (D,T) of organization (for a detailed discussion of this point, see Kreps 1991b).

In considering these examples and the other case discussed earlier, three methodological issues need to be highlighted. The first concerns the empirical independence of the elements as individually necessary conditions of organization. The descriptive power of the code—and by implication our archival analyses—becomes highly suspect if that independence cannot be shown. Tasks and domains have been termed the *ends* of organization. It is essential to show that these elements are not mere abstractions of structural means (R and A); and that the presence of either (D) or (T) does not necessitate the presence of the other (Turner 1989). Recall that the presence of either (D) or (T) must be documented in explicit communications and not inferred from tacit understandings that may or may not have been operating. This quite conservative measurement strategy ensures the empirical independence of (D) and (T) from each other and the remaining elements. Notice with the one-element example (a T form) that a division of labor between the two units is explicitly represented in communications, but there is no representation of a single entity in communications, and no evidence of mobilized resources or conjoined action.

Resources and activities have been termed the *means* of organization. In parallel fashion, it is essential to show that (R) and (A) are independent of each other; and also that the copresence of either with (D) or (T) is always uncertain rather than given. The two-element example (R-A form) illustrates both of these points nicely. The independence of (R) and (A) is evidenced by their temporal discontinuity. There is considerable uncertainty, however, about the presence of (T). To repeat: the existence of tasks must be evidenced by explicit communications which specify a division of labor. The evidence is clearly lacking in this example. To argue that there are always tacit understandings—that is, that this example must include (T) as well—is only to beg questions about what tacit understandings are and how they come into being (Collins 1985 on Garfinkle). The fact that tasks and domains must be observed in communications is an important strength of the methodology, not a weakness.

A second issue relates to determining existence thresholds for each of the four elements. With respect to *ends* (D, T), in the first two examples references to (D) and (T) in communications are explicit, but our use of the phrase "collective representations" from Durkheim (1938) begs the question of how widespread they have to be. The emergency periods of natural disasters are like simple societies in the sense that communications within and between social units are often highly circumscribed. This means that the boundaries of collective representations are identified more easily in the events we studied than would be the case with regional or national disasters.

With respect to *means* (R, A), any transfer or exchange of resources between network members meets an (R) threshold. Judgments are not based on *our* assessment of the importance of the mobilization to the network as a whole or to its member units. Instead, we assume that any transfer or exchange of resources mentioned by interviewees was deemed by *them* to be important. This nontrivial mobilization of resources points to a threshold where the origins of new organizations from existing ones becomes a genuine possibility. The temporal and spatial criteria of (A) are not standardized from one case to the next and, therefore, existence thresholds are less precise for this element as opposed to the others. It is clear from the examples, however, that our judgments about temporal and spatial boundaries of conjoined action are consistently narrow.

A precise numerical criterion of conjoined action is perhaps the most difficult one to identify and use. We do not consider a meeting involving single representatives of network members conjoined action. Such a meeting is below a threshold of pooled activities (Thompson 1967). Sustained involvement (in delimited time and space as noted earlier) of larger numbers or proportions of people from network members may be conjoined action, but we think a criterion of proportional involvement applied to member activities themselves is a better criterion than the sheer number of people engaged. The two-element example (R-A form) highlights the difference we have in mind. If a criterion of proportional numbers is applied, (A) is not present because the state agency was a large organization and only a few of its personnel went to the disaster area. However, if a criterion of predominant disaster related activities of member units is applied, (A) is present because the state agency was not involved in the disaster in any other way.

A final methodological issue relates to the fact that our studies of the origins of organizations and social networks, respectively, involved two different operationalizations of the structural code. In analyzing the DRC archives initially, I first searched for and located organized responses, and then reconstructed how they were created. But in studying social networks, no broader encompassing organization was assumed to exist. Instead, we sorted social networks of existing organizations into dyadic pairs, and then tried to find evidence of organizing as specified by the code. I uncovered many instances of organization (four-element forms) with the first approach. We uncovered no instance of organization with the second approach, even though we had hoped to do so. Although we do not think the absence of four-element forms of association identified from the data on dyadic social networks is an artifact of methodology, that possibility certainly exists. Studying organization as entity and process at the same time is fraught with some methodological and broader epistemological dilemmas that need to be given sustained attention (Enoch 1988).

The Restructuring of Organization

A third aspect of our work with the DRC archives involves using the code to describe changes in existing organizations during and after disasters. Data production has seemed less difficult here, compared to the previous applications (perhaps because in these cases the existence of organization is not at issue). Moreover, the findings have been more detailed because the archives yielded new measurements on temporal properties of change (see Linn and Kreps 1989; Farmer 1989; Kreps, Crooks, and Linn 1987). In what follows, I discuss the questions being asked and the key empirical dimensions of restructuring being measured. In so doing, I highlight (from Warriner 1989) what my colleagues and I think is the key issue being raised by this particular application of the code.

Two structural questions are at issue: First, when an organization restructures, what changes and how does it happen? Second, is there any patterning of changes over time? The first question is relatively easy to answer. If core properties of any organization are the four elements of the code, then we should be able to describe organizational changes in terms of that code. The following two examples are shortened summaries of the first and seventh restructurings by a mental health agency. The seven restructurings occurred over three months. The first one is characterized as an A-T-R form of restructuring and took place during the immediate emergency period. It was followed by five additional restructurings that, although not described here because of space limitations, were of forms designated as A-T-R-D, T-R, T-R-A, R, and R-A. The seventh and last restructuring took place about three months later and resulted in the merger of the agency with another local mental health unit.

First Restructuring (A-T-R)

The agency begins responding to needs for disaster related information shortly after impact (one hour). By chance, it is one of the few units in the impact area that still has telecommunications. That circumstance, combined with public knowledge of its hotline service, results in the agency responding to phone calls related to the disaster (A). The volume of calls is quite high, and there is initial uncertainty about what to do about them. A decision is made to suspend all routine operations and to develop new procedures related to monitoring calls, referring requests or offers of assistance to the proper authorities, and transmitting information by phone or by messenger (A-T). The focus of the initial message service relates to maintaining linkages among mental health agencies. Members of the unit then recruit additional volunteers because they are needed to provide the expanded service (A-T-R). The development of what amounts to a small scale message center and referral service during the emergency period takes less than one full day. The service remains in operation for about two weeks, at which time most local telecommunications have been restored.

Followed Sequentially by These Five Restructurings

A-T-R-D, T-R, T-R-A, R, R-A

Seventh Restructuring (D-T-R-A)

Approximately six months prior to the disaster, the agency had begun negotiations with another local mental health unit in the hope of merging. Some tentative plans had been made for this, but the proposed date of the merger was over a year away. The catalyst for this initiative was criticism of the agency by its local oversight board. While the criticism subsides noticeably following the disaster, it reappears about three months after the event. As a result of its favorable evaluation of the agency's disaster related activities, the leadership of the proposed partner to the merger suggests that the consolidation should be pushed ahead. A formal meeting between representatives of the two potentially merging units takes place one day after the criticism of the oversight board re-surfaces. During that meeting, the general framework under which that new unit would operate is determined (D). Then over a period of about 14-15 days job descriptions and job assignments are negotiated and agreed upon (D-T). Late in the negotiations (last few hours), final agreements are reached on the location of the facility for the merger and a necessary slight reduction in personnel (D-T-R). The merger is created without conflict between the two units and with the approval of the oversight board. The new agency begins operations about 17 days later (D-T-R-A).

Note that with these seven restructurings all possible degrees of change (one- to four-element changes as depicted in Table l) are documented, with the last restructuring describing the absorption of two existing organizations into a completely new one.

In archival studies of two events (a tornado and flood), we have identified 167 restructurings by 42 organizations. For 27 of these 42 organizations, we were able to document chains of restructurings such as the one previously described. The average number of forms in the chains is about five, and the range is two to eighteen. These chains are particularly important because they provide the opportunity to describe the patterning of organizational changes over time. Although our efforts are quite preliminary, we have isolated four temporal properties of social structure in our archival studies (Wallace 1983).

Two of these properties are relevant for every discrete episode of change we have documented. *Order* is defined as the number (one to four) and type of element changes (see Table 1). *Timing* is defined as the elapsed time from an identifiable catalyst (Dubin 1978) to change in the first appearing element, and to change in any subsequently appearing elements. The remaining two properties come into play when a chain (or series) of restructurings has been documented (Collins 1981). *Rhythm* is defined as the regularity in the *order* of element changes across a chain of restructurings by the same organization. Of particular interest to us here is possible patterns in the degree of change (number of elements implicated) as well as possible complementarities between changes in ends (D, T) and means (R, A) of organization. *Periodicity* is defined as regularity in the *timing* of element changes across a chain of restructurings by the same organization. Of particular interest to us here is the actual pacing of change. Are there greater time lags between catalysts and initial element changes (perhaps because of possible internal resistance) than those that transform later (perhaps because a kind of structural momentum builds once change is underway)?

These and many other questions can be addressed once the aforementioned temporal properties have been measured. My purpose here is not to summarize what we have found thus far in studies of disasters (see Linn and Kreps 1989; Kreps, Crooks, and Linn 1987). Rather, it is simply to emphasize that the archives often can yield the kind of data required to unravel the patterning of social structure. As Warriner (1989) points out, the real challenge for sociologists is to discover regularity in a constant process of social change rather than simply assuming that regularity from the outset.

In a sense, the emergency periods of disasters are poor contexts for measuring such concepts as rhythm and periodicity because everything happens so quickly. That is why I believe that the conceptualization and measurement of organizational change proposed in our work may be easier to exploit in historical studies of larger structures (for an empirical example, see Farmer 1989). The documentation of change—at least for more recent history—is very good. Consensus can be achieved about what are significant changes. And given the generally broad time periods of historical transformation, the feasibility of identifying longer chains of restructuring is quite high. Once that is accomplished with respect to this or any other structural code, the prospect of discovering intriguing relationships between and among the above four properties will be excellent. We are not at all sure, for example, that what we have called chains are not more appropriately termed *sequences*, only parts of which have chainlike properties. We also think that with longer chains or sequences of restructurings, it is possible to identify rhythmic cycles of various types. Perhaps some of these cycles exhibit periodicity, but others do not. These matters are fascinating to ponder, but there can be no substitute for the major descriptive spadework that needs to be done.

Organizing and Role Enactment

The final, ongoing aspect of our work with the DRC archives actually began prior to the aforementioned studies of restructuring and social networks.

Susan Bosworth and I began a collaborative effort to bridge two levels of analysis. Specifically, we wanted to see how patterns of organizing, as described by the code, were related to role enactment, measured through new archival data production. Some preliminary work on thirty-eight organized responses resulted in a publication (Bosworth and Kreps 1986) that offered an empirical strategy for measuring role making and role playing during an emergency, and also suggested that our perspective on organizing became more powerful when the concept of role was added to it.

A constructive dialogue on the article between Ralph Turner and ourselves then followed (see Turner 1989; Kreps and Bosworth 1989). The result was a follow-up study and new data production involving Bosworth, myself, and three graduate students (Mooney 1989; Russell 1989; Myers 1990). That study is just now being completed. As was true with the other aspects of our research program, my purpose in this chapter is not to summarize findings from our role studies. Instead, I want to communicate how role enactment is being examined and what this examination says about the strengths and limitations of the archives for research on social structure.

Our conception and measurement of role enactment has benefited much from the guidance provided by Turner's article titled "Strategy for an Integrated Role Theory" (1980). As in all of his work, Turner attempts to reconcile a flexible conception of social structure with Meadian principles (Blumer 1969; Stryker 1980; Turner 1962, 1968, 1985). In the article, Turner highlights three processes he believes must be taken into account in any theorizing about social structure and role enactment: role allocation, role complementarity, and role differentiation. We are specifying them empirically as the status-role nexus, role linkages, and role performance. Our conception of the three processes is discussed briefly, and then a case example shows how we measured them.

Role Allocation (Status-Role Nexus)

Turner (1980) defined *role allocation* as the assignment of particular persons or categories of people to particular roles. The theoretical issue for us is how one system of roles (an organized response) evidences stability or change of role allocation vis-à-vis the more inclusive system of roles (an impacted community and broader society) of which it is a part (Bates and Harvey 1975). We specify the role-allocation process as consistency versus inconsistency of the status-role nexus from pre- to postdisaster time periods.

A *status* is a social position that has expectations of behavior attached to it. These expectations are referred to as roles (Merton 1957). The question we address is as follows: Is allocation of a postdisaster role to a given person expected given one or more positions in that individual's predisaster status set? Where allocation is expected (e.g., a mayor taking command at an emergency operations center), there is consistency of the status-role nexus; where it is not expected (e.g., a local doctor doing the same thing), there is inconsistency of the status-role nexus.

Accordingly, we initially must identify and label a primary postdisaster role for each participant in an organized response selected for study. We accomplish this by *naming* a role as closely as possible to what the incumbent is actually doing (e.g., search and rescue leader as opposed to search and rescue worker). We then comb the archives for information on the incumbent's predisaster status set to identify that individual's *occupation*. Next we must determine whether or not the postdisaster role would be expected of someone having that occupation (e.g., a county sheriff serving as a search and rescue leader in a rural area).

Obviously occupation is only one of many social positions in any individual's status set, but we focus on it for several compelling reasons. First, there is no question that occupation is a major status for the kinds of people chronicled in the archives. It is equally clear that such occupations link these individuals pointedly to at least a broader community role system. Second, although the archives virtually never provide a complete accounting of an individual's status set, they very regularly include information on type of employment or specific job title. The potential to standardize measurement from the archives is therefore fairly good. Third, if a participant selected for analysis is not employed for pay, we usually are able to identify another, analytically comparable primary status (e.g., student, housewife) that will allow for some judgment about consistency or inconsistency of the status-role nexus. Absent data on either postdisaster role or predisaster status (occupation in the vast majority of cases), the case is considered uncertain on this dimension for a given participant.

Even assuming there is information on occupation or some other equivalent status, measurement error is a concern because our judgments might simply be wrong (e.g., would you expect a pathologist to be providing medical treatment?). Where respondents do not provide their own assessments of status-role expectations, or when validity does not appear to be obvious, we contact people in our own community having the same job titles as the participants and ask them if the given postdisaster role we have observed would be expected of them. We use the same strategy for measuring role complementarity. Obviously the kind of problem we face with the archives could be largely overcome in the field simply by asking incumbents the same direct questions.

Role Complementarity (Role Linkages)

Similar to role allocation, some conception of a broader role system must be assumed in discussing what Turner refers to as *role complementarity*. By role complementarity he means that every role in a system is formed in such a way that it is interdependent with at least one other (alter) role (Bates and Harvey 1975). The theoretical issue for us is the stability or change of role complementarity from pre- to postdisaster time periods. Any time we observe occupationally based continuity of role linkages between selected participants and those they interact with during the emergency period, the complementarity of the role system from pre- to postdisaster times remains stable (e.g., a mayor and a police chief). However, any time we observe occupationally based discontinuity of role linkages, the complementarity of the role system changes in response to these same circumstances (e.g., a medical technician and a telephone company executive).

For any selected participant in an organized response, we make judgments about continuity or discontinuity of role links whenever there is evidence of direct contact between the participant and someone else during the emergency period. The contact can be within or outside the boundaries of the organized response. Any identified direct contacts are broken down into four general categories: *intra-role* (links with people performing the same role as the selected participant, and within the same organized response); *inter-role* (links with people performing different roles than the participant but within the same organized response); *intra-role boundary spanning* (links with people performing the same role as the participant, but not within the same organized response); and fourth, *inter-role boundary spanning* (links with people performing different roles than the participant, and not within the same organized response).

All judgments about role complementarity are based strictly on whether predisaster occupational statuses of any incumbent pair are or are not linked routinely in everyday circumstances. The fact that role incumbents may or may not know each other as individuals is irrelevant. Each identified link is judged continuous, discontinuous, or uncertain depending on the availability of occupational data on the incumbent pair.

Role Differentiation (Role Performance)

Turner (1980) considered role differentiation (and accretion) as perhaps the most basic social process. By that he means that in any situation where people interact there is a natural tendency for their behaviors, sentiments, and motives to become distinguishable as discrete entities called roles. Once roles become differentiated, it logically follows that the behavior, sentiments, and motives that appear subsequently in similar situations will become patterned (i.e., they will become assigned to now existing roles). The theoretical issue for us is stability or change of role differentiation from pre- to post-disaster periods. We focus on the actual performance of postdisaster roles, which is quite different from their allocation to particular categories of individuals (status-role nexus).

Two issues must be addressed in measuring postdisaster role performance: *incumbency*, which can be either existing or new; and level of *improvising*, which can vary from low to high for either existing or new incumbents. We determine incumbency initially for any participant selected for role analysis: if the named postdisaster role is represented anywhere in the person's current predisaster status set (e.g., a formal Salvation Army volunteer staffing a food kitchen), incumbency is termed existing; but if it is not represented anywhere (e.g., a merchant next door dropping in and providing the same service), incumbency is termed new. That determined, we then must make some judgment about the level of improvising.

We have developed the six-fold typology depicted in Fig. 1 for examining role performance. The terminology is consistent with our previous dialogue with Turner (1989). The three types on the left side of the figure depict varying degrees of improvising by an existing incumbent. The three types on the right side of the figure depict parallel degrees of improvising by a new incumbent. Social structure and role invention are exhibited as the twin foundations of role differentiation. The six types of role performance are specified as follows:

(first level of improvising)
Formal Role Enactment: existing incumbent, conventional performance
Role Prototyping: new incumbent, conventional performance

(second level of improvising)
Working Role Enactment: existing incumbent, improvised performance
Role Redefinition: new incumbent, improvised performance

(third level of improvising)
Radical Role Transformation: existing incumbent, fundamental change in performance
Radical Role Redefinition: new incumbent, fundamental change in performance

The lowest level of improvising is represented by *formal role enactment* and *role prototyping*. Both types are characterized by familiarity with role expectations and conformity to them during the disaster. The middle level of improvising is represented by *working role enactment* and *role redefinition*. Both types are characterized by some deviation from the conventional enactment of an existing role that is based on incumbent interpretations of the role or the special circumstances in which it is being enacted. The highest level of improvising is represented by *radical role transformation* and *radical role redefinition*. Both types are characterized by major deviation from the conventional enactment of a given role that is based, once again, on incumbent interpretations of the role or the circumstances in which it is being enacted.

Figure 1
Role Differentiation: Types of Role Performance

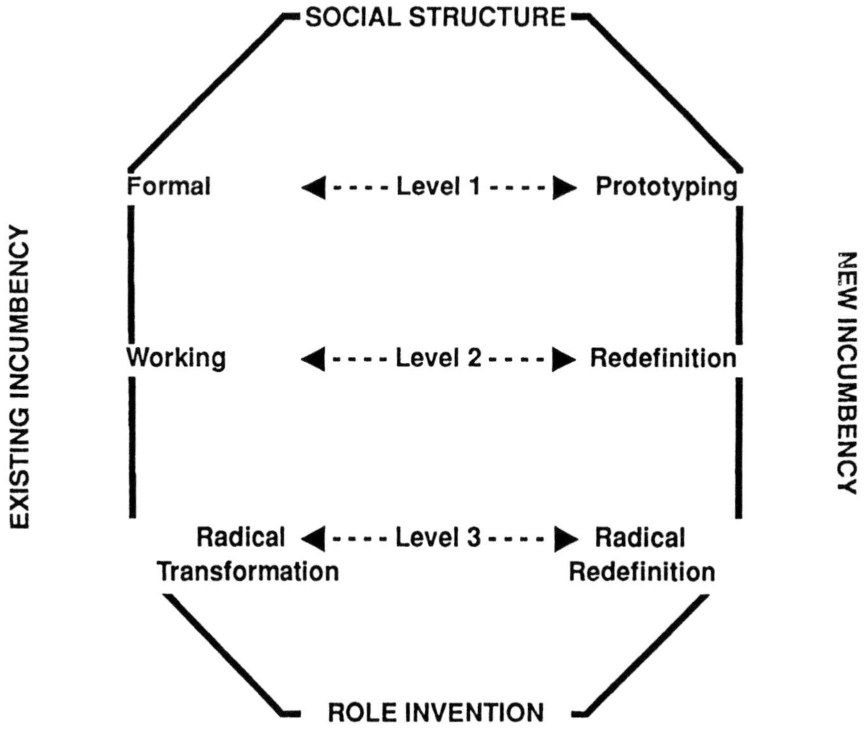

Basic to the typology depicted in Fig. 1 is an assumption that although *no* role performance takes place in a social vacuum, *all* role performance involves some degree of improvising. Both social structure (order) and role invention (action) are thereby affirmed (Alexander 1982). Keeping that symmetry in mind, formal role enactment by an existing incumbent and role prototyping by a new incumbent are analytically comparable in the sense that role differentiation has already happened. Working role enactment by an existing incumbent and role redefinition by a new incumbent are analytically comparable also, but now in the sense that redifferentiation is occurring. Radical role transformation by an existing incumbent and radical role redefinition by a new incumbent are comparable as well, but in the sense that the redifferentiation now is quite extreme.

The following case description illustrates how we implemented the aforementioned conception and measurement of role enactment. Raw data on this particular case were extracted from six interviews. The case describes the actions of a structural engineer who becomes the leader of a disaster operations unit called "disaster control" following an earthquake. With respect to the structural code, the engineer plays an important role in an organizing process that begins with an elemental division of labor (T), followed by the rapid mobilization of resources (T-R) and performance of activities (T-R-A). The actual naming of the organization and the establishment of its domain (T-R-A-D) later completes the origins of this short-lived but very important emergent organization. The organization performs critically needed coordination activities during the five days of its existence.

> A state's largest city is impacted by an earthquake, the worst in the state's history. The superintendent of the city's building construction and maintenance department, a structural engineer and architect by training and experience (we will refer to him as SEA), becomes a key community leader in the immediate emergency period. Prior to the disaster, SEA is in charge of the maintenance of eighty-six buildings in the city and has three engineers, six mechanics, and one administrative assistant working directly under him. His primary job is to oversee the inspection and repair of these buildings. He also spends time reviewing architectural plans for new buildings, makes corrections as necessary, and plans the division budget. SEA's unit is only one of several in the city's Department of Public Works.
>
> The earthquake hits at approximately 5:30 P.M. on a Friday afternoon, a time when many people, including SEA, are on their way home from work. When the earthquake subsides, SEA again heads for his home and passes a friend with whom he had planned to have dinner. He learns from the friend that his home and his wife are safe, and immediately turns around and heads back toward town. He says that as a superintendent in the public works department, his services will be needed.
>
> SEA arrives at the Public Safety Building at about 5:45 P.M. and finds what he calls mass confusion. He goes to a desk and suggests that people be assembled into teams that might be useful for search and rescue or damage control. The teams are then formed and made up of volunteers, police and fire department personnel, city public works personnel, as well as national guardsmen. Within an hour of the earthquake, these teams are out in the impact area performing predesignated tasks. At about this time, a central desk in the building is being manned by members of the police department. People are asking questions that these policemen cannot answer. SEA goes to the central desk and starts answering questions, giving specific directions, and making specific decisions.
>
> The organization works through the night and becomes identified as "disaster control" by Saturday afternoon. It is characterized frequently as the organization most involved in the disaster response. As national guard and army rescue teams become involved, they choose to receive their instructions from SEA and his orga-

nization. For the next five days SEA leads what is described as an effective response to the earthquake. In addition to forming search and rescue and damage assessment teams, the organization becomes involved in creating a filing system to catalogue all of its volunteers, commandeers fuel and other supplies, installs temporary city generators in public buildings, makes temporary toilets, opens shelters, provides security guards, and performs many other tasks. All of the disaster-related activities of "disaster control" are led by SEA in the name of the city, and without any official authority.

The legitimacy of this emergent organization continues to be evidenced by its role in coordinating the activities of the military, state and local government agencies, and private voluntary agencies. Moreover, several members of the city government, such as the city manager, city attorney, and members of the police department work out of "disaster control" for several days. It is also noteworthy that SEA, who has never previously attended city council meetings, is asked to be present at every emergency meeting following the earthquake. He plays a prominent role in these meetings, even though he has no place on the council and no real connection with it.

By late Tuesday the council decides formally that the city's emergency response should be handled by the various city agencies and departments that were intended to handle such problems. They decide also that "disaster control" should begin fading out of operations. By Wednesday afternoon "disaster control" is sending all requests for assistance to various city departments. The transition is difficult at first because local people have become familiar with its existence and coordination domain.

With respect to *role allocation*, we judge that SEA has an inconsistent status-role nexus because neither expectations tied to his predisaster occupation nor his specific job title as building maintenance superintendent include being the leader of this emergent unit following the disaster. Although certainly SEA is expected to be involved in some capacity because of possible damage to municipal buildings, the organization he heads performs a quite wide-ranging community coordination function that goes far beyond the mandate of his office. Had the mayor, city manager, or civil defense director played this leadership role, our judgment would have been different. But this case points to change rather than stability of role allocation. An individual not expected to enact a role, in effect, has risen to the occasion. The reallocation appears to be functional, although the archives indicate that there is some opposition to SEA's leadership. The reallocation seems to be tenable in the sense that SEA expresses intrinsic satisfaction about what he is doing.

With respect to *role complementarity* (second dimension), the worksheet presented in Table 2 classifies our judgments about SEA's predisaster status links by categories of his postdisaster role links. All of the city employees listed in the table enact the same status roles pre- and postdisaster. Of those remaining on the list, the union business agent directs the secretarial staff of "disaster control," the bill collecting agent serves as its resources coordinator, the executive secretary (wife of the bill collecting agent) does clerical work, the local business owner (a former civil defense director) serves as local

Table 2. Predisaster Status Links by Category of Postdisaster Role Links for Structural Engineer/Architect (SEA)

Predisaster Status Links	Link Continuous
WITHIN ORGANIZED RESPONSE	
Intra-role	
none	---
Inter-role	
Union business agent	no
Bill collecting agent	no
Police captain	yes
Police chief	yes
City attorney	yes
City planner	yes
Public works building construction and maintenance foreman	yes
Public works plumbing and heating foreman	yes
Assistant to SEA (engineering side)	yes
Public works maintenance person A*	yes
Public works maintenance person B*	yes
Public works maintenance person C*	yes
Building inspector, public works	yes
Chief of surveys, public works	yes
Executive secretary	no
BOUNDARY SPANNING	
Intra-role	
none	---
Inter-role	
Occupation unknown	unc
Business owner	unc
Mayor	yes
City manager	yes
Administrative assistant to city manager	yes
Army colonel	no
National guard member A*	unc
National guard member B*	unc
Public works street and sewer general foreman	yes
Public works supervisor of design section, engineering division	yes
Civil engineer, public works	yes
Professor	no

	Number of postdisaster incumbents	N=27
	Number of postdisaster roles represented	N=24

Incumbent Links:	Continuous	67%	(18)
	Discontinuous	18%	(5)
	Uncertain	15%	(4)
		100%	(27)

*Indicates that the incumbent played the same postdisaster role.

civil defense director, the army colonel provides security guards, the two national guardsmen do security work, the college professor serves as a search and rescue coordinator, and the person whose occupation is not identified (occupation unknown) is a member of a mountain rescue team.

Overall, SEA has a minimum of twenty-seven direct links with post-disaster role incumbents, and there are many more (e.g., several city councilmen) for whom we could not specify precisely the extent of direct contact. Some fifteen of the identified direct contacts are interrole links within the organized response, and twelve are boundary spanning interrole links. We judge eighteen of the twenty-seven interrole links to be continuous (67 percent), five to be discontinuous (18 percent), and four to be uncertain (15 percent). The relatively high proportion of continuous links is clearly job related, with many links coming from the department of public works itself, and others from SEA's specific responsibility for maintaining municipal buildings.

Of the five discontinuous links, the status role of secretary clearly is the most difficult to judge because it can be interpreted so broadly. In this instance, the person is an executive secretary for a federal aviation agency. Although all people in supervisory positions have regular contact with secretaries, we judged that routine contact between a local department of public works superintendent and an executive secretary for a federal aviation agency is highly unlikely. Of the four uncertain contacts, three result from the lack of information on employment (two national guardsmen and occupation unknown). Finally, the local business owner link is considered uncertain because we have no information on the type of business he operates. As part of his job, SEA would have regular contact with some vendors. But in the absence of specific information, we thought it inappropriate to define regular contact with local business owners so inclusively.

Although this case evidences a facilitating mix of continuous and discontinuous role linkages, the general pattern is one of stability rather than change of role complementarity. Regardless of one's judgment (earlier) about role reallocation, there is no question that the stability of role complementarity facilitates greatly the activities of this emergent organization. Although new relationships are being improvised on the spot, they extend from established relationships that are longstanding. In this sense, the role dynamics are both functional for "disaster control" and quite tenable for its leader.

With respect to *role differentiation* (third dimension), we judged SEA to be engaged in role redefinition (new incumbent, improvised performance). He states explicitly in the archives that being the leader of a community coordination unit like disaster control was a new role for him (incumbency). As described, his role performance requires common sense and decisive action in the absence of authority. Therein lies the rationale for our characterization of his role performance. SEA orders specific actions (e.g., commandeering fuel and other supplies, controlling access to impacted areas) for which even the city government may not have authority at the time. Thus, although SEA's role enactment may or may not entail radical role redefinition, there is no question that his level of improvising is far greater than role prototyping. From the standpoint of disaster control, at least, the role redifferentiation that occurs seems to be both efficient and effective (functional). From SEA's own standpoint, there appears to be no question that his role redifferentiation is essential (and personally tenable) for getting the job done.

In conclusion, this case reveals the uniqueness of the three dimensions of role enactment we have isolated in our research. In Turner's terms, the evidence for both role reallocation and role redifferentiation is quite compelling. But so also is the evidence for relative stability of role complementarity. In reconstructing this and many other cases from the archives, the difficulty of predicting one dimension of role enactment from knowledge of the others becomes very apparent. Given our intense concern about isolating unique properties, we are quite encouraged by what such cases seem to be saying about role dynamics. However, the measurement requirements related to these dimensions are quite demanding. An explicit postdisaster role for each participant must be determined and labeled. The predisaster occupation or other central role of the participant must be identified and, ideally, so must his or her broader role repertoire. All postdisaster role networks of the participants must be located, and the occupation or other central status of counterpart incumbents must be specified. Finally, relatively detailed descriptions of postdisaster role performance are required so that judgments about incumbency and improvisation can be made.

Can the archives yield all of the above? The short answer is not in any complete sense. That is why we have yielded thus far only 257 cases with more or less adequate data on role enactment. But those cases have provided a wealth of information on the dynamics of role during the emergency period of disasters. They also imply an effective strategy for much more structured data collection in the future. Simply put, we are now in a position to measure with considerable specificity stability and change of role enactment in field studies of emergency functioning. The DRC archives have pushed us in the right direction.

Conclusion

Studying role performance, organizational change, social networks, and the origins of organization during emergencies and disasters involves use of basic sociological concepts. When all is said and done, how helpful have the archives been in conceiving and measuring these concepts and the social processes implied by them? There are obvious limitations to the archives and the field research from which they have been produced. These limitations concerned me when I worked at the DRC, and they were key points of criticism when I was asked several years later to evaluate the major strengths and weaknesses of the DRC research (Kreps 1981).

I noted at a conference in 1979 (Wright and Rossi 1981) that DRC's work was unabashedly exploratory. By that I meant that the DRC had been more inclined to break new ground and empirically illustrate ideas than to test theoretical models. To that end, its research had relied predominantly on unstructured and semistructured data collection techniques. My depiction of DRC in 1979 is as apt today as it was earlier. Since its inception, then, the primary approach of DRC research has been to study a wide range of organizational and social network response patterns in different types of disasters. Research results generally have taken the form of empirically based descriptions of different types of organizations, global conceptions of the actions of various types of organizations, and general discussions of organizational and community problems related to disaster response. On the applied side, a key result has been a coherent planning perspective, one which defines improvisation and modest preparedness as central tools of emergency management (see Kreps 1991a for an elaboration of the DRC-based perspective). On the theoretical side, an important result has been incorporation of disaster studies into the mainstream of collective behavior and (to a far lesser extent) organizational theories.

The obvious weaknesses of DRC research must never be overlooked (see Kreps 1981). First and foremost, the quality of descriptions of disaster events has been very uneven, and this probably has resulted from the DRC's desire (and often funding requirement) to study as many events as possible. With some exceptions, thoroughly detailed case studies have been sacrificed so that a greater number of events, communities, and organizations can be studied. A second and related methodological point is that links among conclusions, findings, and data collection are somewhat vague in many DRC publications, particularly those containing general discussions of organizational problems and what to do about them. To discuss general patterns and problems is to raise questions about their marginal distributions over the large number of events studied. Finally, the DRC's research has evidenced only a small number of attempts at quantitative analysis. In part, this relates to the difficulty of measuring key analytical properties of organization in other than qualitative terms. But it also results from the DRC emphasis on exploration and description rather than hypothesis testing.

Ironically, my own researches with data produced by the DRC field studies provide effective rejoinders to each of the concerns that I am on record as having raised. With respect to the first criticism, although the quality of descriptions is certainly uneven, my work has shown that a good number of rich descriptions can be extracted from the archives. With respect to the second criticism, links between conception and measurement have become increasingly precise in our archival work over time; and comparisons of events, communities, and organizations have become more feasible as our respective numbers of cases continue to increase. With respect to the third and last criticism, we have engaged in several model-building efforts (e.g., Bosworth and Kreps 1986; Saunders and Kreps 1987) that merge qualitative descriptions of the content of organizing with quantitative analyses of its forms.

So my feelings about the overall value of the DRC data archives and field research methods that generated them are very positive. There have been important theoretical and practical yields from DRC research, and a large number of people have received sociological training in the process. There is no question that, overall, the field research could have been executed with more highly structured data production. Had that been done, however, it is a quite open question whether or not the theoretical and practical yield would have been any greater. It is important to note that my colleagues and I have the luxury of knowing what to look for in the archives. The original DRC field teams were often scrambling as best they could to document what was happening in real time. It must also be remembered that the emergency periods of disasters are not easy to chronicle in the best of circumstances, and those doing the fieldwork at DRC were and remain primarily graduate students. Yet in spite of these limitations, hundreds to thousands of very useful descriptions of what active social units and participants are doing have been obtained. Such is the stuff of empirically grounded theory.

Public Risk Communication

COLLEEN FITZPATRICK and DENNIS S. MILETI

The communication of climatological, geological, and technological hazards risk information to the public is a continual effort nationally. Such risks include, for example, hurricanes, floods, volcanic eruptions, earthquakes, and transported or stored hazardous materials. Although these risks vary in character, they are similar in that they can result in low-probability/high-consequence disaster events.

Efforts to communicate information generally fall into two categories. The first, hazards communication, is principally concerned with informing citizens and communities about the nature of the risk associated with, for example, living near a hazardous waste site or along the hurricane vulnerable gulf coast. The second category of risk communication addresses risk in times of impending emergencies and disasters and is best cast as warnings communication. Such information efforts communicate warnings to an endangered public to elicit public protective actions. Central to both warnings communication and hazards communication are the perceptions that the public holds and forms about the risk.

Enough research evidence exists to suggest that both warnings and hazards communication are needed to help the public perceive hazardous situations accurately (Kunreuther 1978; Mileti, Hutton, and Sorensen 1981; Turner, Nigg, Paz, and Young 1981). Although extensive research on warnings communication has been accomplished, much work has yet to be fulfilled on hazards communication. Few formal empirical attempts to measure the impact of nonemergency public hazard education on risk perception and behavior have been performed, and the conclusions that can be drawn from existing studies are, at best, unclear. Roder (1961), for example, investigated the impact of risk education on perception by distributing floodplain maps to residents of Topeka followed by measuring their perceptions of risk related to residing in or near floodplains. The study concluded that the maps had no effect on public hazard awareness. The effect of a tsunami hazard public education campaign was examined by Haas and Trainer (1974), who concluded that no significant changes occurred either in public knowledge about the hazard or in risk response as a result of the campaign. They did conclude, however, that public risk perception was elevated as a result of mass media and personal contact. Waterstone (1978), in assessing the effectiveness of a public flood brochure, arrived at a different conclusion, noting that the brochure increased hazard knowledge and awareness as well as encouraged the adoption of family emergency plans. In contrast, Ruch and Christensen (1980) determined that the effect of a public hurricane awareness program was actually to decrease the risk that the public perceived. Palm (1981), in a study of the impacts of the 1977 Alquist-Priolo Act disclosure requirements, concluded that earthquake hazards information had little impact on public perception of risk. In contract, Baumann (1983) found that public flood education not only increased hazard awareness, but also increased public flood mitigation behavior. This set of contradictory research findings on the impact of public hazard education programs suggests that risk education can enhance risk perception in some limited ways, decrease risk perception, or have no effect. Additionally, results concerning public behavior in response to hazards communication is also mixed, ranging from no effect to encouraging mitigation and preparedness.

In contrast, studies on emergency communications and public warnings of impending disasters are more numerous and conclusive. Research on public response to warnings began in the 1950s as part of a research program in the National Academy of Sciences (NAS) and focused on both natural and technological emergencies. Descriptive research was continued in the 1960s by individual social scientists. Later, investigations of public warning response placed less emphasis on describing the response, and focus was placed more on discovering how demographic factors (e.g., sex or age) covaried with public risk perception and behavior. Most current studies attempt to model the impact of complex sets of factors and their interaction on risk perception and response to risk communicated in disaster warnings. This body of sociological research suggests an underlying general model for effective public risk communication. It is the purpose of this chapter to synthesize the results of research on the warning process.

Warning as a Social Process

A major conclusion from this body of work is that response to warnings of impending disaster is not well characterized by a stimulus-response model; for example, it would be an oversimplification to suppose that if one person is generally fearful of a hazard, while another is not, the former would "over-respond" to risk communicated in warnings, and the latter would "under-respond." Instead, people who receive warnings attempt to form personal definitions about the risk and what to do about it. These definitions are shaped in several phases in which the individual: (1) hears, (2) understands, (3) believes, (4) personalizes, and (5) decides and responds. Factors that affect his process are discussed next.

The first stage of public response to hazard warnings is hearing that there is an emergency, typically through a public warning like a siren or a worded message (Mileti and Sorensen 1987). It cannot be assumed, however, that when a public emergency warning is broadcast, everyone will hear it. Even when it is physically possible for people to hear the warning, various factors may inhibit a message from actually being heard. People may fail to listen because of habituation (for example, they never really listen to television) or selective perception (for example, they hear only what they want to hear because a predisposed bias to maintain a routine in their lives). Regardless of the reason, the failure to hear a warning generally precludes or at least delays self-protective action (cf., Anderson 1969; Turner et al. 1981; Lardry and Rogers 1982; Perry and Lindell 1986; Bellamy 1987; Tierney 1987).

Second, once heard, the emergency information (or warning) must be understood. This does not mean understanding what is heard, but rather personally attaching meaning to the message. Meaning or understanding can vary among people, and these varied understandings may or may not conform to the meaning intended by those who issued the warning (cf., Haas, Cochrane, and Eddy 1977; Foster 1980; Perry, Lindell, and Greene 1981; Lehto and Miller 1986). For example, one person may understand a flood warning as referring to a high wall of inundating water, whereas another may visualize ankle-high runoff. Ashfall may be construed as a suffocating, blanketing coverage, or as a light dusting of powder. A 50 percent probability of occurrence may be interpreted as certainty by some or unlikely by others. A singular warning message might be understood as involving no risk by some, but as a threat to human life by others. In each of these examples, people attach meaning to a message. Depending on individual perception of risk, however, each person understands the message to mean something different.

In addition, understanding is connected to people's frames of knowledge and reference. It may be difficult for people to understand a natural hazard warning when they do not understand much about the hazard. In this sense, understanding also defines and bounds perception of risk and what to do about it. Emergency information must thus be able to provide the public with accurate and common understandings of the situation in an emergency. The differing understandings that might emerge among people if an emergency occurred can be avoided with proper planning.

Third, appropriate action is also encouraged if people believe that the warning is real and that the contents of the message are accurate. As with hearing, belief in the warning message may vary among a public at risk (cf. Clifford 1956; Wallace 1956; Demerath 1957; Williams 1957; Fritz 1961; Moore, Bates, Layman and Parenton 1963; Drabek 1969; Mileti 1975; Mileti, Drabek, and Haas 1975; Quarantelli 1980; Yamamoto and Quarantelli 1982).

Fourth, people also consider the implications of warnings for themselves and their social group (usually their family). If people do not think that emergency information was meant for them (the "it-can't-happen-to-me" syndrome), they likely will ignore it. If people think they are the intended targets of emergency information (the "it-must-be-me" syndrome), they may act accordingly. Personalization can lead to both underresponse and overresponse in emergencies (Perry, Greene, and Mushkatel 1983; Nigg 1987).

Finally, when a person has heard the emergency information, formed an understanding of what is being said, defined a level of belief in what is being said, and determined a level of risk personalization, then behavior results based on the perceptions formed (cf., McLuckie 1970; Mileti et al. 1975; Baker 1979; Flynn and Chalmers 1980; Quarantelli 1980; Nigg 1987; Perry 1981).

A person typically goes through these phases each time new warning information is received. In addition, during the emergency warning period, people do not passively await the arrival of more information. Instead, most people actively seek out additional information. When warning information is received, most people try to verify what they heard by seeking out information in another warning message or from another warning source or person. This seeking of new information to confirm prior information, or receiving new information that confirms prior information, has typically been referred to as the warning confirmation process (Danzig, Thayer, and Galater 1958; Drabek 1969; Drabek and Stephenson 1971; Mileti et al. 1975; Quarantelli 1984).

Confirmation is a main reason that telephone lines become busy after a public emergency warning is issued; people call friends and relatives to get their interpretation of the event and to find out what they are going to do. Confirmation occurs because people are information-hungry following receipt of warnings. Rarely are people overwhelmed by information in a warning context. Instead, there is an information void caused by uncertainty, particularly when rare or unfamiliar events are about to occur. This void typically creates a public demand for more information than is being disseminated in the warning message. Confirmation, then, plays an important role in the warning process and is a part of each stage in the social process; it facilitates understanding, belief, personalization, and decision making (Mileti and Sorensen 1990).

To reiterate, the warning-response process begins when the warning is heard. Although hearing a warning necessarily precedes response, hearing is often insufficient by itself to make people take action. The next stage is understanding the warning. Then people must come to believe that the warning is true and accurate. Next, people must personalize the message to make it relevant to themselves. And finally, they must decide to take action and overcome constraints to taking that course of action. People usually follow their decision about responding unless constraints prevent them from doing so (for example, lacking an automobile in which to evacuate). These responder and situational characteristics vary widely among members of a public in any one warning circumstance, as well as between different events. But, as we discuss in the next section, what happens at each step or link in the chain of factors is also the result of interaction between those who send emergency information (the "senders") and those in the public who receive it (the "receivers").

The Determinants of Risk Perception and Behavior

The long-established principle that people behave in response to hazard warnings in ways consistent with their situational perceptions of risk has been researched in a variety of natural and technological emergencies. Early works have been followed by literally dozens of investigations on how situational risk perception influences behavior in an emergency. There are many studies of sufficient quality across a variety of emergency types to provide, when viewed collectively, both scholars and planners with sound knowledge about how and why the public comes to respond to hazard warnings. Determinants of risk perception and behavior can readily be grouped in terms of "sender" and "receiver/situational" characteristics in the hazard warning communication process.

Sender Determinants

Sender determinants are largely informational or related to emergency information. Ten dimensions of sender characteristics are discussed in the chapter. First is the *source* of the information. The individual or agency from which the emergency public information or warnings emanate must seem credible and reliable to the people receiving the warnings. Warnings are most believable if they come from a mixed set of persons, because people have different views about who is credible and who is not. Warning messages that contain endorsement by a mix of scientists, organizations, and officials serve to alleviate the possibility that any one source could be deemed noncredible (cf. Drabek 1969; Drabek and Stephenson 1971; Mileti et al. 1981). For example, a warning message from a mixed set of sources could say, "The mayor and the head of civil defense have just conferred with scientists from our local university and the National Weather Service as well as with the head of our local Red Cross chapter, and we now wish to warn you that. . . ."

Second, message *consistency* is also a determinant of understanding, belief, and personalization. Early documentation of this relationship was provided in a study of the Rio Grande flood (Clifford 1956) which found that inconsistent information caused confusion, and people were less likely to understand or believe that the flood was going to occur. Fritz (1961) reached the same finding in a study of warning response in a range of disasters. The message must also be consistent in the way it conveys information about the level of risk.

A warning message is best if it contains consistency in the information it gives and the tone in which it is given (cf. Drabek 1969; Foster 1980; Perry and Greene 1982; Sorensen 1982 and 1985; Quarantelli 1984; Rogers 1985). Inconsistency in the tone or information in a message creates confusion and uncertainty among recipients. It is not consistent to tell a public to evacuate but that their children will be kept in neighborhood schools. Nor is it consistent for a message to imply "Something bad is happening but there is no cause for concern."

Unfortunately, in most emergencies there are numerous inconsistencies across different warnings as more is learned about the impending event and updates are issued. For example, inconsistencies can appear as new information reveals that the hazard has decreased or increased, the number of people at risk has become larger or smaller, and so on. Frequent updating of the situation can help to overcome such inconsistencies.

A third dimension is the *accuracy* of what is said in an emergency message. A warning message must contain accurate, timely, and complete data. Accuracy is the extent to which the contents of a message about risk, location, and what to do is or is not fully factual. If people learn or suspect that they are not receiving the "whole truth," they are likely to ignore instructions about how to respond, and instead respond in ways consistent with their suspicions. Errors in past warnings have been found to cause people not to believe subsequent warnings (Mileti et al. 1975). Accuracy is enhanced simply by being fully open and honest with the public about a hazard. In addition, accuracy is important in parts of the warning that may be viewed by officials as being trivial. For example, calling Broad Street "Board" Street by mistake may send a signal to the public that other essential information is also incorrect even though message recipients can correct the error on the basis of personal knowledge.

Fourth is the *clarity* of the emergency information. A warning message must be worded clearly and in simple language that can be understood so that people know what is happening and what they should do about it (Quarantelli 1984, 104). Lack of clarity in a message can lead to people misunderstanding the message (cf., Williams 1957; McLuckie 1970; Drabek and Stephenson 1971). For example, "A possible transient excursion of the reactor resulting in a sudden relocation of the core materials outside the containment vessel" might better be stated as, "Some radiation may escape from a hole in the nuclear reactor."

Fifth, a message must convey a high level of *certainty* about the events taking place and what people should do. Even if there is a low probability or an ambiguous situation, the message about it should be stated with certainty (even certainty about the ambiguity). A message characterized by certainty might say, for example, "There is no way for us to know if there really is a bomb in the skyscraper, or that it will actually go off at 3 P.M. if there is, but we have decided to recommend that the building be evacuated now, and that we all act as if the bomb threat is a real one." Certainty determines the level of belief in a warning and affects decision making (cf. Turner et al. 1979; Perry, Lindell, and Greene 1982). In a study of response to earthquake prediction, for example, Mileti et al. (1981) found that warnings become more believable as the probabilities attached to them become greater. Certainty in warning messages extends beyond message content to include the tone with which it is delivered to the public. The warning should be spoken by the person delivering it as if he or she believes or is certain about what is being said. If warnings are certain, decisions to respond are more likely.

Sixth, *sufficient information* should be given in a message so that the public has an idea of exactly what is happening. Not knowing, or feeling that one has insufficient information creates confusion, uncertainty, and anxiety. Whereas too much detail in the message may overwhelm people, the message must contain enough information so that the public's first response is not to fill the information void with uninformed misperceptions or fears. The amount of information provided affects understanding, personalization, and decision making. A study of family response to hurricane and flood warnings by Leik, Carter, and Clark (1981), for example, found that general and vague warnings caused people not to take protective actions. In a study of response to the Mt. St. Helen's eruption it was found that more detailed information led to higher levels of perceived risk, and, therefore, to taking protective action (Perry and Greene 1983).

A warning message, then, must provide the public with sufficient information about the impending hazard by describing the event that may occur and how it poses a danger to people. It is insufficient, for example, for a warning to simply state that a dam may break. This warning must also describe the height and speed of impact of the floodwater that will ensue, and the size and location of the areas that could be effected. A warning could describe "a wall of water 20 feet high moving at 40 miles per hour," "an explosion hotter than the inside of the sun covering half of the county," or "a seismic shake severe enough to bring down half the unreinforced brick buildings in the city." If a hazard is well described, people are better able to understand the logic of protective actions and are provided with a rationale for subsequent behavior. Hazards should be described with sufficient enough detail so that all members of the public understand the character of the disaster agent from which they are to protect themselves. Informing the public about the physical characteristics of the hazard will reduce the number of people in an endangered public who misperceive the hazard and then make poor response decisions because of those misperceptions.

Seventh, a message must contain a clear statement of *guidance and specifics* about what people should do about the event being described, and how long a time they have in which to act. It cannot be assumed that members of the public will know what constitutes an appropriate protective action; the protective action must be described. Public understanding of communicated emergency risk information is enhanced if it is specific regarding the risk, the hazard, what the public should do, and how much time is available before impact (cf. Drabek and Boggs 1968; Mileti and Beck 1975; Flynn 1979; Dynes, Purcell, Wenger, Stern, Stallings, and Johnson 1979; Carter 1980; Perry and Greene 1982; Perry, Greene, and Mushkatel 1983; Quarantelli 1984; Nigg 1987). A study of the Big Thompson Canyon flood found, for example, that people who received warnings during the flood were not necessarily advised on what to do (Gruntfest 1977). As a consequence, many who were warned incorrectly attempted to drive out of the canyon and were killed.

This point is not as obvious as it seems. For example, warnings must do more than tell people that they should "get to high ground." High ground for some may be the low ground for others. High ground must be defined, for example, "ground higher than the top of city hall," or specify areas to which people should evacuate.

Eighth, *frequency* or the number of times the warning message is delivered affects hearing, understanding, believing, and deciding, and is thus important at most stages of response. Numerous studies underscore the importance of repeated hearing of a warning as a condition for response (cf. Fritz and Marks 1954; Drabek and Boggs 1968; Mileti and Beck 1975; Baker 1979; Turner 1983; Mikami and Ikeda 1985; Rogers 1985). The frequency of public messages should be referenced so that people will be informed about when they will hear the message, or a new message, again. This information can reduce anxiety created by not knowing when one can confirm what is happening or learn more details. Frequent messages can also help reduce the effect of misinformation and misperceptions.

There are some potential advantages of frequently recurring warning messages. Frequently recurring warnings (e.g., "This message will be repeated over this same station every fifteen minutes, unless new information updates are available") focus people on official warnings, reduce rumors, and increase public confidence in the validity of the warnings.

There is no magic formula for specifying how frequently a warning message should be repeated, but some guidelines can be established on the basis of knowledge about how the public processes warning information. In part, dissemination frequency is geared to the dynamics of the emerging risk and its severity, as well as being influenced by increased or changed knowledge about it. Frequency is best dictated by the needs of the public at risk. It is difficult to provide people at risk with too many warnings. People want updates of information even when there is little change in the content of the warning. In protracted emergencies, however, there is a point of diminishing returns after which messages that contain no new information may be counterproductive.

Ninth is the specification of *location* effected by the message. Identifying a location is important in determining belief and personalizing a warning. The emergency warning information should clearly state the areas effected or potentially impacted by the event. People must be told if they are the intended recipients of the message or not. For example, Diggory (1956) showed that the closer the respondent's proximity to a threatened area, the more likely that person will believe the message. Other studies show that more location-specific messages lead to greater levels of personalized risk (Perry and Greene 1983; Nigg 1987).

Detailing the location of risk is best done in ways readily understood by the public. For example, a flash flood warning could say: "The area of town that will flood will be between Second and Fifth Streets, from Elm Avenue to Magnolia Boulevard." If there is reason to be concerned that people who are safe could think that they are unsafe, then the warning should address them—for example, "People who live in other parts of the city will not experience flooding"—but the warning should then explain why they are safe. This is usually necessary because a wider audience than those at risk will hear the warning message.

And, tenth, the *channel of information* plays an important role in warning response. Risk information communicated over multiple channels, such as printed and electronic media or personally delivered, has been shown to enhance hearing, understanding, belief, and response by a public at risk (cf. Mileti and Beck 1975; Flynn 1979; Turner et al. 1979; Carter 1980; Hiroi, Mikami, and Miyata 1985; Rogers 1985; Nigg 1987). Warnings can be issued to the public in a variety of ways, for example, by voice, electronic signals, or printed media. Voices can be direct or broadcast over loudspeakers, public address systems, telephone, radio, or television. Signals include sirens, alarms, whistles, signs, and lights. Leaflets, brochures, or video can be used to distribute graphic information and printed messages. Effective warnings use a range of possible channels instead of a single channel. This helps reach as many people as possible in a short time.

Receiver/Situational Determinants

Actions in response to impending natural or technological disasters have also been shown to covary with the situational and personal characteristics of those who receive hazard warnings. These receiver/situational characteristics fall into six general categories.

First, *environmental cues* or physical characteristics of an emergency setting can interact with sender factors to convey information to receivers. The visibility of the hazard is a physical characteristic. It is difficult (or at least more difficult), for example, for the public to believe a flood warning on a sunny day or to heed an evacuation warning if neighbors are not seen evacuating. People often wait to evacuate in a hurricane until they see the weather change. Such environmental cues are important for understanding, believing, personalizing, and confirming the threat, as well as responding to the threat (cf. Drabek 1969; Mileti et al. 1975; Flynn 1979; Quarantelli 1980; Cutter and Barnes 1982; Saarinen and Sell 1985; Bellamy 1987; Rogers and Nehnevajsa 1987; Tierney 1987). It is, therefore, important when no cues exist to use an artificial means to break a routine-appearing environment. Sirens can help accomplish this. Location of the risk or geographical proximity of those at risk to the impending threat is another type of physical factor that affects belief and personalization (cf. Diggory 1956; Flynn and Chalmers 1980; Perry and Lindell 1986).

Second, *social setting* factors characterize the context in which the emergency information is received. Such factors include whether or not the family is united when the warning is delivered, what activities are being performed at that time, and what others are doing to respond. Social setting factors affect beliefs, decisions, and response (cf. Clifford 1956; Moore et al. 1963; Dynes and Quarantelli 1968; Drabek and Stephenson 1971; Flynn 1979; Cutter 1987). Mack and Baker (1961), for example, reported that family unity at the time of a warning increases the likelihood of belief. Drabek and Stephenson (1971) noted that families who are united at the time of a warning are more likely to respond to the warning. The importance of family unity in evacuation decisions in human-made emergencies is substantiated by the Three Mile Island (TMI) accident in which only 10.6 percent of the sample of households had some members evacuating and some members staying. This means almost 90 percent of the households behaved as family units. In addition, neighbors and friends evacuating are major influences in decisions to evacuate. In an article by Cutter and Barnes (1982), it was noted that, for some people at TMI, evacuation behavior was associated with knowing a neighbor had evacuated.

Third, recipients' *social ties* can affect decisions to respond to warnings (cf. Clifford 1956; Gruntfest 1977; Worth and McLuckie 1977; Mileti et al. 1981; Rogers 1985; Bellamy 1987). Perry (1979), for example, found that as family cohesion increased, the likelihood of evacuating in response to a flood warning increased. This was also found for nuclear emergencies where Sorensen and Richardson (1983) found that knowing someone who worked at Metropolitan Edison was related to the decision not to evacuate during TMI.

Fourth, *sociodemographic* characteristics of the receiver, such as resources, gender, and socioeconomic class, can influence hearing, understanding, believing, personalizing, and responding (cf. Friedsam 1962; Flynn 1979; Turner et al. 1979; Perry, Lindell and Greene 1981; Yamamoto and Quarantelli 1982; Rogers 1985; Nigg 1987; Perry 1987). For example, older people were less likely to have heard the Rapid City flood warning regardless of the source of the warning (Mileti 1975). And, in an analysis of the TMI accident, Sorensen and Richardson (1983) found that older people were less likely to evacuate, which is consistent with other studies of the TMI evacuation. Gender has also been found to be related to warning beliefs in that women are more likely to believe a warning than men (cf. Drabek 1969; Turner et al. 1981; Yamamoto and Quarantelli 1982). The basis for this association, however, is not well understood.

Fifth, *psychological* characteristics of the receiver, for example, cognitive abilities, personality, or attitudes can also influence reception of a warning. Limitations in cognitive abilities to process information are a constraint to all people receiving warnings. Variations in that ability influence understanding and deciding (cf. Sims and Baumann 1972; Flynn 1979; Turner et al. 1981; Quarantelli 1980; Perry 1987).

Personality traits are also related to decision and behavior. The personality factor most investigated in the context of warning response is *locus of control*. Simply stated, people with an *internal* locus of control are very self-determined; people with an *external* locus of control are characterized by fatalistic views of the world (cf. Rotter and Mulry 1965; Rotter 1966; Strickland 1965; Davis and Phares 1967; Lefcourt 1976). *Internal* people tend to feel they have control over their behavior, and *external* people feel their fate is in the hands of others. People with an internal locus of control are more likely to hear, believe, personalize, and respond to a warning than people with an external locus of control (cf. Dynes et al. 1979; Flynn 1979; Turner et al. 1981; Lardry and Rogers 1982).

And, sixth, *prewarning perceptions* play a role in hearing and decision making. The concept of *selective perception* refers to the tendency to filter information to conform with existing views. Without adequate emergency information, people may disregard warnings if their risk perceptions are already biased. Such prewarning perceptions may also influence decision making (cf. Baker 1979; Flynn 1979; Livermore and Wilson 1981; Mileti et al. 1981).

A Theoretical Synthesis

Research over the last several decades has addressed public warning response in a wide array of climatological, geological, and technological events.

Studies have been of various types; some have been descriptive, whereas others tested hypotheses. Some have used sophisticated multivariate analysis, whereas most have been content to explore the character of a few hypotheses based on simple statistical tests of correlation and significance. Enough evidence exists to conclude that it is imprudent to presume that all members of a public hear a warning just because one is issued, and the literature clearly shows that hearing a warning is influenced by both sender and receiver/situational factors.

The research record points out the characteristics of warnings that maximize the probability that they will be correctly understood, believed, personalized, and acted on. The most effective warnings are those that are specific about impact location, protective actions to take, the time to impact, and the character of risk. Additionally, the most effective warnings are consistent and certain, address why they should be acted on, delivered through multiple channels of communication, repeated frequently, and labeled as coming from a panel of officials, scientists, and experts credible for everyone.

People respond to warnings through a social psychological process. Planning for a sound public response to future emergencies means that this social and psychological process must be understood and addressed. Which persons in an endangered public do and do not hear, understand, believe, personalize, and respond to emergency warnings is not the result of chance. The sequential steps in the warning response process are the consequences of the effects of the risk message and the personal/situational characteristics grouped into the categories of receiver and sender factors.

Clearly, then, risk communication is a process and not an act. Communications that result in more accurate public perceptions of risk and public behavior in proportion to the risk that is faced typically have been comprised of multiple communications, arranged in a programmatic format, that take a variety of communication variables or factors into account, for example, source, consistency, accuracy, clarity, certainty, guidance, frequency, location, communication channel, and so on. Public warning response is best understood and planned for if it is viewed as a series of related sequential factors: hearing warnings, understanding what is said, believing what is heard, personalizing what is believed as may be appropriate, deciding what to do, and then engaging in response behavior.

On the basis of the research record, we have developed a causal model, shown in Fig. 2, that posits the following relationships: (1) the odds of public compliance with warnings are enhanced if warnings are personalized by those who should personalize them and not personalized by those who are not at risk; (2) the probability of effective warning is enhanced by the level of belief elicited by emergency warnings; (3) belief of emergency warnings can have its best effect with respect to personalization if it is preceded by accurate public understanding of what is being said in a warning; and (4) understand-

Figure 2
A Model of the Theory of Public Risk Communication

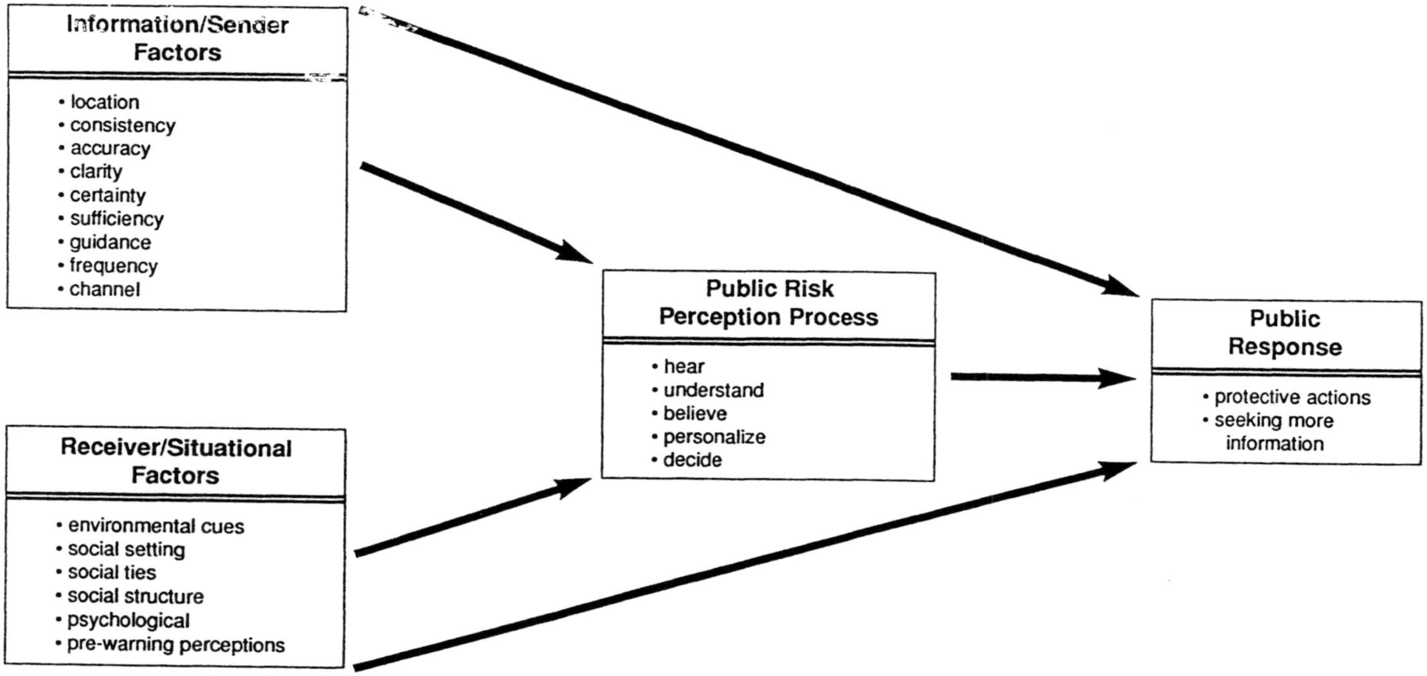

ing the contents of a warning presumes that warnings are heard by the public. This model still needs future empirical test.

Conclusions

Although the empirical evidence on risk education remains ambiguous and does not lend itself well to drawing conclusions, the research record on public risk communications in emergencies clearly suggest a general conclusion that could inform future risk communication programs. Receiver characteristics vary widely among members of a public in any one warning circumstance, as well as between different events. In warning events that provide convincing and reasonable emergency warning information to the public, the understanding, belief, personalization, and response of the public can be sound. The effects of receiver determinants on warning process outcomes are not unchangeable laws of nature. It is possible to design a warning system with sender characteristics that maximize the probability of sound public response and also minimize the negative impacts of some receiver characteristics. Communications that result in more accurate public perceptions of risk and public behavior in proportion to the risk that is faced typically have been comprised of multiple communications, arranged in a programmatic format, that take a variety of communication variables or factors into account. This theory of communication of public risk has much empirical support from different hazard types and study locations. Many agencies at all levels of government have also applied these ideas successfully.

A Model of Evacuation Compliance Behavior

RONALD W. PERRY

Warning response behavior represents one of the most studied phenomena in disaster research. Furthermore, there have been many studies of a particular form of warning response: evacuation. It is most interesting, however, that much of this literature consists of single site, single disaster-type accounts of whether or not (and in what ways) citizens complied with a warning to evacuate a dangerous place. Even in the case of this "well studied" phenomenon, there is a paucity of theoretical frameworks through which one could begin to connect the plentiful empirical efforts into a meaningful gestalt for understanding evacuation behavior or its components.

Indeed, only one general theory of evacuation has been proposed (Quarantelli, 1980) that addresses several levels of analysis (individual, groups, organizations, etc.) and deals with the full range of related phenomena including threat characteristics, community context, and social processes. There are also a handful of narrower models of evacuation decision making among individuals in the literature. Some of these are theoretical models comprised solely or largely of demographic variables (Nelson et al. 1989). Others draw on social or psychological approaches, but were originally designed to deal with the larger phenomenon of warning response (Mileti 1974; Perry et al. 1981) and adapted to predict evacuation compliance.

The purpose of this chapter is to develop a theoretical model specifically for predicting evacuation compliance behavior. Such a model can not only serve as a basis for understanding the calculus citizens use in deciding to comply with an evacuation warning, but may also serve as a framework for organizing and ordering the plethora of existing descriptive studies. The model will be tested on evacuation compliance data from three different types of disaster event: a flood, a volcanic eruption and a hazardous materials incident.

The Modeling Process

The term model has been used in a variety of ways by natural, physical, and social scientists. Indeed, there is considerable variation even among social scientists about the referent for this term. Recently Hartman (1988, 142–50) catalogued a number of ways in which the term model has been used. His treatment involved attempting to abstract general meanings of the term that might serve as classifications under which a variety of uses might be grouped. This lead to the identification of analogical models—an analogy of the phenomena to be explained—and theoretical models—an interpretation of some theoretical system.

The type of model being constructed here best fits into the category of theoretical models. Thus, these models seek to duplicate the systems of relationships among concepts (based on definitions and linking statements) that are specified by theories. This is distinct from analogical models that are usually less complex, but approximate the systemic operation of the theory (cf. Hartman 1988, 143).

The focus here will be on a particular type of theoretical model: the causal model. In the tradition of Simon-Blalock causal thinking (cf. Blalock 1961), the causal model is usually represented in two forms: a graphic form summarizing direct and indirect relationships, and a verbal form composed of propositional statements that explicitly link the concepts which compose the model. A collection of variables and operationalizations (and the rules of correspondence that connect them to concepts) may also be seen as a component of causal models (Costner 1969). Finally, the concepts and variables and their interrelationships identified in the model all reflect a theoretical perspective or logic that is usually elaborated as the basis for the model. In this way, the model contains both a structural and theoretical component.

As the popularity of causal modeling increased among social scientists, so has the size and complexity of the models themselves. As models became larger (more variables and more stages), the algebraic bases of structural equation models proved easily expandable. Thus, causal modelers began to relish the challenge of even larger and more complex models (cf. Berk 1988, 155; Heise 1975, 27).

Recently, however, serious questions have been raised about the virtues of such large, complex models and the logics that create them. Interestingly, the scholars raising these questions are people who were prominent in the development of causal modeling in the first place (cf. Duncan 1975, viii; Duncan 1984). Stanley Lieberson (1986) reminded researchers that the objective of modeling is explanation and not necessarily the construction of elaborate edifices. Lieberson suggested that a basic tenet of causal thinking is parsimony; one should create a model with the smallest number of concepts and assumptions with the greatest explanatory power. In fact, Blalock (1984) reemphasized the importance of the simplification process in causal modeling.

Interestingly, however, many model builders have more carefully attended to a different assumption required in the causal modeling process: that all "relevant" variables or causes are included. For some modelers, meeting this assumption has involved including greater numbers of variables rather than the adoption of a careful conceptualization process that might have resulted in "cleaner" concepts to be operationalized (cf. Blalock 1982, 27).

Although Lieberson's critics focused on quantitative aspects of his arguments (Arminger and Bohrnstedt 1987), his work seems to have more implications for assembling models than for the statistics involved in testing. Lieberson is not arguing per se against the creation of complex or multistaged models. Instead, he is questioning the scope and causal logic used to construct models. He suggested that model builders start with a different assumption (Lieberson 1986, 186):

> the dependent variable in a typical setting is actually responding to a small number of causes.... A complicated network of causal relations will also appear, but these are largely superficial causes that do not have any impact on the dependent variable, which is actually affected by a very small number of basic causes.

Lieberson's charge goes beyond seeking parsimony and urges that causes (variables) be chosen carefully—on both empirical and theoretical grounds—such that only the most crucial appear in models. For Lieberson, models should be composed of "basic" causes and not "surface" causes. He argues that the focus of model building should be understanding (explanation) of a target concept, rather than on the elaboration of a larger causal network that might be associated with that concept or its antecedents.

There is much intuitive appeal to the idea of using a small number of Lieberson's "basic" causes in model construction. What is needed, however, is a clear way of identifying (or distinguishing) basic and surface causes. His critics point out, and Lieberson admits, that this doesn't currently exist and may be some time in coming. Further, Lieberson argues that such distinctions are theoretical in nature and may be transitory in that as theoretical logics change, so will the determination of basic and surface causes. Interestingly, he argues that one key to identifying basic causes will be their performance in models tested over time and across events (Lieberson 1986, 188).

In constructing an evacuation compliance model, an attempt is made to work within the framework outlined by Lieberson. Thus, we focus on creating a single-stage model using variables that serve as indicators of basic causes. In the absence of an established logic for identifying basic causes, selection of variables for an evacuation compliance model is based on both theoretical and empirical grounds. That is, variables are identified initially on theoretical grounds and ultimately included only when there is also research support for their importance in evacuation decision making.

An Evacuation Compliance Model

The principal theoretical perspective to be used here draws on Ralph Turner's work (1964) on emergent norm theory. Emergent norm theory was originally adapted to the general warning response problem (Perry, 1979). Our primary modeling goal is to understand how a citizen makes a decision to comply with an evacuation warning. From an emergent norm theory standpoint, the problem here is to understand how an individual evaluates and acts upon a message. The evaluation aspect implies that at least four issues (representing concepts) must be resolved by the warning recipient. The first of these is the initial message itself: what environmental danger is being described (message content)? The second issue relates to assessing the accuracy of the message: is it correct or did I hear right (warning confirmation, warning source credibility)? Third, the individual must determine how much personal danger is involved: will this event negatively affect me (perceived risk)? Finally, warning recipients must decide if they are able to effect some protection from the pending event: Do I have a plan that can minimize what I believe will be the negative effects of the disaster agent (Plan)?

It should be noted that a considerably more intricate model can be developed using the same emergent norm assumptions. For example, in working with the general warning response problem, Perry (1985, 65–94) developed a multistage model that involved using different indicators for some of the emergent norm-based concepts and at the same time included elements of a bounded-rationality approach to decision making. The resultant model was based on a variety of fairly precise assumptions about the processes through which people make decisions. The approach described here makes many fewer assumptions about decision processes. Indeed, there are only two: (1) people will try to verify messages, and (2) when in danger people will attempt to protect themselves, Quarantelli's (1980, 87–88) work provided the impetus for using an approach with a minimum of assumptions:

> In conclusion, it should be noted that a focus on decision making may imply a more conscious and deliberative act than might often be the case.... Decisions to evacuate may actually be less a weighing of alternatives and deciding, than the development of an informal consensus or implicit understanding about what should be done.

Quarantelli's observation is consistent with Lieberson's concern for developing parsimonious models reflecting the most basic influences on the dependent variable.

The emergent norm approach focuses on understanding the individual's "definition of the situation" as a function of receiving an evacuation warning. The modeler is alerted to five variables that represent the four issues described above: warning confirmation, warning source credibility, warning content, perceived risk, and possession of an adaptive plan. A sixth variable should be included in any evacuation compliance model on empirical grounds: family context in which the warning is received. The following paragraphs briefly review the research literature that describes the relationship of each of these variables to evacuation compliance.

Family context is used to refer to the extent to which family members are safely accounted for at the time of warning receipt. The importance of this variable stems from a long research tradition indicating that families evacuate as a group or only when missing members have been determined to be safe (cf. Killian 1952; Quarantelli 1960; Drabek and Boggs 1968). Thus, one proposition in our model asserts that unless family members are accounted for, citizens will not comply with an evacuation warning.

The possession of an adpative plan refers how much the individual has formulated a specific tactic for accomplishing an evacuation. Such plans can be conceived as based on information derived from the warning message itself, from past disaster experiences, from knowledge of the threat environment, and/or from personal disaster planning activities (cf. Perry et al. 1981, 40–43). Without regard to where the plan information was derived, several empirical studies have suggested that evacuation compliance is facilitated by possession of an adaptive plan (Windham et al. 1977, 15; Leonard 1973: Perry and Greene 1983). A second proposition for the evacuation compliance model is that the more precise the adaptive plan, the greater the likelihood of evacuation compliance. For this analysis, adaptive plan will be operationalized as a series of four ordered categories ranging from no notion of a plan, to "a general idea of how to evacuate", through knowledge of a safe departure route only, to knowledge of a safe route and a safe destination.

The importance of personal risk in both warning and evacuation compliance has been documented at least since the first NORC studies of disaster response (Fritz and Marks 1954). Reviews of the empirical literature (cf. Perry 1985, 78–80; Drabek 1986, 71) reveal considerable support for the notion that evacuation compliance is greatly enhanced if the warning recipient believes that the pending disaster will result in direct and personal harm. Conversely, people who believe themselves outside the likely impact area or who believe that the personal consequences of impact will be slight, tend not to comply with evacuation warnings. Thus, a third proposition in our model may be stated as the higher the level of perceived personal risk, the greater the probability of compliance with an evacuation warning. Personal risk will here be operationalized as a four category ordinal scale, where respondents designate themselves as outside the impact area, certain to experience slight damage, certain to experience moderate damage, or certain to experience severe damage.

Warning confirmation refers to the warning recipient's ability to confirm the accuracy of an evacuation warning with some source or sources. Successful warning confirmation appears to create confidence in the veracity of the warning and to enhance the probability of evacuation compliance (cf. Drabek 1969; Drabek and Stephenson 1971). Interestingly, warning confirmation has sometimes been used as having an indirect relationship to warning compliance, affecting that variable through warning belief (definition of situational danger) or warning content (cf. Mileti et al. 1975, 51). In the present model, confirmation is assumed to have a direct impact on compliance for a theoretical reason. That is, emergent norm thinking suggests that developing a definition of the situation is a collective (or interactive) endeavor, and this has a direct effect on compliance behavior. This interactive component is achieved through the confirmation process, wherein information received may be confirmed, or new information may be introduced from the source, or other information may be identified as inaccurate. Thus, another proposition for our model may be stated as: confirmation of an evacuation warning increases the likelihood of compliance. Confirmation was here operationalized as a dichotomy; the respondents either reported confirming the warning with any source or failed to do so.

When a citizen is warned by a source that is considered to be credible—that is, reliable and trustworthy—the likelihood of evacuation compliance is increased (cf. Perry 1985, 44–47). In keeping with our theoretical perspective, a message received from a credible source deals with the issue of the quality of warning information rather than specific content itself. Thus, warning from a credible source enhances the probability that the recipient will develop confidence in the accuracy of the message, the completeness of the message, the timely nature of the message, and the appropriateness of suggested protective measures. A fifth proposition in our model is that persons warned by a source perceived to be credible are more likely to evacuate. Credibility will be operationalized as the respondents' characterization of warning source as credible or not credible.

Finally, warning content has long been known to be related to warning response, particularly evacuation (cf. Mileti 1974; Perry 1985, 78–79). Much work has been done on the desirable content of warning messages, essentially concluding that they should contain basic decision information such as the likely magnitude and timing of the event, location of the impact area, and concrete suggestions about what actions would be protective (cf. Williams 1957, 15–19; Drabek and Boggs 1968, 445–47; Mogil and Groper 1977). In this context we may derive the last proposition for our model from flood evacuation studies that showed that the more specific the warning content, the more likely the warning recipient is to comply with an evacuation warning (Perry et al. 1981, 96). Content was operationalized here as a series of ordinal categories representing increasing specificity, beginning with the general content that danger was present, to indicating that a disaster will strike a particular place, to specifying that a disaster will strike the individuals current location, to the statement that the disaster will strike the individuals location with a description of one or more appropriate protective actions.

Data Sets and Analysis

In the preceeding section, a single-stage model of evacuation compliance behavior was developed that expresses compliance as a function of six variables: family context, perceived risk, warning content, warning confirmation, source credibility, and possession of an adaptive plan. This model is fitted to data in terms of a particular dimension of comparative study: event type. From a modeling perspective, one expects the same variables to control evacuation compliance (that is, the model "fit") in any event that shares the context established by the assumptions under which the model was constructed. Warning response models of all types must hold as a principle assumption that the speed of event onset and level of social preparedness are such that (1) time for forewarning exists, and (2) that a warning message was communicated to recipients prior to actual impact, and (3) that there was sufficient time after warning receipt (response time) for the individual to undertake some response if deemed necessary. Although other characteristics of the disaster event itself—absolute amount of response time, scope of impact, duration of impact, secondary impacts, and familiarity of threat, for example—may vary, the three assumptions above are critical for meaningful analysis of evacuation response behavior.

Secondary data is used to assess the utility of the evacuation compliance model across three disaster events: a riverine flood, a train derailment, and a volcanic eruption. Because of the specific nature of these three disasters, one proposition in our model is dropped from the analysis. All of the events occurred in a time frame such that none of the respondents reported that any family members were away from home or unaccounted for at the time of warning. Thus, family context becomes a constant—rather than a variable—in the data sets, and must be removed from the analysis. It remains theoretically important, however, and is consequently retained in the general model itself. The following paragraphs are devoted to a brief description of each data set.

The flood data represent an event that occurred in Abilene, Texas. The genesis of the flooding was the overflow of two creeks following heavy rain, which resulted in the inundation of a low-lying segment of the community (cf. Perry and Greene 1983). Flooding in Abilene is a seasonal event, therefore constituting a relatively familiar threat for the local populace. Because the creeks rose over a period of hours, local authorities had time to formulate and deliver an evacuation warning with ample forewarning for citizens.

The train derailment took place in Mt. Vernon, Washington. It arose when a switching accident knocked a tank car carrying twenty-five thousand gallons of propane off the tracks near downtown. Although no breach of containment was visible, local authorities decided that the threat was sufficient to require evacuation of the surrounding downtown area, including a contiguous residential area. The evacuation was to be maintained until the arrival of a hazardous materials response team and the replacement of the tank car on the tracks. Authorities issued an evacuation warning that advised citizens to leave immediately (very short forewarning). It should also be noted that a propane gas threat was very unfamiliar to local residents; the most common disaster threat in the area was seasonal flooding.

The third data set is drawn from Toutle, Washington and deals with the response to the 18 May 1980 cataclysmic eruption of Mt. St. Helens volcano. Toutle is located near the north fork of the Toutle River, on the northern flank of the volcano. At the time of the eruption, the principle threat to the town was flooding associated with volcanic mudflows. Local authorities issued a preimpact evacuation warning, with instructions to leave immediately (very short forewarning). Toutle is subject to periodic seasonal flooding, and eruptive activity had been occurring for a number of weeks. Thus, although volcanic mudflows were themselves not a familiar threat, most Toutle residents were aware of the volcanicity and felt that some protective action may be necessary at some point (Greene et al. 1981, 49).

In each community a probability sample of warning recipients was drawn. Personal interviews were conducted, with completion rates of 92.3 percent in Abilene (n = 185), 90.4 percent in Mt. Vernon (n = 147) and 90.0 percent in Toutle (n = 90). A slightly modified version of the same questionnaire instrument was used in each community. Table 3 shows the zero-order product moment correlations for each independent variable with evacuation compliance. These coefficients indicate that all variables, as hypothesized, show a positive relationship with the dependent variable. Furthermore, all coefficients are statistically significant except for the possession of an adaptive plan in Mt. Vernon and Toutle. It is possible to speculate that this may be the case because evacuation is not a complex adaptive action and can be achieved with a minimum of preplanning. If this is the case, however, it would not be appropriate to infer that adaptive planning is not important for models predicting more general warning response behaviors.

To more precisely assess the relationships hypothesized in the evacuation compliance model, ordinary least squares regression analysis was employed. Table 4 shows the standardized partial regression coefficients (called beta coefficients) and the squared multiple correlation coefficients for each data base. Although the numerical value of the beta coefficient is not a measure of explained variance, these values are used to assess the relative importance of independent variables in the same data set (cf. Blalock 1979, 481).

Table 3. Zero Order Correlations of the Independent Variables with Evacuation Compliance

	Toutle	Mt. Vernon	Abilene
Confirm	.31*	.56*	.59*
Credibility	.22*	.59*	.37*
Plan	.01	.05	.35*
Content	.46*	.53*	.29*
Risk	.74*	.75*	.67*

*$p < .05$

In Toutle, perceived personal risk is by far the most important variable in understanding evacuation compliance. This is consistent with the research literature that documents that warning recipients are most likely to undertake any protective action when they believe themselves to be in serious danger. Risk is followed in importance by specificity of warning content and source credibility. Warning confirmation shows a slightly smaller beta, and adaptive plan is the least important of the variables in the model (neither of these two betas are statistically significant).

The R^2 indicates that the overall fit of the model to the data is good; the five variables acting together explain 61 percent of the variance in evacuation compliance. Statistically, in Toutle—a disaster marked by short forewarning and a moderately familiar threat-agent—perceived risk, warning content, and source credibility were the most important variables for understanding evacuation behavior. The lower relative importance of having an adaptive plan may indicate that evacuation in Toutle does not depend on extensive planning. There are essentially only two roads leading away from town. The lower relative importance of warning confirmation may be a function of the short period of forewarning and citizen familiarity with the threat. Short forewarning cuts the time available for confirmation in general. Because citizens were aware that Mt. St. Helens had erupted before and was likely to do so again, one would not anticipate that there would be a pressing perceived need for confirmation. Interestingly, the warning from a credible source and message content were statistically significant in this context. Apparently, when forewarning is short and the threat moderately familiar, the combination of specific warning information delivered by a credible source is sufficient to create evacuation compliance.

The fit of the model to the Mt. Vernon data was also good, yielding an R^2 of .74 (indicating 74 percent of the variance in compliance is explained). Furthermore, inspection of the beta coefficients indicates that the relative importance of the variables is almost identical to that observed for Toutle. Again, perceived risk shows the largest beta by far. This is followed by source credibility and message content, which show very similar magnitude beta coefficients. Warning confirmation and possession of an adaptive plan again rank at the bottom of the list, although confirmation is statistically significant here.

In general, the similar patterns observed in these two communities are probably a function of similar threat settings. In both cases we are dealing with very short forewarning that left limited time for confirmation behavior (although confirmation was statistically significant in Mt. Vernon, its beta weight still places it next to last among the five independent variables). At the same time, warning content and source credibility were statistically significant and follow risk in importance. This would be consistent in a situation where the threat was unfamiliar, because a specific message from a trustworthy source would tend to elicit quick action on the part of a warning recipient.

The evacuation compliance model also fits the Abilene flood data well; the R^2 indicates that 64 percent of the variance in evacuation compliance is explained by the joint action of the independent variables. These data represent a longer forewarning event with which the local citizens were very familiar, and show a different pattern of importance among the independent variables. The consistent point is that once again, perceived risk shows the largest beta by far.

Following risk, warning confirmation and possession of an adaptive plan show the next highest beta coefficients. Interestingly, warning content and source credibility show the smallest betas (and are not statistically significant). The increased importance of warning confirmation is consistent with our earlier hypothesis that longer forewarning provides the opportunity for confirmation behavior. It is possible that in a slow developing, familiar disaster threat, warning recipients are less concerned with either the content of the warning (because familiarity would permit the belief that one can anticipate the nature of the event) or source credibility (because familiarity would permit the belief that one has valuable personal judgment and evaluative skills about the threat). Under such circumstances heightened confirmation behavior (in the form of information gathering) like that observed in Abilene in expected.

The increased importance of possessing an adaptive plan may stem from

Table 4. Standardized Partial Regression Coefficients (Beta)

	Toutle	Mt. Vernon	Abilene
Confirm	.10	.13*	.37*
Credibility	.14*	.25*	.05
Plan	.03	.09	.11*
Content	.17*	.24*	.09
Risk	.63*	.53*	.49*
R^2	.61*	.74*	.64*

*p < .05

the nature of the threat in this community. Under most circumstances, accomplishing a minimally effective evacuation (i.e., clearing an area) requires little planning. Of course, the effectiveness can be greatly enhanced if the person knows appropriate routes, appropriate safe destinations, takes important papers and medications, and so on. But minimal performance can be attained with a minimal plan. This logic can become complicated, however, when either route or destination information is not clear or obvious. In Abilene, route information was a potential problem. Although flooding was a common threat, multiple flood sites from multiple creeks was less common and raised the question of what routes of egress were safe and appropriate. In this case, citizens with an adaptive plan had the advantage of being more likely to have devised alternate evacuation routes, thereby increasing the importance of planning for evacuation compliance.

Summary and Conclusions

In the preceeding analysis, we have drawn a variety of conclusions. First was the idea that the evacuation compliance model fit each data set well, as measured by level of explained variance. This allows us to conclude that within each event, the variables in our model do indeed control much of the variance in evacuation behavior. One might infer, based on such promising analyses, that the concepts represented by the variables approach Lieberson's idea of "basic causes." Of course, the definition of basic causes is a cumulative matter, and such a designation for any concept must await lengthy comparative testing. Empirical issues demand attention as well because, for example, the model tested here could benefit from more precise measurement. But a reasonable beginning has been achieved.

Standardized partial regression coefficients (betas) were used to examine the relative importance of variables *within* each data set. Such coefficients cannot technically be compared across data bases because each coefficient is standardized in terms of the standard deviation in the data set at hand. At this point it is important to consider comparisons of the role of specific variables across the data sets. To do this appropriately, one can examine unstandardized partial regression coefficients (usually symbolized as *b*).

The *b* statistic represents the amount of change in the dependent variable associated with a unit change in a selected independent variable when all other independent variables are held constant. The unit of measurement (and, hence, change) is represented by whatever units were used to measure the variable originally. For example, the variable risk would be expressed in terms of our four-unit risk scale. Since *b*s represent the same measuring units, they may be compared across data bases, but only for the same variable. Each variable in the model, when expressed as *b*, will have different measuring units and therefore may not be meaningfully compared with other *b*s in the same data set.

Table 5 shows the unstandardized partial regression coefficients for each variable across the three data bases. We already know from earlier analysis that perceived risk is by far the most important variable within the model. That this is generalizable across each of the data sets is supported by the similar magnitude (and statistically significant) coefficients for risk in Table 5.

It was also argued that source credibility and warning content is expected to be more important when forewarning was short and event familiarity was low to moderate. Table 3 shows that the largest *b* for credibility is found in Mt. Vernon (short forewarning, low familiarity), followed by Toutle (short forewarning, moderate familiarity) and lowest in Abilene (longer forewarning, high familiarity). Similarly, the *b* for warning content is highest in Mt. Vernon, followed by Toutle, and lowest for Abilene.

The earlier analysis also suggested that in longer forewarning, high-familiarity events, warning confirmation takes on increasing importance. This is supported in the data in that the *b* for Abilene is more than two times larger than that for Mt. Vernon, whereas the *b* for Toutle is even lower and not statistically significant. The role of adaptive planning across the data sets remains unclear. The *b* is smallest for Toutle, indicating that particularly in that setting an adaptive plan was not crucial to evacuation behavior. The *b*

Table 5. Unstandardized Partial Regression Coefficients (b)

	Toutle	Mt. Vernon	Abilene
Confirm	.39	.63	1.46*
Credibility	.54*	1.36*	.30
Plan	.06*	.17	.16*
Content	.34*	.66*	.13
Risk	.88*	1.09*	.90*

*p < .05

for planning is of similar magnitude in the other two events, however. The larger and statistically significant *b* for Abilene is consistent with the earlier argument that complicated threat impact characteristics may force citizens to rely more on previous adaptive planning. Mt. Vernon, a situation without any particular impact complications, also shows a numerically large *b*, although it is not statistically significant. Unfortunately, within the three data sets available, it is not possible further test the role of planning. Additional data from communities with varying impact complications will be necessary to resolve the issue.

In closing, the most important conclusion to be drawn is that the evacuation compliance model fits each of the three different events well, with explained variance from 61 to 74 percent. The idea that the same model will predict evacuation behavior in connection with a flood, a volcanic erruption, and a hazardous materials incident calls into serious question the popular strategy of partitioning research on disaster events in terms of the type of event (e.g., "natural" versus "man-made"). This approach represents an attempt to accumulate knowledge about disasters by classifying those events phenotypically; that is, on the basis of surface characteristics rather than in terms of more theoretical distinctions.

The three events studied here represent comparable events in that each involved some forewarning and was characterized by definable speed of onset, duration of impact and scope of impact. These particular dimensions were drawn from Kreps's (1989a) classification system for disaster events—a theoretically based system grouping disasters in terms of physical and temporal dimensions. The success obtained here in fitting the same model to events that appear different on the surface, but would be similarly classified in a disaster typology, suggests that development of classification systems through taxonomy is an important enterprise for the growth of knowledge about human and organizational behavior in disasters (cf. Kreps 1989b; Perry 1989). In terms of the growth of modeling in disaster research, two paths need to be pursued: Models need to be tested comparatively on multiple data bases, and attention should be given to selecting and classifying data bases in terms of the growing numbers of typologies and classification schemes for disasters. We have apparently reached a point in scientific development where the empirical side of research (model building and testing, for example) must proceed parallel with the theoretical side (taxonomy, for example) to effectively accumulate knowledge.

EMS in Halifax after the 6 December 1917 Explosion: Testing Quarantelli's Theories with Historical Data

JOSEPH SCANLON

> Maimed and senseless, with blood-soaked bandages, with dangling limbs and powder-blackened faces, they were carried or, those who could walk, staggered in. Motor cars, carriages, grocery waggons, coal carts and every other conceivable variety of vehicle, hastened to the hospital doors....
>
> The beds were soon filled to overflowing. Mattresses were dragged out of stores and placed upon the floors, in halls, in offices, in every available space. The wounded or dying lay in helpless misery upon bed or mattress or sat in listless waiting upon the floor. Already, dozens of surgeons and nurses were at work staunching the blood or stitching up the gaping wounds. Young women from the St. John Ambulance Brigade and other first aid workers arrived every minute and threw themselves into the breach.... They worked feverishly ... but still the never ceasing stream of patients came, and still the tramp, tramp of their bearers filled the halls.
>
> The operating room had been destroyed, but new ones were quickly improvised in the midst of beds ... for every room was filled, and still on, on they came.... some had eyes gouged out, some had ears or noses torn off, hundreds had their faces gashed or beaten out of possible recognition. As fast as one patient was off the table, another was laid onto it, and the sickening odor of chloroform, ether and fresh blood intermingled, filled every room.
>
> Bell, *The Romance of the Halifax Disaster*

The aforementioned quotes do not sound like romance. That is because, although they are from a novel, they reflect the personal experience of the author, F. McKelvey Bell, a lieutentant colonel with the Canadian Army Medical Corps, a key figure in the medical response to North America's worst disaster, the Halifax harbor explosion of 1917.

The story of Halifax has been told many times. About 8:45 A.M. 6 December 1917, a French munitions ship, the *Mont Blanc*, in from New York to join a Canadian convoy, caught fire after colliding with a Norwegian relief ship, the *Imo*. The *Mont Blanc* was loaded with explosives and at 9:05 A.M., twenty minutes after the collision, it blew up, right in the heart of the city's inner harbor. The result was that everything within 800 meters was destroyed, and there was massive damage for another 800 meters (Bronson 1918). The incident left 1,963 dead, and up to 10,000 injured, many mutilated or blinded by flying glass. It devastated the north end of Halifax and parts of neighboring Dartmouth, across the harbor.

> Large portions of the North End had been totally destroyed: houses, schools, churches and businesses had been completely flattened or burned ... several factories were destroyed.... For approximately one square mile, the devastation was complete with only an occasional wall standing. (Metson 1978)

> The glass and iron frame roof ... of the city's only railway station, dropped in upon the platforms and tracks. The Richmond freight terminal, its sidings and marshalling yards were a mass of shattered railway cars and debris. The Dockyard and Wellington barracks were shattered. The tall brick bulk of the Acadian Sugar Refinery was snapped off like a carrot.... The factory of the Dominion Textile Company was burned to a shell. (Raddall 1948)

The blast created a tidal wave that threw a tugboat on top of a pier, carried some persons up a hillside, and even exposed the harbor bottom. It also created a shock wave that shattered windows and sent debris flying through the air. Many injuries were caused by splinters of glass going through walls, clothing, flesh, and worst of all, into eyes. Finally, because many buildings were of frame construction, heated by wood-burning stoves, the blast was followed by widespread fires and, eventually, a fire storm. Many who died were trapped in the burning wreckage. Even today, the blast is considered impressive: At 2.5 kilotons, its force was one sixth the power of the bomb used at Hiroshima.

Impact on Medical Services

At the main provincial hospital, the Victoria General, glass from shattered windows flew about the nurses' residence. The skylight in the operating room fell onto the head of the surgeon. The chief surgeon was hit in the jugular by flying glass. Other hospitals were also hit (Bell 1918B):

Cogswell St. Military Hospital—all the windows and, in some cases, the casements of same, were blown into the wards, the outer doors were torn down, parts of the ceiling of several of the wards fell, and the ceiling of the operating room was destroyed.

Camp Hill Hospital—slight damage, a few of the windows and some small portions of the ceiling being effected.

Nova Scotia Infectious Hospital, Dartmouth—serious damage, all of the windows and the ceiling of one of the wards very shattered.

Damage was not confined to the hospitals. When a physician's home was badly damaged, the physician suffered three broken ribs and a bȧdly cut right wrist. His wife had a severe gash on her right arm and cuts on her chin and to her lips.

Despite these problems, the medical community carried on. At the Cogswell Hospital, which was worst damaged and closest to the blast, four to five hundred patients were treated the first day. At the physicians' home—it was also an office—the physician and his wife carried on.

Literature Review

There is an official history of the explosion by Archibald MacMechan, a professor at Dalhousie University (Metson 1978). There is Smith's book, *Heartthrobs of the Halifax Horror* and a fairly new book, *Shattered City*, by Janet Kitz. In addition, two articles appeared in the *Canadian Medical Association Journal* (Tooke 1918; "The Halifax Disaster" 1918), and two others appeared in foreign medical journals, one published shortly after the incident, one not published until 1976 (Foster 1917–18; Connelly 1987). All of these, the books and even the articles in medical journals, only touch on the overall medical response.

On that topic, possibly the best published accounts are in three books: Bell's novel; another novel by Hugh MacLennan; and a third book, this one nonfiction, written by another novelist, Thomas Raddall. All three were writing from personal experience. Bell, as already mentioned, is a physician and was involved in the initial response. MacLennan was a youngster at the time but he spent much of 7 December in the blast area, and his father was a physician at Camp Hill. His well-known novel, *Barometer Rising* (1958), reflects his own and his father's experiences. Raddall, also a youngster at the time, was at school but escaped injury when the blast occurred. His version appears in *Halifax, Warden of the North* (1948), a history of Halifax.

But none of these published accounts, fiction or nonfiction, match the detail available in the Public Archives of Nova Scotia and the Public Archives of Canada. In Nova Scotia, the best source is an unpublished medical history by David Fraser Harris. In Ottawa, it is the records of the Canadian Army Medical Corps that were, until recently, kept secret. By 1917, Canada had been at war for three years, and Halifax was the key North American port for the conflict in Europe. Canada did not want its enemies to know 15 military personnel had died, 464 had been injured, 39 severely, or that the city of Halifax had been substantially destroyed ("Synopsis").

There is, of course, no direct link between this material, the Halifax explosion, and Henry Quarantelli. He wasn't born when the explosion happened or when Samuel Henry Prince's study of the response became the first sociological study of disaster (Prince 1920; Scanlon 1988A; Scanlon 1988B). But because this material has only recently become available, it offers a rare opportunity to test Quarantelli's theories about the response of emergency medical services (EMS) in disaster against what actually happened in a major medical disaster, even though that disaster happened long before Quarantelli developed his theories.

Initial Medical Situation

What the records show is that the initial EMS response was informal. Family members dug their relatives out of the wreckage. Those with first aid skills used them. A youngster remembers her mother providing the only treatment she ever got: "I still have a scar on my left cheek from a sliver of glass which Mom pulled out when I came home" (Gormley 1977). Another child remembers his eye hanging out on his chin until it was put back in by a nurse he happened to meet on the way to the hospital.

Other survivors found medical help elsewhere. Some were treated by uninjured passengers and by a physician who happened to be on an incoming train (Harris):

> All the passengers turned doctor. The first one I treated was a little girl. . . . Her eyes were filled with glass. I got the glass out but she couldn't see. She was blind.
> The women passengers got to making bandages. They used the towels and sheets belonging to the Pullman then tore up their underclothes. The water was run off from the engines to bathe the wounds.
> Fortunately, Major DeWitt had his hypodermic case and morphia in his pocket and the sheets from the sleeping car were used for bandages.

Still others went or were taken to surviving ships. One U.S. warship, the *Old Colony*, acted as an emergency hospital for days. Hayes recalls (1917):

> I found that fifty-four persons had been brought on board seventeen of whom had died before receiving medical treatment, the living were given first aid and put to bed on board ship.

Many others victims went directly to the office of the nearest physician.[1]

When we reached Uncle Al's (Dr. Allison Payment) I began to realize something very serious had happened and that people as well as houses had been hit. The house seemed filled with people and nurses and doctors were scurrying to and fro. (Baker, n.d)

In an instant, walls, ceilings, plaster, glass, furniture, were in turmoil, and wounded from the whole neighborhood were rushing into the house looking for my husband, who is a doctor. He was not at home, telephones were out of order, the house was filled with injured (McIntosh, n.d).

Finally, many victims—including most of the seriously injured—went to the hospital. And they went by the hundreds. Patients were helped in on foot, carried in, brought in on makeshift stretchers or, sometimes, brought in on carts, horse-drawn wagons, motor cars, or by ambulance. On arrival, they would be placed anywhere there was room, on a table, a floor, in a corridor or, sometimes, despite the cold, even outside on the steps. The situation was exactly as described in Bell's novel (Harris, n.d).

The ground in front was jammed with autos, waggons, and every conveyance capable of carrying the sufferers. The hallways and offices and every bit of floor space was filled with human beings.

Almost immediately, people in all conditions of injury began to arrive at the institution [Victoria General]. They were brought in in autos, ambulances, cabs, delivery wagons, flat wagons, in any and all kinds of vehicles capable of carrying a person. This, coupled with the rush of ambulatory cases, wounded, bleeding, and fleeing to us for help, created in a short time, an almost innumerable company seeking aid within our doors. Floors of wards, halls, corridors, offices, the dispensary ... became literally covered with men, women and children, some dead, some dying, others variously injured.... The predominance of facial injuries rendered the scene all the more ghastly because of the freely flowing blood.

During normal emergencies, hospitals often report they are overcrowded and can receive no more patients. This did not happen in Halifax. Writing about the acting superintendent of the Victoria General, Charles Puttner,[2] Aldrich reports:

He turned no one away, and every bed was full, and where no were available, the wounded and the dying were laid gently on the floors of the wards, the halls, the offices, and even the cellar.

Although no one was turned away, some victims, seeing the extent of the problem, may have left on their own. Murphy recalls seeing injured persons arrive at a hospital, notice how many others were there with injuries more serious than their own, then walk away (Harris, n.d.).

The decision on who went where was often a matter of chance. A fourteen-year-old delivery boy for a grocery store found himself carrying a cartload of victims from place to place. He wanted to go home to his family, but an armed soldier ordered him to take the persons to hospital. After finding no one at the Cogswell Street Military Hospital, no room at Camp Hill, he ended up at Victoria General, also full. He started to cry. Someone—he assumes it was a physician—told him all right, the hospital would take his patients. The cart was unloaded—this was a cold, December day—and the patients were put on the only place left, the hospital steps (interview with survivor).

Treatment

The medical staff available did what they could, but it was a very difficult task, partly because of the lack of space and skilled help. Volunteers performed tasks normally performed by medical personnel. One of them, J. C. McKeen, later chairman of the board of the Nova Scotia Light and Power Company, Limited, reported he worked first as a rescue worker, later as a volunteer ambulance driver, finally as an anesthetist (McKeen 1967):

> During the course of the second day in our extremely busy operating room ... the young medical student ... passed out from sheer exhaustion after 40 hours of work. Doctor Cox sent for someone to replace him and I was selected.
> Finally, after some hours when the nurse administering anesthetic could carry on no longer, the job of giving anesthetic fell to me.
> It is horrible to relate, but with one patient waiting to go on the table, one on the table and one being taken out, a huge surgical bucket had become filled with enucleated eyes.

The accounts of physicians were similar if more clinical (Harris):

> The face, head, hands and arms fared the worst. One frequently found flaps cut as regularly as if they had been done by a scalpel. Many of these wounds showed contusions. A dark, bruised area with many cuts scattered all over it was a common enough observation. The deep cuts showed, as a rule, much laceration. Glass was found in the muscles, fascia, bones, tendon sheathes, in fact everywhere.
> The simple fractures require no special mention unless it be to say ... they were not so numerous as the compound. These fractures were quite the worst I had ever seen and they all went septic and no wonder. When one finds a compound, comminuted, lacerated fracture with bits of old wall plaster, stones, laths, almost everything—driven in and among the fragments and the soft parts, he will do his best to clean up the wreck, but he is almost bound to have sepsis.

Statistics compiled later showed 226 fractures. It is not clear if this covers all those seen in hospital, all in one hospital, or all for whom there are records. The records show 34 amputations: leg, 11; arm, 13; foot 4; fingers, 4; toes, 2; hand, 2.

A physician who arrived on 10 December discovered that a physician from New Glasgow had been operating twenty-four hours at a time, stopping for a few hours' rest, then going back to the operating table (Tooke 1918).

> Most of the night he had been working in the kitchen and operating on the floor. . . . He was manifestly exhausted and he told me he had done so much work that his instruments would no longer cut.

As McKeen indicates, the most horrifying cases were the eye problems. The data shows 333 survivors were treated for eye injuries. Of these, 41 were blind, 87 had one eye removed, and 61 were still being observed early in the new year (Bell 1918B).

One physician came by train from Ottawa, together with his own operating team comprised of a general practitioner from Maine as assistant, a medical student as anesthetist, a nurse from the Royal Victoria Hospital in Montreal, and a retired army person as orderly. They dealt with forty-eight eye cases, eighteen involving removal of one or both eyes (Tooke 1918).

> It must be remembered that the eye injury was frequently only an associated condition, of the most ghastly incised wounds of the head, face and neck. These wounds were invariably caused by glass and could only be described as hideous. . . . Practically every face wound was septic; nay more, each was welling out with a copious purient discharge while other wounds appeared almost to be gangrenous.

Overall Response

During the first hours, this activity was uncoordinated. Each physician and each hospital was alone. But, less than two and one-half hours after the explosion, Halifax began putting together a formal, coordinated response.

At 11:30 A.M., just two hours and twenty-five minutes after the explosion, the deputy mayor, H. S. Colwell, convened a meeting, in "the only room in the building not so badly wrecked by the explosion as to be unfit for the purpose." Present were some members of city council, some prominent citizens, and the lieutenant governor of Nova Scotia, head of state for the province. Colwell announced he had already sent out a request for outside medical assistance. And, he stated, he, the city clerk, and the chief of the police had asked the military to assist with the response.

Colwell's decision to seek military aid gave the city access to enormous resources. Halifax was Canada's main east coast port, the place from which military supplies were flowing to Europe. At any time, thousands of able-bodied soldiers were on hand. Equally important, Halifax was also the main port for reception of war wounded, which meant the city was home to two military convalescent hospitals and their staffs. Military involvement meant military transportation was available for the injured. It meant—once the meeting authorized this—soldiers could commandeer civilian vehicles for emergency use. And it meant the military hospitals would remain open to civilians.

The first meeting not only approved Colwell's initiatives, but also it decided to establish a formal relief committee. The chairperson was the lieutenant governor, McCallum Grant, a man well liked and widely respected. Colwell was secretary. The meeting also set up a series of other committees to deal with matters like search and rescue, transportation, and the dead. By 2:30 P.M., the morgue committee had decided to use Chebucto School as a morgue because the funeral homes were overcrowded. The structure created at that first meeting was modified on several occasions and new personnel were put in charge of some activities, but the basic management pattern persisted for more than a month. It was finally abandoned only when the federal government, using emergency powers under the War Measures Act, took over.

An EMS System Emerges

The initial EMS response was managed by Colonel Paul Weatherbe, an engineer. He and some associates looked after billeting of incoming U.S. medical personnel, established some dressing stations, and created a central dispensary. They also arranged for physicians on standby to respond to persons needing treatment at home. (About one hundred persons a day were treated during home visits.)

As the numbers of incoming medical personnel increased, the medical community, especially the medical military community, began to be concerned about this arrangement, involving, as it did, nonmedical control of medical operations. Bell, accompanied by W. H. Hattie, another army corps physician, and Fraser Harris (the man who wrote the unpublished history), called on the Halifax Relief Committee to present their view that emergency medical response should be run by emergency *medical* personnel. It was essential, they argued, that such tasks as the creation and inspection of emergency hospitals be under medical supervision.

Bell went further. He argued the incident that led to the emergency was a result of the war and that medical control should be in military rather than civilian hands. The committee agreed and on 9 December put EMS under Bell's control. From then on, Bell ran an EMS system consisting of five key elements: dressing stations, a quick-response team, a medical supply system, emergency hospitals, and, of course, the regular hospitals, the ones that had existed prior to the blast. These EMS components are discussed next.

Dressing Stations

The initial dressing stations were informal. They operated wherever the injured congregated or where medical supplies could be found. They were usually staffed by soldiers.

Later, a series of dressing stations were set up. Though they were usually placed in makeshift locations—one was in a veterinarian's office—they were staffed by nurses and other skilled personnel.

> All persons injured or ill, if able to walk, were requested to call at the nearest (dressing) station during the day, if unable to walk, a doctor or trained nurse once the case was reported was sent off. (Johnstone 1919)

Among the places providing treatment for the injured survivors were: St. Mary's Hall, Convent of the Sacred Heart, Halifax Dispensary, St. Paul's Hall, Ladies' College, St. Mary's College, Bellevue Hospital, Knights of Columbus Hall, and five other places, identified only by location. (Although the street names and numbers have changed since 1917, a review of old maps shows them in a neat semicircle around the main impact area.)

Standby Team

As long as the search for survivors continued, a team of physicians and nurses were ready, at any time, to respond to calls for medical aid. Over the five days starting the day after the blast, physicians were dispatched to deal with more than four hundred cases (Harris, n.d.).

> all cases requiring medical aid—seventy-five or eighty a day—were, without exception, attended to within one half hour at the latest from the time the call was received

This team, working out of city hall, kept going even when a severe blizzard disrupted transportation to Halifax and made movement in the city difficult.

Central Supply System

But treating so many injured in so many locations wasn't just a matter of staff. It also required supplies. A central medical supply center was set up at what is now TUNS, the Technical University of Nova Scotia. It was staffed by forty-four persons, four of whom were druggists. The supply center provided all kinds of supplies, including drugs, to the hospitals, the dressing stations and other places looking after the injured. It was directed by a visiting pharmacist and functioned as an arm of the Red Cross (Harris, n.d.).

> ... took possession of the whole of the technical college, had a staff ... working day and night, of which the Canadian Commercial Travellers' Association formed a large part, commandeered the supplies of the National Drug and Chemical Company.
> The hospitals were visited twice daily by two members of our staff, and all their requisitions were filled the same day.... We found some slight difficulty in securing small surgical instruments which, we found, were greatly needed.
> Kept two others employed delivering material to all the hospitals and dressing stations, maintaining a large prescription department, employing three registered pharmacists and other help, and was the chief basis of medical supplies.

Emergency Hospitals

The fourth component in EMS was a set of emergency hospitals, some on board ships like the *U.S.S. Colony*, most in hastily refurbished buildings.

Although some staff came from Halifax or nearby communities in Nova Scotia, most were from the New England states. During the days after the explosion, medical staff came from Boston, Massachusetts (32 physicians, 73 nurses, 13 others), Providence, Rhode Island (52 physicians, 52 nurses, 3 others), New York (supplies, 4 staff), and Maine (12 physicians, 13 nurses, and 11 others) (Smith).

These new hospitals were not put in place immediately. It took time to turn damaged buildings into hospitals. It also took time for the staffs to arrive. The blizzards had made train travel slow-to-impossible. The speed at which the new hospitals were created was, however, incredible. The first medical relief team from Boston arrived Saturday morning, forty-eight hours after the explosion. It was to take over Bellevue, an officers' mess (Bell 1918B).

> At eleven A.M., the officers were ordered out, and the engineers started work repairing the building, the windows and doors of which were shattered. The A.M.C. (Army Medical Corps) orderlies and about twenty Jackies (sailors) from the American ship, "Old Colony," worked with the nurses and doctors to clean and equip the building. All stores were drawn and in place and at 9 P.M. the same evening i.e. in ten hours from the commencement of the work of repair, the hospital completely equipped from kitchen to operating room, stocked with food and medicines, was handed over to the Massachusetts unit as a hospital ready for patients.

This activity prompted an American to write home:

> It would stir your blood old boy if you could come down Spring Garden Road and see the Stars and Stripes of Old Glory waving over ... Bellevue, that old House where the British general always lived. (Letter is in files of Public Archives of Nova Scotia. It is dated 18 December 1917.)

The work at Bellevue was repeated as other buildings were turned into hospitals, and as dressing stations were opened in locations near the impact area. On one day, 9 December, three hospitals were established: 127 beds at the YMCA, 159 beds at the Ladies College, and 130 beds at St. Mary's College (now St. Mary's University).

When the speed at which these hospitals were created was reported in the American media, it was apparently accompanied by a comment on "American efficiency." That irritated the Canadians. The hospitals were staffed by Americans but were put together mainly by staff from the Canadian Army Medical Corps. Bell commented, "It is only fair to the Engineers, A.M.C. and Ordnance Corps to state emphatically that this work ... was entirely done by local units" (Bell 1918B).

That wasn't the only news report that caused irritation. On 2 February, the Halifax paper carried a story about the medical response, which stated that local doctors "notwithstanding that they were called upon to treat their regular patients, also helped materially with the wounded." The physicians replied (Harris):

> The fact is that they and certain colleagues from outside Halifax bore the whole burden of surgical relief work until the various hospital units from the United States were installed in their various buildings ... the Halifax doctors instead of helping materially with the wounded, virtually performed all the vast work that was to be done ... before the American surgeons had time to arrive on the scene.

Regular Hospitals

The speed and efficiency with which the new hospitals were created and staffed should not obscure the fact that the key element in the initial EMS system was the existing hospitals. They carried the burden of initial response and were where the majority of emergency surgery was performed.

Camp Hill, for example, with its 250 beds, had taken in one thousand four hundred patients in a few hours. There and at the Victoria General surgery had gone on nonstop for twenty-four to forty-eight hours despite the fact that Camp Hill, a convalescent hospital, had no operating room. In contrast, the largest of the new hospitals had 130 beds and accepted that many patients; and at Bellevue, the first emergency hospital, physicians performed just ten operations.

Facilities Expanding

By creating the emergency hospitals, and by expanding the bed space in existing hospitals (Camp Hill, 400 extra beds; Cogswell, 200 extra beds;

Victoria General, 350 extra beds), the Medical Relief Committee had managed to find two thousand beds by 14 December, one week after the explosion. Bell was thanked and congratulated for excellent work.

The situation was also slightly relieved by sending 150 casualties by train to Truro the day of the explosion and sixty more to New Glasgow a day later. At New Glasgow, where there was time to prepare for the arrival of victims, the citizens turned a school into a 160-bed emergency hospital completely equipped (Harris):

> this hospital contained an operating room with four tables, and operating equipment, a dispensary with full stores, a kitchen with two ranges, not to speak of a supply of hot water and several baths both on the main and upper floors. The whole hospital was wired for electric light and electric heating.... the excellent work (was) done due to the energy and sustained efforts of the men and women, boys and girls of the town, who were assisted by the various churches and the staff of the Aberdeen Hospital.

Other regional communities also offered help. Parrsboro offered to take 50 injured. Pictou said it could take 40. Antigonish offered to handle 25. Sackville said it could take 50. Louisburg offered to take care of 30. These offers were not accepted. The survivors wanted to stay in Halifax and refused to move.

Moving out of town was not the only option rejected. When tents were set up as emergency shelters, survivors declined to use them. By that time, other emergency shelter, including space in private homes, was becoming increasingly available.

Survey of Medical Needs

Given this organization it might have seemed all those needing care would be receiving it, but the American doctors were not satisfied. One of them, Richard Giddings of the U.S. Army Medical Corps, argued large numbers of wounded were not getting proper care.

> As a result of observations made on the 8th by various members of this unit, who had visited many people in their homes, we were able to suggest the mapping of the city into districts, with the recommendation that a house to house canvas be made first by social workers, who would report as to whether medical or surgical help was needed, the case then to be seen by a doctor or a nurse. This suggestion was made because our doctors found that large numbers of injured people requiring surgical aid had sought the shelter of buildings near the devastated area, where

> they were content to stay. So dazed were they by the disaster that they did not realize that help would come to them for the asking. Also the members of our staff had found that many people could not leave their places of refuge because they had literally lost all their clothes.

The Canadians were not pleased by these comments but agreed a canvas should be made and that it be supervised by the latest arrivals, the team of physicians and nurses from Rhode Island. The findings brought Bell some satisfaction (Bell 1918B):

> This canvas was ordered because of the persistent reports ... that in these districts there were numbers of persons seriously injured who had received no medical or surgical attention. We found, however ... this area had been pretty well cared for; everybody had at least received first aid, and while there were a considerable number of serious eye cases, still no instance of neglect could be found.

Medical Relief Committee

As time passed, the system began to change. First, some of the American ships had to leave. Then some of the dressing stations closed. By 13 December, just six days after the explosion, the American medical teams began to leave as well. (To thank them, the lieutenant governor held a farewell dinner.)

As the American medical staff left, some emergency hospitals started to empty and, eventually, shut down. On 19 December, there had been 749 patients in hospital. By 12 January, the number had dropped to 344 ("Daily Consolidated report, December 19, 1917;" "Weekly Consolidated Report, January 12, 1918").

There was pressure for other changes. The Victoria General began to argue that its disaster victims should be moved into one of the emergency hospitals or into one of the many shelters. The Victoria General felt its role as a provincial hospital was being undermined because it could not service physicians throughout the province.

The local physicians were also starting to worry about being paid. Canadian Army Medical Corps personnel were paid regular salaries whatever they were doing: responding to a disaster did not effect their income. Local physicians, however, needed medical fees to survive.

The relief committee first maintained that all physicians should provide emergency services for nothing. The physicians argued this was all right for a while but an emergency had to end. The issue was resolved after a special committee of the Medical Society met with the rehabilitation committee and made clear the physicians were not trying to charge for what they had done during the actual emergency. It was agreed, after discussion, that the emergency period had ended 20 December. After that, medical services would be charged for (Minutes).

That meant all medical services by physicians were free for the first fourteen days, including 7 December, the day of the explosion. Eventually 69 physicians received an honorarium, usually $300 (six got $500), for the period immediately following the end of the two-week emergency period. A total of $23,630 was paid to eighty-four physicians by the end of June 1919, most of it to practitioners from Halifax-Dartmouth, but some from Truro, which had received a trainload of casualties.

Third, the Canadian military bureaucracy started to demand its supplies be located and returned. It required officers and lesser ranks to explain what had happened to medical supplies. And it placed advertisements in the newspapers warning civilians they could be prosecuted if they did not return the many missing army blankets.

Rehabilitation

Some problems lingered on. There were postincident emotional problems. Harris compiled examples of some cases where it appeared the impact of the trauma of disaster was somewhat delayed. (Because he later suffered a breakdown, he may have been among them.) He included in his list a physician who had committed suicide, apparently depressed by his experiences during the tragedy. And, for many of the victims, treatment would continue for years, even for life.

Others had no place to go, even if their medical problems had been resolved (Harris).

> there were persons in hospital ready to be discharged. the Emergency Shelter Committee should look after all such persons needing shelter.

There was follow up by trained medical social workers who kept track of the survivors after they left hospital and made every attempt to get them back into the work force. They were assisted by federal government staff who were involved in helping soldiers adjust to similar disabilities.

Finally, as Prince suggested in his thesis (Prince 1920), there were some positive, permanent developments. Some of the money donated by Massachusetts was used to establish an outpatient clinic at the Victoria General. The clinic was run by Radiology and included radiation therapy (Scammell).

Innovative EMS System

What happened in Halifax was not based on careful planning. In fact, Prince suggested there were no plans (Prince 1920).

Strangely, this may have been an advantage. Much medical planning for mass casualty situations assumes that there will be a disaster site, that some sort of triage can be done on site, and that the casualties can be directed in a controlled flow—worst first—with the load spread evenly among the hospitals. Halifax is one of many incidents that shows such systems cannot work when the damage is so widespread there is no site and when, inevitably, victims make their own decisions about where to go for medical help, even whether to go to a hospital (Scanlon 1989).

What happened in Halifax was that an informal system—makeshift field-dressing stations, treatment in physicians' offices—turned into a formal system. Casualty treatment centers were established in a semicircle around that same area. Supplies were taken to where they were needed. Emergency hospitals were created. Physicians canvassed the impacted area. The system, in short, reached out to provide service where it was needed.

Not only that, Bell, the man who ran the system, suggested in a final report to the federal government that such a decentralized system could be put in place by planning. He proposed hospital equipment be standardized, buildings be identified in advance as potential emergency hospitals, supplies be stockpiled for emergency use, and a scheme for first aid posts be created with the voluntary agencies getting together to plan to service such a system.

What Bell suggested—based on his experience with the worst mass casualty incident in North American history—is, in essence, what Quarantelli suggested in his book on EMS (Quarantelli 1983). Working without the benefit of experience of a major incident but with the knowledge of a social scientist, Quarantelli argued that it could be more effective to deal with mass casualties with a system of what amounts to M*A*S*H-type field hospitals, casualty clearing stations, similar, perhaps, to that which is done for rock concerts, for the Mardi Gras, or, in Canada, as was developed for the visit by the Pope. He suggested, in short, bringing the system to the people rather than trying to force the people to adapt to the system.

What happened in Halifax suggests, therefore, that Quarantelli's theories—based on a knowledge of human and organizational behavior—not only could work but did work. Furthermore, they worked so well that Bell saw them as applicable to other incidents, not just to Halifax. The nature of the medical response in Halifax—not available to Quarantelli when he wrote his book—constitutes solid evidence that Quarantelli's approach, constructed without the benefit of a major mass casualty incident, probably could work very well indeed.

Notes

1. The fact some injured were treated in physician's offices helped to reduce the strain on the hospitals. But it also meant physicians who would normally respond to a hospital emergency, were not available.

2. Puttner, incidentally, was not a physician. His Ph.D. was in pharmacy. Because of this, he had also become a radiologist. He is revered as a founder of the field of pharmacy in Nova Scotia.

Postdisaster Sheltering and Housing: Social Processes in Response and Recovery

ROBERT BOLIN

This chapter is based on a taxonomic model and conceptual analysis presented by Henry Quarantelli in a 1982 monograph. I have chosen to highlight this monograph because it has influenced my approach to the study of disaster recovery as a social process. Quarantelli's (1982) sequential, four-phase taxonomic model of the types of postdisaster housing captures the essential features of the complex social processes underlying sheltering and housing. In this chapter I review Quarantelli's model and highlight its significance in the study of household and community recovery from natural disaster. To that end, I present findings from a broad range of recent housing and recovery research and discuss them in the context of Quarantelli's 1982 analysis.

Background

The sheltering and housing of victims after a communitywide natural disaster is a social process whose shape and outcome are determined by a complex set of factors. It involves, in the first instance, characteristics of the households impacted by a natural disaster. These include factors such as household social and fiscal resources, loss levels sustained in a disaster, housing needs and preferences of the victims, knowledge of and availability of shelter and housing alternatives, victim willingness to relocate, availability of help from support networks, and demographic characteristics of the household (Bolin 1989; Oliver-Smith 1990).

The actions of household members are not the outcomes of individual choice but rather are influenced and constrained by social, cultural, ecological, historical, and political-economic forces. These factors include the avail-

---

Research supported by National Science Foundation Grant # CES-8803188 and BCS-8902584. Conclusions and recommendations are those of the author and do not necessarily represent the views of the National Science Foundation.

ability of nonhazardous areas for temporary housing facilities, historically prevalent community housing practices, governmental housing and aid programs, relocation of employment sources, postdisaster land use changes, and hazard mitigation efforts. In addition, political factionalism and the promotion of specific class interests in the reconstruction process can influence postdisaster sheltering and housing (Bates 1982; Caporale 1989; Oliver-Smith 1991; Quarantelli and Dynes 1989).

Quarantelli's (1982) taxonomic classification offered the most useful approach to discussing and analyzing postdisaster sheltering processes. In order of occurrence from disaster impact, the four types (phases) of sheltering identified by Quarantelli are emergency shelter, temporary shelter, temporary housing, and permanent housing. Quarantelli (1985) suggested that such a classificatory scheme is necessary to properly reflect the heterogeneous nature of postdisaster shelter arrangements.

Emergency shelter consists of the often unplanned and spontaneously sought provisional shelter arrangements. Such shelter is typically sought in the immediate pre- and postimpact phases of a disaster and is frequently based on convenience or opportunity, rather than on predisaster planning. The exception to this is in the case of Red Cross shelters that are planned for and provisioned as part of routine emergency preparedness.

Often areas become de facto emergency shelters simply because disaster refugees congregate there. As Quarantelli noted (1982), victims will tolerate primitive conditions in these sheltering arrangements because of the assumed short-term nature of their stays. In the United States, emergency shelters are often mass sheltering arrangements in public or other large buildings (schools, churches, etc.) and, as just noted, are planned for and managed by the Red Cross. There is ample evidence in the literature that mass shelters are avoided by the public in the United States if they have alternative sheltering arrangements to turn to (Quarantelli 1982; Drabek 1986). In other instances, emergency shelters are established by victims themselves on an impromptu basis, often in yards or parks near their homes (e.g., Bolin and Stanford 1991), and are typically occupied only until victims feel it is safe to return to their homes.

When victims seek shelter away from their predisaster residence for a period extending beyond the height of the emergency phase, they may be said to be in temporary shelter (Quarantelli 1985). Although this shelter is often an extension of emergency shelter, particularly in the case of Red Cross mass care facilities, victims usually seek other more private arrangements, often with relatives or friends (e.g., Bolin and Bolton 1986).

The distinguishing feature of the temporary shelter phase is that stays are long enough to require the provision of food and sleeping facilities (Quarantelli 1982). Consequently, there is a need for organizational planning and the provision of resources by central authorities. This phase of sheltering also requires the presence of emergency services personnel and the availability of medical services in the case of publicly provided mass shelters (Quarantelli 1985). Thus Red Cross emergency shelters may be considered both as emergency and temporary shelter, their function changing with the victims who use them.

In situations where a portion of the community escapes damage, victims may be "absorbed" as evacuees into the homes of nonvictims, although this is dependent on the existence of social support networks (Bolin 1988). In many U.S. disasters where there is residential dislocation, mass shelters provide both emergency and temporary shelter. However, in most instances such shelters are underused by victims if they have access to other sheltering alternatives such as the homes of friends or relatives (Quarantelli 1982).

As Quarantelli noted (1982), it is difficult for emergency planners to anticipate how many victims may go to the homes of kin as their temporary shelter rather than go to mass care facilities. It is also virtually impossible to predict how many victims may actually come to temporary shelter facilities after a disaster that heavily impacts housing stocks. Thus, such planning has to have flexibility to respond to unanticipated and often fluctuating demands over the emergency period (Quarantelli 1985).

The third phase, temporary housing, "involves the reestablishment of household routines but with the understanding that more permanent quarters will be obtained eventually" (Quarantelli 1985, 130). In the United States temporary housing is most often provided for by FEMA, usually in the form of grants to allow victims to rent living units until they can reestablish permanent housing. In some instances mobile homes are provided when existing housing rental stock is not sufficiently plentiful to house disaster victims. There is sufficient evidence that mobile homes are a particularly problematic form of temporary housing (cf. Bolin 1982; Golec 1983; Erikson 1976).

Temporary housing can become permanent housing, particularly in underdeveloped countries and other areas with a shortage of suitable housing (Bates and Peacock 1987; Bolton 1979; Oliver-Smith 1986). However, temporary housing facilities are seldom planned for or situated with that eventuality in mind (Haas et al. 1977). Concerns over temporary housing becoming permanent can result in delays in the development of temporary housing facilities (e.g., Bolin and Stanford 1990). As Quarantelli noted (1982), permanent housing is the least planned for of all housing phases. The privatized approach to housing in the United States generally leaves it up to the victims to establish new permanent housing, sometimes with the assistance of government funded grants and loans such as from FEMA and the SBA (e.g., Bolin 1982). How permanent housing is reestablished often reflects and is determined by predisaster social inequalities in the community (Bates 1982; Oliver-Smith 1986).

Given the heterogeneous nature of postdisaster housing, all four types of disaster housing/shelter may occur simultaneously in a community (Quarantelli 1985). That is, some victims may be still in emergency or temporary shelter, whereas others may have already established permanent housing again (Bolin 1986). As Quarantelli (1985) noted, this can create a wide range of demands on various organizations involved in the provision and management of the various phases of housing.

The variability in the speed at which victims are able to return to permanent housing reflects the underlying class structure of the society. Lower SES victims typically have fewer resources available to speed their return to permanent housing (Bolin 1982). Thus, postdisaster housing processes tend to mirror preexisting social inequalities.

Although temporary sheltering and housing arrangements have not been studied extensively (Quarantelli 1985), some researchers have discussed problematic features that may occur. As discussed by Quarantelli (1982), temporary shelter and temporary housing are the targets of the most complaints by victims. These complaints often reflect underlying difficulties in both the housing itself and the organization and administration of the housing programs. Among the reported difficulties are indecisiveness by local planners as to where to situate temporary housing facilities, organizational delays in the acquisition and disbursement of temporary housing, inconsistent standards of victim eligibility, housing inappropriate to the cultural preferences of the victim community, poorly situated temporary shelter and housing camps, inaccessibility of employment opportunities for victims, transportation difficulties for dislocated victims, conflicts among residents of housing camps, and the hazardousness of the housing itself (Bolin 1982; Bates 1982; Davis 1975; Caporale 1989; Loizos 1977; Raphael 1986). As Quarantelli (1985) noted, it is not clear if many of the reported problems associated with temporary shelter arrangements are the result of the characteristics of the disaster event, the physical nature of the housing, or organizational problems in providing and managing the temporary housing programs.

The literature identifies some temporary shelter and housing alternatives that are frequently used in disasters in which there is either a threat to, or actual destruction of, housing stock in a community. In instances where there is large scale proactive evacuation, such as in hurricanes, mass care emergency and temporary shelters are typically established in large buildings such as armories, schools, and churches (Quarantelli 1984). Common forms of temporary housing include victims living in damaged homes while repairs are made, the acquisition of apartments or rental homes to live in until permanent housing is reestablished, living in mobile homes while rebuilding a home, and "doubling up" with a host family (Bolin 1982; Raphael 1986; Quarantelli 1985). The latter, although documented in some studies (Drabek

1986), appears a viable and sustainable temporary housing alternative mostly in societies with strong kinship orientation. Although the host-victim relationship has not been studied extensively in the United States, Bolin (1982) found that beyond a period of approximately one month, relations can become strained within the confines of a single family dwelling.

Housing and Recovery

Beyond merely describing the variety of postdisaster housing arrangements, researchers have, in the last fifteen years, begun to focus on the dynamics of household recovery as a social process. In keeping with the notion that sheltering and housing are components of a social process (Quarantelli 1985), housing has been studied as an indicator of the postdisaster recovery process for families and households. Initial work in this area came under the rubric of "family recovery" (Bolin 1982) and entailed multidimensional modeling of the processes whereby families reestablish housing circumstances comparable to their predisaster state. These models are founded on a longitudinal and processual conceptualization of housing recovery including some assessment of exogenous factors (Drabek and Key 1984).

Bates (1982; see also Peacock et al. 1987) developed a "domestic assets index" that is a quantitative scale used to measure the value of a household's physical assets. It is based on assets used to perform vital household functions or activities and appears useful as a measure of household recovery after disaster (Peacock et al. 1987). The intent of the use of the domestic assets index is to assess the impact of disaster on households in comparison to nonimpacted households. In other words, household assets are continuously changing, and a disaster may be expected to have variable impacts superimposed on the existing household change vectors. As Peacock et al. (1987, 83) concluded about household recovery after the Guatemalan earthquake of 1976:

> Housing programs that were designed by agency personnel and planners to provide temporary housing proved to have long-term negative effects on recovery. . . . Temporary houses simply became semipermanent dwellings, perhaps due to economic and social factors associated with the "normal" conditions of underdevelopment in Guatemala. Permanent housing programs . . . had exactly the opposite effect in that they not only produced recovery on an average, but resulted in positive increases in . . . domestic assets well above recovery levels because they provided participants with long-term housing that was superior to predisaster housing in most cases.

This research provides empirical support for the linkage between housing processes as discussed by Quarantelli (1985) and overall recovery. Further, because of the unequal distribution of permanent housing aid, the net outcome of the earthquake recovery process was an increase in societal inequality (Peacock et al. 1987; Bates 1982). Thus, social class was a determinant of the type of housing aid received and the levels of recovery attained (cf. Bolin and Bolton 1986; Wijkman and Timberlake 1988).

Hogg's research on housing recovery after the Venzone, Italy, earthquake also illustrated the complexities of housing processes after disaster. Hogg wrote (1980, 182) that in Venzone the reconstruction:

> radically altered the character of the community as a whole ... [F]ollowing relocation some wage earners have to spend more time and money commuting ... to work, or some people find their new neigbors incompatible. The houses tend to be small, identical and closely spaced in any one area of prefabricated dwellings and this can have adverse psychological effects on the inhabitants.

Hogg noted that the cultural dynamics in Italy were such that inhabitants of the Friuli area were less worried about the speed of reconstruction (cf. Haas et al. 1977) and more concerned about maintaining architectural integrity of style when homes and public buildings were rebuilt. The desire to properly rebuild centuries old and historically significant buildings slowed the pace of reconstruction significantly.

Patterns of postdisaster shelter and housing are bound up in the larger social processes of recovery and reconstruction. Both housing and overall recovery processes are strongly influenced by the social dynamics already present in the predisaster social structure (Oliver-Smith 1991). Recovery, as used here, refers to a condition in which a community achieves a state after a disaster that it would have achieved if the disaster had not occurred at all (Bates and Peacock 1987). Recovery, in this sense, incorporates the notion that disasters interrupt social processes, including those involved in housing, often either accelerating or decelerating them.

Case Studies of Sheltering, Housing, and Recovery

To illustrate the use of Quarantelli's housing process typology, in this section I review postearthquake sheltering, housing, and reconstruction processes after three California earthquakes. Each case study is used to illustrate aspects of postdisaster shelter and housing. The three quakes considered below are Coalinga (1983), the Whittier-Narrows earthquake (Los Angeles County, 1987), and the Loma Prieta earthquake (Santa Cruz County, 1989).

The Coalinga earthquake of 1983 was of moderate intensity (Richter magnitude 6.7), but it did significant damage to the small and geographically isolated community of seven thousand in central California. The quake caused an estimated $31 million in damages, most of it to unreinforced masonry buildings in the central business district and to surrounding residential areas. More than 950 residential units (35 percent of the total) were heavily damaged or destroyed by the temblor (French et al. 1984).

Temporary housing took two primary forms. Because some of the homes were uninhabitable, the Federal Emergency Management Agency (FEMA) brought in two hundred mobile homes as part of their temporary housing program. The other main component of FEMA's response was through grants to qualified victims to pay for rent while in temporary housing facilities. The rent subsidy program's effectiveness is limited by the availability of surplus housing stock in disaster stricken communities. In Coalings and other earthquakes of the 1980s in California, the lack of surplus low-income housing has repeatedly created temporary housing difficulties for victims with limited economic resources (e.g., Bolin and Bolton 1986; Bolin and Stanford 1991).

Patterns of temporary shelter use and household recovery tended to follow class lines in Coalinga. Victims with lowest incomes experienced the greatest proportional losses to their housing. A disproportionate number of these victims were Hispanics (Mexican-Americans). As Bolin and Bolton (1986, 207) noted, Hispanic victims experienced more difficulties in reestablishing permanent housing than did Anglo victims because of their socioeconomic resources:

> Hispanics lived in the oldest and most damaged housing; they generally rented; and the type of housing that they occupied was either unlikely to be repaired or, if it was fixed up, it was likely to be priced out of the market it once occupied.

Additionally, Hispanic victims had fewer financial resources, no insurance coverage, and higher losses than Anglo victims. They also had to rely more on federal aid and housing programs.

Federal housing aid in the United States provides greater resources to homeowners than renters, further complicating the recovery of lower income victims who are forced to rent because of their income level. Hispanic victims were more likely to use mobile homes as a temporary housing facility than were Anglos, reflecting the fewer housing options available to them (Bolin and Bolton 1986).

Because few households in Coalinga, either Anglo or Latino, had earthquake insurance, and because federal and state aid was inadequate, less than one fifth of the residents had all of their disaster losses covered by aid programs and insurance. The remainder fell back on alternative strategies to deal with losses, including borrowing money or doing without certain amenities that were previously available to them. Hispanics were more likely to experience a decline in standard of living than were Anglos after the earthquake (Bolin and Bolton 1986).

Whereas the Coalinga earthquake impacted a relatively isolated town with an economic base derived from both oil and agriculture, the Whittier Narrows earthquake of 1 October 1987 struck wide areas of the heavily urbanized Los Angeles metropolitan area. The quake, a moderate magnitude 5.9 event, did widespread but limited damage throughout Los Angeles County (Weber 1987). The greatest concentration of damage was in Whittier (population seventy-five thousand), east of the city of Los Angeles. The 1 October quake was followed by a significant magnitude 5.3 aftershock three days later that did additional damage to buildings in Whittier.

Some four thousand buildings in Los Angeles County were heavily damaged or destroyed, including 61 buildings in an eight-square-block area of Whittier's central business district. An additional 750 residential units in Whittier sustained heavy damage, and approximately half were condemned as unsafe in the two-week period following the main shock (Bolin 1989). Total losses from the earthquake were placed at about $360 million. As with Coalinga, impacts of the quake and the resultant recovery processes in Los Angeles County coincided with existing social class and ethnic divisions (cf. Tubbesing 1989).

Because of the uneven distribution of earthquake impacts, certain areas of Los Angeles County had concentrated damages and loss of housing stock. Some of these neighborhoods were home to significant numbers of ethnic minorities, many recent immigrants from Latin American and Southeast Asia (Rubin and Palm 1987). Latinos, particularly those recently from Mexico, were reluctant to return to apartments after the main shock because of hundreds of aftershocks that occurred (Bolin, 1989).

Recent experiences of some Los Angeles victims with the Mexico City (1985) and El Salvador (1986) earthquakes were ostensibly responsible for victims' hesitancy to return to built structures. This reluctance was not overcome by building inspectors' assurances that buildings were, in fact, safe for occupancy. At the height of the emergency, several thousand victims in east Los Angeles, primarily Latinos, were camped in yards and city parks, rather than returning to their apartments in the barrios. The Red Cross developed multilingual outreach programs to attempt to get non-English-speaking immigrants to return to their homes. Bolton (1988) found that Latino victims of Los Angeles had strong commitments to their neighborhoods and were reluctant to seek temporary or new permanent housing located away from those areas. The predominance of poverty within the Latino community compounded the difficulties of housing those displaced by the earthquake.

As is typical in many urbanized areas in California, Los Angeles has a pronounced shortage of housing for lower income households. Much of the low-income housing stock in California is in the form of tenements built of seismically vulnerable unreinforced masonry construction. These structures are being torn down or seismically retrofitted and upgraded to be safer. Consequently what was affordable housing, albeit deteriorated and seismically risky, is becoming less available or more expensive. Because the Whittier Narrows quake impacted most heavily on the structures used for low-income housing, those displaced by the quake were from precisely that part of the class structure least able to find new housing at prices they could afford. Thus, the earthquake increased the marginalization of a segment of the population already on the periphery of the wealthy urban core.

In Whittier proper, the quake damaged a wide range of housing, including large older single-family dwellings as well as apartment houses. Although homeowners in Whittier relied on FEMA temporary housing programs to pay for transitional facilities, it appears that apartment dwellers often opted to simply leave the area rather than go through the process of attempting to procure available FEMA temporary housing aid.

The return to permanent housing for homeowners in Whittier was delayed in some cases by the amount of time it took to receive SBA loans to rebuild homes. Additionally, those few victims with earthquake insurance, in some instances, took more than a year to achieve a settlement with their insurers, again delaying the reconstruction of their homes. Such delays in acquiring recovery resources resulted in some victims still being in temporary shelter whereas others had already returned to permanent housing (Bolin 1989; cf. Quarantelli 1985). Although renters were less likely to use aid programs, their mobility improved their access to new permanent housing compared to homeowners with heavily damaged property.

As rebuilding began in the most heavily damaged areas of Whittier, conflicts soon emerged over the types and location of new, permanent housing as well as over redevelopment plans for the central business district. At issue was changing patterns of land use regarding single family dwellings and apartment/condominium complexes. Long-term residents of the area wanted to see a pattern of large older single family dwellings maintained. Although many of these houses were damaged in the quake, largely because they shifted off their foundations, most were reparable if owners could obtain adequate funding for their reconstruction. Developers, on the other hand, sought to buy up damaged homes and convert the properties to multifamily dwellings, thereby changing the character of the neighborhoods.

As a consequence, the reestablishment of permanent housing in Whittier engenedered social conflict. Several neighborhood-based groups fought to maintain land use and housing patterns as they were before the quake. Developers and speculators attempted to use the earthquake as an opportunity to increase housing density and subsequent profits. As Quarantelli noted (1982), the return to permanent housing is rarely planned for as part of emergency planning. The failure to anticipate the types of issues and conflicts that emerged in Whittier resulted in various planning and legal delays in reconstruction and the reestablishment of permanent housing.

Loma Prieta, California's 1989 earthquake, the most powerful to strike an urban area since the 1906 earthquake that destroyed San Francisco, is the third case to be considered. The magnitude 7.1 October 1989 Loma Prieta earthquake produced a complex of shelter and housing issues that are only slowly being addressed or resolved. The most concentrated damage to housing and businesses occurred in Santa Cruz County, south of the Bay Area, and near the epicenter of the quake (Bolin 1990).

Temporary shelter and housing in the cities of Santa Cruz (population 48,650) and Watsonville (population 30,000) became key social and political issues. Problems in sheltering and housing victims in Santa Cruz County were strongly influenced by existing social trends that were intensified by the disaster. Those same trends and conditions are still influencing recovery processes in the area (Bolin and Stanford 1990).

Both Watsonville and Santa Cruz are relatively discrete communities, separated from the urban sprawl around the San Francisco Bay by mountains. Whereas Santa Cruz is a predominantly middle-class, university, and tourist town, Watsonville is a major agricultural center with a large resident Hispanic population. Land and house values in Watsonville and Santa Cruz have been influenced by development and real estate speculation from San Francisco Bay Area-based investors. Affluent commuters to the San Jose/Silicon Valley area are also exerting upward pressures on home values by living in Santa Cruz County and working in the Bay Area.

Prior to the earthquake, Santa Cruz County evidenced sharp class divisions that, although not openly antagonistic, manifested themselves in increasing difficulties for low-wage agricultural and service workers in trying to find adequate housing. In Watsonville, where the majority of victims were Latino agricultural workers, the earthquake exacerbated the shortage of low-income housing to a point of crisis. In Santa Cruz, postquake sheltering was also complicated by a lack of available and affordable housing for the displaced. In Santa Cruz those with the greatest problems in finding new housing included elders displaced from residential hotels, lower income Hispanic service workers, and members of the sizable population that were already homeless before the quake.

Of the estimated 13,000 persons in the affected area left homeless from the quake, 8,000 resided in Santa Cruz County (Bolin 1990). FEMA received more than 4,000 applications for temporary housing in the county, a good indicator of the demand for suitable housing. The housing problems in Santa Cruz County resulted in the Red Cross keeping its mass care emergency shelters open sixty-six days, more than four times as long as in the Whittier Narrows earthquake (Bolin and Stanford 1990). Even when Red Cross shelters were finally closed, a total of twenty-one families were still staying in one Watsonville shelter and had to be placed in motels.

FEMA's initial response to the transitional housing of victims was through its normal temporary housing program that provided cash grants to qualified victims. Amounts of these grants were determined by "prevailing rates" for available rental units, in this case $960 per month. For renters, grants were available to cover two months rent and for homeowners, three months. Victims who could demonstrate that the money, once granted, was actually used for temporary housing, could be eligible for further assistance, if a longer stay in temporary housing was needed.

For many victims in the county who received the temporary housing grants, it became apparent that no housing of an appropriate sort and in an appropriate location was available for short-term occupancy. After significant political pressure by a local congressional representative, FEMA agreed to provide mobile homes to victims unable to find rental housing. The trailers, located in special camps prepared by the County, were to be available for a maximum of eighteen months, with an eligibility review period every two months. Persons could remain in the trailers contingent on either currently rebuilding a damaged home or actively searching for new permanent housing, a requirement likely to be the source of significant future conflict (cf. Bolin 1982; Golec 1983).

In Watsonville the provision of temporary shelter and housing was bound up in a complex set of class and ethnic antagonisms that have. at different times, involved the Red Cross, FEMA, the Department of Justice, as well as city and county officials. These conflicts emerged over preexisting social inequalities, and focused on issues of substandard housing and the lack of adequate housing for low-income residents in the county. By destroying nine hundred housing units in Watsonville, the earthquake made housing the subject of public debate. The earthquake was used by community activists as an opportunity to throw open the political agenda in Watsonville and to attack existing social inequalities.

That those most in need of housing after the quake were low-income Mexican-Americans employed as laborers in the agribusiness sector added complexity to housing and recovery. Hispanics in the area have been inextricably bound to agricultural production and its attendant low wages. On the other hand, local land values and rents are rising from Anglo land owners and speculators engaged in real estate development for an upper–middle-class clientele. As a result, the labor needed to drive the multimillion dollar agricultural production of the county is being priced out of housing by those who variously consume or profit from that production. The earthquake, by creating a crisis in low-income housing, made these conflicts visible.

Emblematic of the political struggle that the earthquake occasioned in Watsonville was a "tent city" that Mexican-American victims established in a park in that city. Because the Red Cross had established five temporary shelter facilities elsewhere, they were reluctant to recognize this impromptu tent camp or to provide food or medical services. Those in the tent camp refused to go to official shelters because of fear of aftershocks and because of an expressed desire to be near their old neighborhoods.

What began as a relatively straightforward process of providing temporary shelter evolved rapidly into a politicized series of social actions. These included charges of racism and cultural insensitivity by Latino activists against the Red Cross and their shelter operations; a Justice Department investigation into allegations of discrimination against Hispanics by the Watsonville city government; and a legal complaint filed against FEMA by twenty citizen groups and community organizations charging that FEMA systematically discriminated against low-income and non-English-speaking victims.

In response to these conflicts some efforts were undertaken by various coalitions of community activist groups, governmental agencies, and private organizations to provide low-income housing as part of the planned reconstruction of the cities in question. More immediate, the need existed to provide temporary housing for low-income victims until recovery and reconstruction moved beyond the planning stage.

One solution to the temporary housing crisis was the use of FEMA trailers. Although FEMA initially did not intend to bring in trailers to Santa Cruz or Watsonville, the trailers became the major source of low-income temporary housing. Whereas middle-class victims avoid using trailers as temporary housing (e.g., Bolin 1982), for many poor Hispanics in California, the trailers represented a significant step upward in terms of housing quality (Bolin and Stanford 1990). Because many victims saw the trailers as a permanent solution to the housing crisis, this view is likely to have complicated FEMA's efforts to manage the trailer camps and to withdraw them after the eighteen-month period of availability has expired.

Shelter and housing in Santa Cruz involved similar concerns to those in Watsonville to the extent that sheltering and recovery were likewise complicated by a lack of affordable housing for lower-income victims. The victim population in Santa Cruz was more diversified by class standing, and less unified by ethnicity than Watsonville's victim population. Although the disaster in Santa Cruz also opened the political agenda to address the housing needs of marginalized groups, including elders, Latinos, and the chronically homeless, it was not characterized by the rancorous and polarizing conflict seen in Watsonville.

The provision of temporary housing in Santa Cruz was complicated to an extent by a number of predisaster homeless who used Red Cross shelters and mass care services. Although these nondisaster homeless were allowed to stay in Red Cross shelters, they were not eligible for longer term temporary housing programs as provided by FEMA. The earthquake did furnish the impetus to address the lack of low-income housing in the city and to explore ways to accommodate the needs of the permanent homeless population in Santa Cruz.

Because of the destruction of several residential hotels in downtown Santa Cruz, five hundred low-income elders were displaced by the quake, and three hundred of those needed new housing that met their specific needs. FEMA worked with local social service agencies to place homeless elders in suitable temporary facilities, although thus far no new permanent housing arrangements have been made (Bolin and Stanford 1990).

Unlike Watsonville, where reconstruction planning responded to the collective demands of previously marginalized populations, in Santa Cruz steps to provide permanent solutions to housing difficulties were initiated "from above"—by city and county government and other political influentials. Six months postimpact both communities were only beginning to formulate reconstruction plans for destroyed business districts and for providing low-income housing for those made homeless by the earthquake.

Officials in both communities have spoken optimistically about a rapid reconstruction and recovery process. However, their planning process, by trying to accommodate the often incompatible demands of private citizens and citizens' organizations on the one hand, and those of business and development interests on the other, is unlikely to proceed rapidly. To expect political consensus on housing and land use in the highly diversified multicultural setting of Santa Cruz County with its increasingly activist marginalized groups ignores existing tendencies. The case of Santa Cruz County illustrates well Quarantelli's assertion that postdisaster housing is a heterogenous social process. A full spectrum of postdisaster housing was being used simultaneously in the county, with new permanent housing being the key issue in recovery.

Concluding Observations

These brief case studies illustrate the types of social processes highlighted in Quarantelli's model of postdisaster shelter and housing. Quarantelli provided a useful taxonomy and heuristic for conceptualizing a key postimpact social process. Although not explicitly tied to recovery in its initial formulation (Quarantelli 1982), the model clearly articulates a process of return to housing facilities that victims consider permanent. That new permanent housing may differ from the housing that existed prior to the event is contingent on the complexities of intermediate housing stages and the types of aid programs available to victims. As the case studies of California earthquakes illustrate, all phases of postdisaster housing can become problematic if adequate planning and resources are not available. The changing ethnic and socioeconomic characteristics of disaster victims in the United States are creating new issues and needs in the provision of temporary shelter and housing that are not always met successfully (Tubbesing 1989). The return to permanent housing clearly receives the least attention from planners (Quarantelli 1985), but it is one of the keys to the overall recovery of families and communities. Quarantelli's contribution to understanding these social processes is important for both researchers and emergency planners.

Stress Coping and Emergency Disaster Volunteers: A Discussion of Some Relevant Factors

NEIL R. BRITTON, CARMEN C. MORAN, and BENJAMIN CORREY

The Permanent Volunteer Within the Emergency Disaster Context

This chapter seeks to identify the principal variables thought to influence individual capacity to cope with stress. The authors, however, are not attempting here to develop a universal model of stress-coping behavior. Rather, interest is directed toward a specific group of workers who operate within a distinctive life-threatening work environment. The environment in question is the large-scale social crisis occasion (i.e., the mass emergency or disaster). Narrowing our focus further, attention is centered on a cohort of unpaid individuals who have a long-term attachment to, and who are recognized as being permanent members of, formally constituted and organized voluntary emergency and disaster service organizations. Within this setting, specific concern is oriented toward identifying individual and social attributes leading to successful stress-coping mechanisms among permanent emergency and disaster volunteer workers. In particular, we look at the volunteer in terms of personality and coping styles, and how much these interact with both the organizational setting and the large-scale social crisis occasion.[1] Within this framework it is useful to consider psychosocial factors associated with the act of volunteering, and then factors associated with appraisal and management of stress in the social crisis situation.

The chapter is speculative in that little practical material is available dealing with volunteers as a specific emergency worker group. Whereas there

---

Funding for this research was provided by a Cumberland College of Health Sciences (University of Sydney) Research Grant. The authors are grateful for comments provided by Mike Smithson (James Cook University of North Queensland, Australia) and Roger Wettenhall (Canberra University, Australia) on an earlier draft.

have been studies on specific types of volunteering behavior such as bystander intervention and voluntary association membership, little attention has been given to the role of the permanent disaster worker in the context of social crisis. Similarly, studies have explored the personality profile of professional emergency workers such as police, ambulance, and fire officers, yet few studies focus directly on the permanent volunteer. Moreover, relatively few studies have been systematically carried out to gauge the influence of organizational factors (both structure and process) on stress capabilities of emergency service personnel; fewer still have concentrated on stress coping within volunteer disaster agencies. Given the reliance of disaster management systems on volunteers (Britton 1990, 1991a; Britton et al. 1990), the current imbalance in the research is inappropriate.

Factors Affecting Volunteer Behavior

There is little systematic information on why individuals become volunteers in emergency and disaster organizations. A considerable body of research on volunteerism focuses on behavior associated with low risk and/or intermittent involvement, and hence is not necessarily generalizable to permanent volunteers.[2] Rushton (1980) proposed two basic types of motives to account for prosocial behavior: empathy related motives (sympathy, concern, role-taking) and personal norm related motives (moral obligation, sense of duty, perceived responsibility). Smithson et al. (1983) noted two kinds of ideologies in people working in helping professions. The first is characteristic of people who are essentially caring and sympathetic to others. The second is a type of "missionary ideology," where people have a mission to change or reform things, rather than help per se. It is possible that permanent volunteers in the disaster context have a similar duality. It is unclear the extent to which delineating these motives can provide information on the personality of those who volunteer, however, as attempts to "locate a single explanatory theory for altruistic behavior at the level of personality have not succeeded" (Kemper 1980, 331). Furthermore, individuals are likely to volunteer for more than one reason, which could reflect different aspects of personality dependent on the situation or context. For example, in a recent U.S. Gallup Poll survey on voluntary action, 52 percent of responses were associated with altruistic motives, but the remaining responses totaled 144 percent, indicating considerable scope for multiple motives for voluntary action (cited in Ferris 1988).

It is important to note that when situational influences are extremely strong, personality characteristics may not be as important in determining actions. Conversely, when situational influences are weak, personality characteristics may be strong determinants of actions. This pattern is supported by observations in the social crisis context: soon after a disaster or a mass emergency (an occasion wherein there are strong pressures for individuals to get involved) many people come forward to volunteer, but in the longer term the number of permanent volunteers remains relatively low at about 3 percent (Britton et al. 1990, 5). Staub and Feinberg (1980) discussed determinants of adult prosocial behavior, suggesting that many personality characteristics contribute to "an inclination to help others in many situations." These include appropriate and nonconflicting values or goals, courage, confidence, ability to take another's role, empathy, awareness of consequences and a tendency to plan, the assumption of responsibility to self for others' welfare, and personal goals such as need for approval or achievement of a desired endstate. Although this personality profile might seem appropriate for permanent volunteers, it does not account for involvement specifically in emergency organizations.

It has been suggested that individuals may be more likely to give assistance to friends and family than they are to strangers (Smithson et al. 1983), although this selective helping proclivity may break down under disaster conditions (Drabek 1986). Amato and associates (1984), for instance, asked thirty people to describe their reasons for providing assistance (money or goods) to victims following the "Ash Wednesday" bushfires in Victoria (Australia) in 1983. Sixty-five percent referred to the victims' extreme and urgent need, and to the fact that they would expect the same sort of aid from the community if placed in a similar situation. This appears to indicate a type of "pragmatic altruism." Wolensky (1979) investigated motives for volunteerism following tropical cyclone Agnes in the northeastern United States and noted distinct types of volunteerism that went beyond simple altruistic motives, for example, egoistic, opportunistic, and communalistic motives. Seen within the social crisis context, these motives could lead to behaviors which generally would be appropriate from a stress management perspective.

In an informal discussion with permanent emergency volunteers in New South Wales (Australia), we found that they usually first offer altruistic explanations for joining volunteer organizations, and later discuss other pragmatic or egoistic reasons. In a recent preliminary study that included an open-ended question on reasons for joining a voluntary emergency organization, responses citing altruistic motives were common (Moran 1989). Respondents tended to give stock answers such as "I wanted to help my fellow man." It is difficult to be confident that altruism is the only motive, especially when the phrasing of the responses is so close to the socially desirable answer. However, the same subjects also scored high on two altruism items embedded in a seventy-two-item questionnaire, so it is possible that these individuals do have higher than average concern for others. But respondents often indicated that other factors such as companionship, social activities, and excitement also explained their volunteer involvement. For instance, in reply to an open-ended question asking "What are the main reasons that you have continued to be a member?" 14.2 percent of a sample of Sydney (Australia) volunteer bushfire fighters indicated "comradeship" was the major motivating factor; and a further 3.3 percent stated that "social benefits" (get-togethers, barbeques, and the like) was the primary consideration for joining (Correy 1989, 114). It is important, therefore, to consider the context of volunteer involvement when attributing motives to individuals. It has been noted, for instance, that

> in many cases, such as rescuing those in need or in emergency situations, direct social rewards are quite likely to result, and it would be naive to think that such rewards never cross the minds of those involved. (Gergen et al. 1972, 117)

Eisenstat and Felner (1983) also noted that with paid human service workers intrinsic factors such as sense of accomplishment and opportunities for personal growth were more likely to explain involvement than were extrinsic factors such as salary. Superior job performance in paid ambulance workers has been correlated with personality variables such as "excitement seeking" and desire for "self-development," suggesting that certain egoistic motives might be highly appropriate for performance in the social crisis context (Dawson 1988). It should not be assumed that because the volunteer does not receive a salary, his or her motives must be solely altruistic. Other intrinsic and extrinsic factors may be operating, and it is possible that these factors may explain volunteers' ability to tolerate aspects of the social crisis event that others would consider aversive and, hence, choose to avoid.

Volunteering as a Stress Management Strategy

The act of volunteering itself may also help individuals deal with stress. An attributional analysis might explain volunteering as providing the individual with a sense of control that mitigates the aversiveness of a stressful event. Because volunteers are aware that they have chosen to be in the crisis situation, they might reframe the aversive situation as one that is, nevertheless, tolerable "since I put myself here." Support for the mitigating effect of volunteering was noted in observations of a Zimbabwe group of sixty-two farmers who voluntarily participated in a residential relocation program (Myambo 1989). Although these farmers experienced considerable stress associated with the relocation, they reported that they could cope with it. They did not differ from a control group on factors associated with "social networks and uplifts and hassles" (Myambo 1989, 84). The farmers were likely to see their stress as situationally based; that is, they made external, causal attributions even though they volunteered for the relocation. Hence, although individuals may volunteer for a stressful activity, they may not adopt internal attributional styles.

There is no a priori reason to expect that permanent volunteers are exposed to a different set of stressors within the social crisis context than paid workers. In Australia, for example, permanent volunteers often respond to the same event as paid permanent rescue workers (Britton 1991a, 1991b). The stressors and aversiveness associated with such events can vary considerably. Furthermore, stressors can occur at the organizational or individual level. Organizational stressors will impinge on individual responses, as the individual functions in and identifies with the disaster organization. For example, much of the emotive reactions of staff and volunteers in the Country Fire Authority in Victoria (Australia) following the Ash Wednesday wildfire was directed at the authority's role and responsibility toward its members after the fires (Symons and Glowaski 1988). It is interesting to note that in the discussion of the firefighters' reactions, and also in the discussion of posttraumatic stress symptoms, no distinction was made between the volunteers and paid workers.

Recent definitions of stress have emphasized the role of appraisal (Lazarus and Folkman 1984). Stress is seen to result from an appraised imbalance between perceived demands of the environment and perceived ability to cope. This approach avoids the problems associated with earlier models that categorized stress either as the stimulus (e.g., smells, sights) or the response (e.g., increased sympathetic adrenomedullary activity) and minimized the role of the individual in a stressful situation. By emphasizing the role of appraisal, the current views of stress place more emphasis on personality and related characteristics. Caution is needed, however, because the idea that there is a simple unitary concept of "personality" is just as difficult to sustain as a simple unitary concept of stress. Smith (1980) grouped numerous personality traits into several clusters considered useful for understanding voluntary involvement in formal and informal associations, but he also acknowledges that situational variables are important in understanding individual social behaviors. In the stress literature, the increasing emphasis placed on the individual's appraisal of the stressful situation has sometimes been at the expense of objective information about the situation. Information about all aspects of a stressful event, the person, and the situation, is obviously important, especially in a social crisis situation.

In the social crisis context, "appraisal" could refer to the assessment of threat. Carr (1974) put forward an equation that has been used to assess perceived threat from aversive events:

$$\text{Threat} = \text{Probability} \times \text{Cost}.$$

Even with the relative simplicity of only two values (Probability × Cost), the behavioral outcomes implied by this formula offer considerable variation. Perceptions of threat and the personal costs that may result vary between individuals. Persons high in anxiety, for example, rate the probability and costs of negative events as higher for themselves than others, and, hence, would perceive more threat in the environment (Butler and Mathews 1983). Certainly, perceptions of threat (including stress and trauma) are crucial to any evaluation of stressors to which rescue workers are exposed.

In a preliminary report of our research it was noted that when permanent volunteers were asked how often in the course of rescue work they were exposed to "something horrific or traumatic" responses varied from "never" to "all the time" (Moran 1989). It seems unlikely that some volunteers escape exposure to trauma in virtually the same environment that others come into constant contact with it. Even allowing for some objective differences in rescue environments, it is likely that the interpretation of the word *trauma* is extremely variable across individuals, and needs to be explicitly defined in further research. Differences in the perception of trauma could then be more confidently explored.

Stress, Resources, and Coping

Perception (or appraisal) alone is not sufficient to account for the aversiveness of stress and the development of subsequent pathology. Other factors such as level of exposure and social support may be associated with different reactions to an event. Vitaliano and associates (1987) recently proposed the following equation to take into account differences in level of exposure to disasters, as well as psychological and social factors that might mitigate response to this exposure:

$$\text{Distress} = \frac{\text{Exposure to stressors} + \text{Vulnerability}}{\text{Psychological Resources} + \text{Social Resources}}$$

This equation is meant to be heuristic rather than mathematical. Moreover, the factors need not be independent. Nevertheless, the equation does allow some predictions. For this chapter, and our current research program, the prediction of interest is that individuals with lower psychological resources will experience higher levels of distress, holding other factors constant.

One type of psychological resource that is relevant in this context is "coping style." There is much anecdotal information that rescue workers use effective coping styles, but little systematic evidence in this regard. Dunning (1985, 126) for example, stated that those involved in emergency work may be "more emotionally stable than the general population, and are less likely than the ordinary citizen to crack under intense pressure" but does not offer any systematic empirical evidence for this. Other authors (e.g., Mitchell 1988) have taken an opposite view, implying the level of coping breakdown is high, although it is not clear how much this impairs functioning at the site of social crisis event. It is likely that both these views have some truth; that is, rescue workers are good at coping at the scene, and for some individuals the cost of coping is high, with extreme wear and tear on psychophysiological resources leading to later breakdown.

Coping may be problem-focused or emotion-focused (Lazarus and Folkman 1984). These two forms of coping are often seen as separate activities, but in the emergency context it is likely that both aspects of coping would be adopted. Many forms of emotion-focused coping exist, one of which is humor. In a recent series of articles and letters in the British journal *The Psychologist*, police and ambulance workers debated the role of humor as a coping strategy. One article centered on the use of humor as a means of saving face around co-workers (Joyce 1989) and saw humor as a means of coping with the social demands of a crisis event. A second article discounted this view and proposed that humor was employed to prevent intense emotional grief from overwhelming the worker thus making the rescue or medical task impossible (Steele 1989). In this case, humor was seen as an effective emotion-coping strategy. In a third article dealing with the cost of coping, the writer argued that not only was humor an effective coping strategy, it was too effective, and as a result many workers failed to come to terms with their true reactions to an emergency scene and became hardened or debilitated in some other way (Kingswood 1989). This interchange of views makes the important point that coping strategies cannot always be seen as unequivocally good or bad. Only when factors such as the context of coping or other indices of personal functioning are included can the appropriateness of a coping strategy begin to be evaluated.

Nevertheless, evaluative labels are applied to coping strategies. Andrews and associates (1989) adopted Vaillant's (1977) terms "immature, mature, and neurotic" for factors on the Defensive Style Questionnaire (DSQ). The defensive or coping styles in the DSQ measure everyday responses to stressors, conceptualized according to DSM-IIIR nomenclature, for example "projection," "humor," and "reaction formation." In our preliminary study, permanent volunteers completing the DSQ did not differ from the norms provided for the "immature" and "mature" defensive styles on the DSQ, and after two altruism items were removed, they did not differ on the "neuroticism" factor. In terms of defensive styles at least, it seems the volunteers were not made up of any extreme type, except that they were likely to be more altruistic (Moran 1989). A second study of ours, although again with a small group, showed similar results, with volunteers scoring close to the DSQ norms (Moran, Britton, and Correy 1992). This evidence is gleaned from preliminary data on small groups, however, and further support for these conclusions is necessary.

Although information on general coping styles describes stable aspects of functioning and can be used to distinguish between anxiety disorders, this does not necessarily help in predicting how an individual will cope in a specific situation, nor how coping will change across various stages of a crisis occasion, such as the movement from threat, impact, and rescue phases of a disaster. Durham and associates (1985) examined coping behaviors reported by paid emergency workers following two social crisis occasions, a building explosion and a tornado (McCammon et al. 1988). Reactions at the site of the events were noted, as well as coping behaviors used following the events. Incidents most likely to have an impact on the respondents at the site were related to death; that is, finding dead bodies, or being at risk themselves. Other factors related to equipment or associated difficulties. Some respondents reported little or no emotional reaction, which the authors interpreted as a reluctance to disclose distress. The most frequently rated items on the Coping Inventory were "remind oneself that things could be worse" and "try to keep a realistic perspective on the situation." The authors interpret these and associated coping behaviors as an attempt to achieve cognitive mastery over the event and ascertain meaning. In the second study only, another item was rated: "remind self providing help"; 78 percent of respondents indicated they used this type of behavior to cope. This item would be especially relevant for permanent volunteers, and it would be interesting to see if they are more or less likely to use particular coping behaviors than are paid emergency workers. Interestingly, in the above studies, between 14 percent and 17 percent of respondents from the two events reported symptom patterns consistent with posttraumatic stress disorder (PTSD) criteria. Unfortunately, the authors did not relate responses on the Coping Inventory to the presence of PTSD symptoms.

A full discussion of PTSD is beyond the scope of this chapter; nevertheless, some points are worth considering. The breakdown of successful coping and the development of PTSD or critical incident stress (CIS) is of considerable importance within any emergency response organization. The potential for PTSD as a result of volunteer emergency and disaster work has considerable additional implications as back up financial, psychological, social, medical, and legal support may not always be readily available. The estimates of PTSD in rescue workers are variable (McFarlane 1988; Mitchell 1988; Raphael et al. 1983; Shepherd and Hodgkinson 1990). Differences in estimates may reflect differences in the crisis occasion under discussion. It seems that most work on PTSD in rescue workers (as distinct from victims) focuses on the small-scale event, so that therapeutic aspects of social support and cohesion found with larger crisis occasions might be missing. There is no doubt that some rescue workers are exposed to horrific scenes and life-threatening events and that some develop PTSD. It is not clear at this stage how much personality characteristics contribute to the development of the disorder, and to what extent virtually anyone could develop it as a result of exposure to the traumatic event.

The diagnostic criteria for PTSD and CIS often explicitly state that the event is the main or sole cause of the disorder, but there is no unequivocal empirical support for this. Even following the Buffalo Creek disaster, which is often cited as an event wherein a high level of psychopathology was recorded in the victim population (estimates were varied), there were still cases of survivors reporting no impairment (Gleser et al. 1981). Unfortunately, some issues related to PTSD may be unresearchable, partly because of the tautology involved in defining PTSD and CIS. Mitchell, for example, defined a critical incident as

> Any situation faced by emergency service personnel that causes them to experience unusually strong emotional reactions which have the potential to interfere with their ability to function either at the scene or later. (Mitchell 1983, 36)

The critical incident is defined in terms of the reactions, and the reactions are seen to result from the critical incident. The same potential for tautology can be seen in discussions of PTSD; for example, the disorder is seen to result from "the experience of a stressor that is likely to elicit a reaction of significant distress in most people" (Durham et al. 1985, 666). There may be no way of avoiding tautological problems with this type of diagnosis.

Not all researchers see the symptoms of PTSD or CIS as warranting a separate diagnostic category, arguing that effects are the same across all stressors, although they may vary in their intensity and be more or less likely to be seen as "symptoms" (Breslau and Davis 1987; Ettedgui and Bridges 1985). It is hoped that empirical, prospective studies will be able to shed more light on this and include information regarding any separate considerations that may be relevant in the case of the permanent volunteer.

Stress and Task Performance

An important aspect of individual differences in the social crisis context is the ability to perform well under stress. Much research on stress and task performance has been discussed under the concept of arousal and the hypothesized "inverted-U" relationship between performance and arousal. The most relevant aspect of this theory for the present chapter is the notion that there is an "optimal level of arousal" that can vary across individuals. That is, some individuals may perform tasks optimally under levels of arousal that would be debilitating for others. Although this theory has some appeal, suggesting rescue workers fit the former category, the concept of arousal has been criticized as failing to distinguish between fearful and nonfearful states (e.g., joy, anger). Neiss (1988) speculated that emotionality and worry, rather than arousal per se, may be the important factors explaining performance decrement. Rescue workers often recall adopting various strategies to control emotionality in the social crisis context such as focusing on the task at hand and deliberately putting the sadness of the situation out of their minds. Specific treatment techniques have been designed to teach individuals to focus on task-relevant responses rather than task-irrelevant responses, which tend to increase with anxiety. Volunteer rescue organizations may be made up disproportionately of individuals who already have this skill. Individuals who can focus on task-relevant responses may be more likely to volunteer for activities that have the potential for high levels of stress and anxiety. A similar logic suggests that these personality characteristics could also be present in paid rescue workers, who might only choose such an occupation because of their own satisfactory past performance under stress.

This does not mean that rescue workers will always be able to attend to the task and never be debilitated by emotion at the rescue site, but it does predict that in general these workers will be able to do so. This seems to fit in with most observations of rescue workers at the site. Rescue workers have reported that factors such as overload, fatigue, familiarity, or identification with victims are variables that sometimes impede the ability to use task-relevant responses or other coping strategies. More information on factors that cause coping strategies to fail could be useful, not only for selection of rescue workers, but also to help prevent stressful secondary reactions (such as depression or anxiety) that can occur when coping is seen to fail, usually unexpectedly. For an individual who has volunteered for emergency work, the breakdown of coping may be particularly unexpected.

Factors Contributing to Stress

Several issues have arisen in the discussion of permanent disaster volunteers. Two important questions have been considered: "Who volunteers and why?" and "Can specific personality or coping strategies be delineated that will provide information on the stress-coping potential of these individuals in the large-scale social crisis context?"

As already noted, there is no single answer to these two questions. Many motives have been noted as contributing to volunteerism, for example, altruism, a personal "mission," and the need for recognition. These factors are not exclusive, and more than one motive may be found in the one person. These observations lead logically to a conclusion of how much motives affect functioning in the disaster context. Although there is no direct information on this issue, there is some indirect evidence suggesting that when motives are frustrated, stress is likely to result in permanent volunteers as well as volunteers in expanded and new roles (see Dynes and Quarantelli 1980, 348). A few examples follow. Berah and associates (1985) noted that after the 1983 Ash Wednesday wildfire in southeast Australia, their volunteer mental health team arrived unasked to counsel victims but were requested to leave (albeit after some time). These researcher-volunteers suggest that conflict over altruistically motivated behavior was the most stressful aspect of the entire wildfire aftermath for some team members. Anecdotal data has suggested that following the 17 October 1989 Loma Prieta earthquake, action-oriented workers in San Francisco showed frustration that there was less activity required than seemed to be the case initially (especially from television reports). Ryan (1989) discussed the case of medical practitioners being denied social recognition for their efforts at disaster sites when there are no survivors—the popular assumption being that there was nothing for them to do if no one survived.

Both motives and coping strategies are complex phenomena, and are themselves likely to be explained and modified by factors outside the disaster context as well as those directly related to it. Factors such as age, gender, cultural backgrounds, and beliefs impinge on disaster workers as much as they do on any social group. Be that as it may, there are no specific predictions regarding direction of influence of these factors because generalizations of this type cannot be supported at this stage. Take for example, the concept of "experience": In their study of the 1979 DC10 air crash on Mt. Erebus, Taylor and Frazer (1982) noted that older emergency workers were less distressed than younger ones when handling and identifying bodies, the assumption being that the older ones had more experience of this aversive activity. In contrast, reports on characteristics of those who develop critical incident stress or a posttraumatic stress disorder often indicate repeated exposure to traumatic events as a variable that has increased rather than decreased vulnerability. In the example of the DC10 air crash, it should be noted that the emergency workers comprised both volunteer mountain rescuers and career police officers. This occasion highlights the multidimensionality of crisis occasions. For instance, the police officers were chosen partly because they "volunteered" to undertake the body retrieval tasks but also because they had experience in body handling (although none of them had previous experience of mass body handling of the type they were subjected to on the slopes of Mt. Erebus). However, none of these officers had experience in working in the harsh physical environment of the Antarctic. On the other hand, although the volunteer mountain rescue climbers were accustomed to functioning in the cold isolation of mountain terrain, they were unfamiliar with mass body incidents.

At this stage, it appears that available knowledge does not allow us to predict which experiences are more important in mitigating the stress associated with large-scale social crises. Likewise, many emergency workers have considerable exposure to small-scale rescue activity (for instance, motor vehicle accidents), but it is not known how much this exposure (which can be extremely stressful or traumatic at times) prepares or sensitizes the individual and how this might affect functioning at a subsequent large-scale event.

Individuals who volunteer for disaster work may be more "hardy" than others to begin with, and this may account for their ability to face potentially severe stressors and trauma on a voluntary basis. Indeed, in the broader area of stress research current concepts such as "hardiness" (Kobasa et al. 1982) and "sense of coherence" (Antonovsky 1987) are being used to explain not only how individuals cope with stress, but also how some actually thrive in its presence. It may be that similar factors will characterize volunteer disaster workers, so that the disaster context is a positive, rather than a negative, challenge. Quarantelli and Dynes (1985) suggested a similar positive aspect to disaster response at the community level. It is unlikely everyone in a

Figure 3
The Stress Wheel

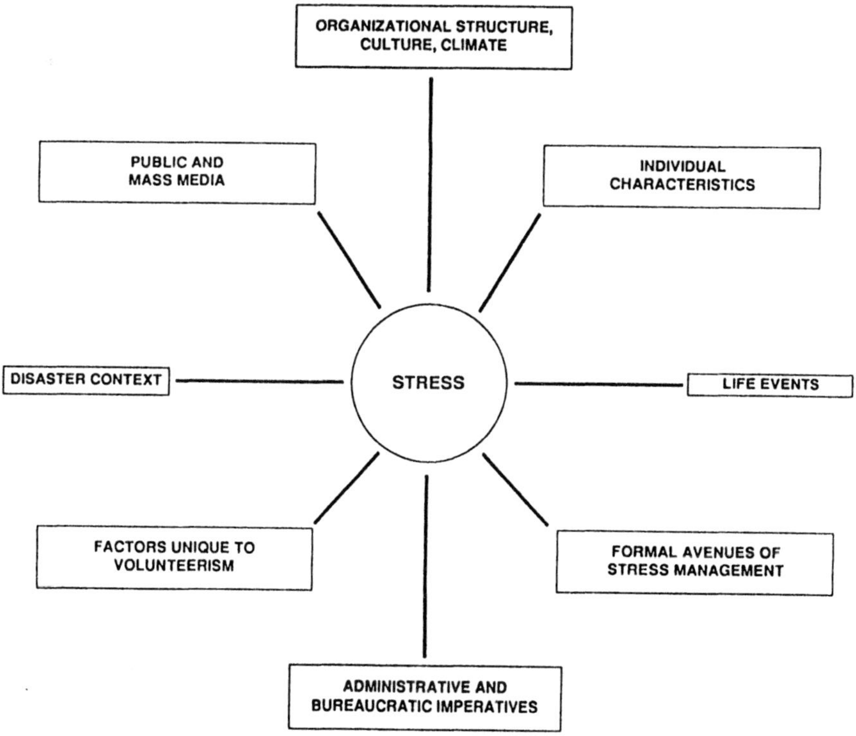

volunteer emergency service will demonstrate the same characteristics, however. We have noted that even with a small group of subjects, scores on subscales of a hardiness questionnaire (Bartone 1989) were bimodal. Furthermore, high scorers on the "committment" subscale were less likely to report psycho-physiological symptoms normally associated with stress (Moran et al. 1992).

Thus, the characteristics of the individual may not only help to mitigate stress, but also in some circumstances may considerably alter its perceived nature. The individual is not isolated of course, and many other factors will also affect stress levels. As noted in Fig. 3, some of these factors are external to the individual. The nature of the disaster or emergency context, for example, may override individual coping strategies in some cases (for instance, where children are involved) or provide additional stressors (for example, poor organizational response). Other factors external to the disaster environment (for instance, family stressors) might also impinge on stress and coping in that context. These factors may decrease or increase stress depending on appropriateness to the disaster context (for example, altruism, social contact). The important point to note here is that these factors interact with each other as well as with the final level of stress. For some individuals, participating in stressful disaster work will enhance functioning and this, combined with other factors, will reduce the subjective level of stress. For others, the stress level will be increased and the potential for impairment high. Personal coping styles will play a part in predicting these different reactions, but other factors will also be involved in determining the overall stress levels associated with the disaster context.

The Permanent Volunteer within the Emergency Organization Context

Questions concerning the role of permanent volunteers in disaster situations have often been raised, usually by career emergency service officers who see their positions as challenged by volunteers. Although it is clear that not all career officers hold this point of view, many do regard permanent volunteers as threats. In the current period of increasing government fiscal restraint and economic rationalization, this may not be quite so unrealistic as it was once thought. More and more governments throughout the world are trying to reduce expenditure (Britton 1991c; Hood 1989; Hood and Jackson 1991), and one method many are turning to is to use volunteers as replacements for permanent positions. Emergency services in particular are turning more and more toward this process of "coproduction" (Estepp 1990; Snook 1990; Sundeen and Siegel 1986; Zech 1982). In Australia, where the national ratio of emergency volunteers to career officers is currently about 10:1, this is a particularly salient issue that is being kept alive by state government actions such as the New South Wales government's decision in 1991 to introduce "community policing" and "shop-front police stations," in which both initiatives rely heavily on volunteer citizens. Further, many Australian career officers express the viewpoint that their lives are endangered because they are forced to operate alongside volunteers who they believe are not as well trained as paid career emergency service people. Career officers readily acknowledge that permanent volunteers undoubtedly possess the all-important qualities of commitment and enthusiasm and are dedicated to offering themselves to their local community. There is, nevertheless, a feeling among career officers that permanent volunteers do not have the prerequisite knowledge, skills, or specific training required to help protect themselves or their colleagues in the field. Not surprisingly perhaps, career officers regard themselves as "professionals" as distinct from the "volunteers," who they perceive to be second-rate emergency workers. As a counter to this, volunteers have questioned the assumption that disaster-relevant skills and knowledge are only associated with full-time employment in functionally specific, emergency-oriented organizations.

To our knowledge, no studies have yet been conducted to establish either the extent of these sentiments or their accuracy. However, a nationwide study conducted by one of the authors between 1983–85 did allow for some quantification. Britton (1985) found that 36 percent of career State or Territory Emergency Service (S/TES) officers who came up through the permanent volunteer ranks indicated that as volunteers they felt career officers displayed attitudes of superiority during operational periods. This rose to 41 percent during nonoperational phases. The same study found attitudinal variations between career personnel in the established emergency services (police, fire, ambulance) toward S/TES officers in general. Specifically, 46 percent of respondents in the S/TES felt the role of their organization was not understood by other agencies, and 50 percent declared the S/TES role was not readily accepted by other disaster-relevant organizations. The significance of these findings is that the S/TES is a predominantly volunteer organization whereas the others are more or less staffed exclusively by salaried full-time officers and administrators.

From the perspective of individuals who comprise Australia's bushfire, S/TES, and similar volunteer search and rescue units, many of the characteristics ordinarily associated with voluntary association membership are absent. Both structurally and functionally, these volunteer agencies typify the classic "machine bureaucracy" outlined by Mintzberg (1979). Further, like government departments such as Australian police, fire, and ambulance services, volunteer emergency agencies are heavily influenced by the dominant military model of "command and control" in civilian social crisis planning and management (Britton 1989, 1991d; Dynes 1983, 1990; Quarantelli 1987). Thus, for the permanent disaster volunteer, a significant source of external stressors (leaving aside the disaster event itself), may come from the disaster-relevant organizational network itself.

This may be surprising at first blush, but if one thinks about specific aspects of organizational culture and organizational climate that permeate the established emergency services, as well as the bureaucratic structure of the disaster-relevant organizational network, it should not be too astounding. Consider the following four propositions, and the ensuing dilemmas.

Proposition 1	Career officers and permanent volunteers are exposed to similar (type and degree) disaster threat and impact consequences.
Proposition 2	The disaster-relevant skills acquired by career officers and permanent disaster volunteers are similar.
Proposition 3	The successful management of large-scale social crisis (i.e., nonroutine) episodes requires an innovative approach.
Proposition 4	Permanent disaster volunteerism is a manifestation of the institutionalized expansion of the disaster citizenship role.

Proposition 1 underscores a general thesis within this chapter. That is, permanent disaster volunteers often perform similar disaster-relevant tasks to career officers. Stresses produced by exposure to the disaster event should therefore be expected to be similar between the two groups.

Proposition 2 suggests that although the focus of training is different, actual skills are similar. The opportunity for career officers to practice their training is greater, although instruction and practice are oriented toward routine accident and emergency roles. Disaster volunteers tend to rehearse more frequently for large-scale crisis although the opportunity to practice is less. Hence, skills training between career officers and permanent volunteers are similar, but the intention differs because of the level at which the two groups are required to operate routinely.

Proposition 3 implies that disaster management is essentially the management of nonroutine episodes. Thus, improvisation is more important than the implementation of routine and standardized responses. The large-scale machine bureaucracies of police, fire, and ambulance services apply rationalistic managerial models that have a tendency to lock organizations into a "disaster as a large-scale accident" viewpoint. This can have unintended consequences for appropriate coordination of interagency response that, in turn, hinders effective organizational and individual learning.

Proposition 4 states that the activities of permanent disaster workers are extensions of the citizenship role. This phenomenon has been well documented in the disaster literature (see Barton 1969; Dynes 1970). Career officers seem not to recognize the significance—or the inevitability—of this, which leads to the nonacceptance of permanent volunteers and emergent groups in the postimpact environment, thereby creating strains between the groups.

What these four propositions suggest is that a key stressor separating the permanent volunteer and the career officer has less to do with disaster as such than with the organizational and institutional contexts within which they both operate. Moreover, one group creates additional stresses for the other, in that the volunteer has to put up with additional stress factors created by career officers, a factor that is enhanced by the bureaucratic nature of the disaster-relevant organizational network.

It should be regarded as a given that in any large-scale event, trained permanent volunteers and nontrained (emergent) volunteers will be part of the disaster impact environment. What appears to be happening, however, is that career emergency officers do not accept this. It is possible that this results, in part, from the specific secondary (occupational) socialization they have received during initial training. This creates a dual dilemma for the career emergency service officer. The first dilemma is that emergency service personnel are socialized into accepting it as their duty to protect the lives and property of fellow citizens. No other social group has this responsibility. However, with respect to disasters, the established disaster-relevant organizational network cannot achieve this responsibility unaided. The second dilemma is that disaster volunteerism is an inevitable and vital component of the disaster impact environment. However, the career emergency service officers do not accept volunteers—either the permanent volunteers or the public-at-large (emergent volunteers).

There is an inherent contradiction with the social responsibility placed on the career emergency officer that becomes manifest during large-scale social crises. This contradiction is fostered by the closed-rank attitude of "professionalism" that pervades established emergency officers, which closes off both the permanent disaster volunteer and the larger citizen group.[3] Ironically however, this attitude may well enhance the career officers' ability to withstand the occupational hazards they face, which includes, of course, disaster-related tasks. This implies that the obvious question of "how does one enact appropriate attitudinal changes (toward permanent disaster volunteers) within the ranks of career officers?" may well be a two-edged sword.

Conclusion

Although many of our comments on stressors, coping strategies, and permanent volunteers remain largely speculative at this stage, it is necessary to consider the implications any relationship between these factors will have on both volunteer disaster work as well as on career officer disaster work. One of the aims of this chapter was to consider successful stress-coping mechanisms, but by doing this we are also likely to identify poor ones. How can this information be used constructively? In a disaster management system that relies on volunteers, the idea of screening out those with "poor coping skills" may be impractical for various reasons. One reason is that volunteerism relies on the goodwill of all involved, and the establishment of a screening system could likely threaten this goodwill. In addition, the number of permanent volunteers is never very high, so that any reduction in numbers needs to be well justified. At this stage it is unlikely that we would be able to delineate relevant parameters with sufficient confidence to justify rejection of volunteers. As Gibbs (1989, 489) noted in her recent review of factors that mediate between disaster and psychopathology, "coping styles appear promising predictive variables," but she has only been able to offer hypotheses rather than conclusions as to these variables. At this stage it may be more fruitful to examine the full range of variables that relate to the stress of the disaster context, not only the personal characteristics of the disaster worker, and then decide which variables would be most feasible to change.

Notes

1. There are, of course, other variables that will relate to the permanent volunteer within the emergency disaster context, for example, gender, cultural background, and beliefs. However, discussion of these variables is beyond the scope of this chapter.

2. Much of this literature has been summarized in Britton et al. (1990, 8–20; see also Britton 1990), where a typology of voluntary action is offered. The seven types portrayed are based on the relative position of each on three dimensions: (a) type of involvement (i.e., "organizational involvement"); (b) degree of commitment (i.e., "exposure regularity"); and (c) inherent danger of the voluntary task (i.e., "cost of helping").

3. It is worth noting that the Australian Bureau of Statistics classifies none of these occupational groups as "professional." Both police and ambulance officers are examples of "Para-Professionals"; and fire officers are located with the "Plant and Machine Operator/Driver" category.

An Ecological Approach to Disasters

FREDERICK L. BATES and CARLO PELANDA

Introduction

Disaster research involves a multidisciplinary effort conducted in fields covering virtually the entire range of physical, biological, and social sciences. Such breadth of coverage is inevitable whenever a complex "natural" event is chosen as an object of study for practical rather than theoretical reasons. In the case of disasters, physical, biological, and human social phenomena operate together in an interacting system to produce an event that unfolds as a multidimensional process in nature having dramatic humanitarian significance.

Because of this multidimensionality, and because of the importance of disasters to human well-being, there is virtually no limit to the scientific perspectives that might be brought fruitfully to bear in disaster research. This richness of scientific opportunity and the diversity of scientific response that has been made to it constitutes, at the same time, the strength and the fundamental weakness of disaster research as a broad field of study. In general it can be said that disaster research has yielded a total body of information that is fragmented by discipline and that lacks overall integration. Even so, a great deal of information of practical value has been accumulated by that uncoordinated effort that has, for the most part, been undertaken to solve practical problems rather than to further theoretical agendas.

Henry Quarantelli and his associates at the DRC, which has long functioned as the "seed bed" for disaster research in the social sciences, have always recognized the complex nature of disasters as objects of scientific study and have resisted developing a specialized independent academic subfield that might be called *disasterology*. Rather, Quarantelli, along with Russell Dynes and others connected directly and indirectly to DRC, have influenced disaster research in sociology by connecting in with the already established subfields of collective behavior and organizational theory. The intellectual strategy of the DRC has, therefore, been to connect disaster research with established research traditions in sociology and in other fields rather than to build an autonomous subfield devoted entirely to disaster research (Dynes, DeMarchi, and Pelanda 1987).

Even though as yet uncodified, however, the theoretical outcome of the work of the DRC and those associated with it across the country and around the world has been to demonstrate that the collective behavior approach should be merged with organizational theory to provide a better theoretical vehicle for dealing with complex unfolding events such as disasters. At the same time the work of DRC seems to point to the conclusion that systemic factors are more relevant to understanding disasters than purely social psychological ones. Thus, the contributions of Quarantelli and his many associates have led to a growing emphasis on the *sociosystemic* nature of disasters and of the various phenomena associated with them.

Any attempt to explore the systemic nature of disasters, however, encounters a general problem related to the theoretical capacity of sociology to offer a systemic approach that can fruitfully be employed as a guide to empirical research. Within sociology the Durkheimian tradition of empiricism still dominates, and when the term *system* is employed, it is almost always used to refer to a "social fact" or "empirical reality" discovered through observation rather than to a social reality constructed through a process of theoretical action in which observations are selected and given meaning by a conceptual apparatus that controls and guides research. As a consequence, "systems theory" remains underdeveloped as a usable tool for organizing empirical observation and constitutes a broad theoretical orientation more or less unsupported by a methodology, especially considering the inappropriateness of survey research and multivariate analysis, the dominant methodologies employed by sociologists. Systems theory in sociology, although not necessarily in all fields of science, is especially underdeveloped in the very area critical to the study of disasters because it has neglected the theoretical placement of human systems in the context of a general ecological field, or active environment, in which human sociocultural systems can be seen as part of larger natural processes. To use systems theory to understand disasters, it is obviously necessary to broaden systems theory by developing a conceptual apparatus that can allow us to construct images of a process which involves organized human action in the environmental context in which it takes place. Not to do so is to fail to take advantage of the central idea embedded in system theory as it applies to sociology.

We agree with Quarantelli concerning the undesirability of making disaster research into a separate discipline. Disciplines should be founded on general theoretical problems defined in terms of a division of labor with respect to phenomenological domains. Because disasters cut across domains (social, biological, physical, etc.), it is inappropriate to set disaster research aside as a separate discipline. However, it could be desirable to identify a broad theoretical orientation that has the capacity of spanning the gaps between disciplines so that communication, especially of scientific results, can flow among specialized fields. Although it is undoubtedly impossible, and for that matter undesirable, to confine disaster research in sociology within a single theoretical perspective, given the complex nature of the phenomena covered by the field, it may be possible to achieve a kind of "loose integration" by developing a perspective within which the physical, biological, and social sciences can identify points of theoretical correspondence, thus allowing each field to use information from others more effectively.

An Integrated Ecological Approach

The theoretical perspective required to move toward the integration of disaster research should be capable of placing human systems within the context of their natural physical and biological environments. It should conceptualize the nature of disasters as total events in such a way that they are thought of as part of a natural process in which human systems and their environments interact, changing in relationship to each other as they impact on each other in a kind of evolutionary process. The theoretical perspective needed should also be capable of conceptualizing the process by which disasters are produced, develop as events and are eventually absorbed as part of the natural evolutionary process of this planet.

It seems apparent that the one existing perspective that can fit these needs is that of general ecology. This term is used here to suggest a new approach that goes beyond the ecology which has been developed in the biological sciences during the past several decades or the human or social ecology that developed in the social sciences. *General ecology*, in contrast to these perspectives, includes human systems as an integral element, self-consciously recognized and accounted for in detail in a conceptualization of a total ecological field.

In general ecology, human societies and their various component elements are recognized as natural systems, having the same status as components of nature as do physical and biological systems. Man and his sociocultural systems are fully recognized as a part of nature, rather than being cast as a kind of outside enemy of the environment or as a kind of artificial "monster" invading the ecosystem and destroying its balance. The founding assumption is that everything, including man and his sociocultural systems, is part of one natural process that needs to be understood as a whole.

The field of disaster research seems ideally suited to supply an intellectual breeding ground for this new kind of "humanized general ecology" because it inevitably deals with the interface between human systems and their natural environments. "Natural disasters" have long been recognized as being the consequence of human systems failing to adapt human populations to the vulnerabilities they face with respect to the environments within which they exist and on which they depend. Technological disasters have also been recognized to be a consequence of the built-in vulnerabilities inherent in the very social, economic, political, and technological systems humans employ to produce and consume goods and services. Because such technological systems develop to provide adaptive mechanisms that relate human populations to their environment by artificial means, these artificial systems must also be regarded as part of a general ecology. Thus, all disasters are recognized to be, by their very nature, ecological events.

Furthermore, it has long been recognized that human systems often exacerbate their vulnerability to disasters by impacting on their environment through their technological systems, thus, exposing themselves to man-produced ecological crises that are clearly the consequence of human interventions, effecting the so-called natural ecological balance.

It is, however, one thing to say that a general ecological approach is an appropriate road to take in integrating the various subfields of disaster research, and quite another to specify how this can be done. In the following pages some beginning thoughts are offered as a basis for encouraging others to join in an effort to develop a more fully specified theoretical apparatus within the field we call general ecology. Because we are sociologists we begin with propositions concerning human systems and extend them to include more detailed consideration of disasters and other environmental issues later.

Some Basic Assumptions of General Ecology

Human beings are dependent on sociocultural systems to provide them with the behavior patterns they employ as a means of drawing resources from their environments and for patterns that adapt them to environmental conditions which can threaten their survival and well being. This sort of sociocultural adaptation is as "natural" to the human species as nest building is for birds or grazing is to cattle. Over tens of thousands of years these sociocultural systems have developed into complex human social systems that involve the organized behavior of huge human populations distributed unevenly over vast territories.

Such complex social systems develop artificial technological systems and build artificial (man-constructed) environments, thus profoundly impacting on the global environment by using vast amounts of natural resources and by converting them into products and services. Historically, human systems have developed in such a way as to consume resources at an ever-increasing rate. This increase in the rate of resource consumption has been accompanied by an ever-burgeoning human population, with rapidly increasing aspirations for richer life-styles and with the concentration of population in huge urban settlements often located in highly threatening environments.

As these trends have developed and industrialization and urbanization have taken place, more or less uncontrolled by concern for the long-range future, unanticipated or unintended side effects have been produced that have negatively impacted on the environment, increasing the probability of future environmental crises and disasters.

This is a "thumbnail" sketch of the historical relationships of human systems to their so-called natural environments. Hidden within this tale, however, is the fact, stated at the beginning, that human systems develop to *provide an adaptation* of a human population to their environments. These systems developed as a means of providing human populations with the resources they needed to survive, reproduce, and protect themselves against threats that could destroy them. As social evolution has unfolded, the aspirations of humans have risen far above the desire for mere physical and biological security to include the desire for comfort and affluence, thus multiplying the demands placed on global resources. Even though aspirations have been generated that go far beyond mere survival and basic environmental adaptation, we must regard human sociocultural systems as being "ecologically" generated because they arise out of the interplay of human populations with specific sets of environmental conditions. In the most basic sense, and more to the point of disaster research, human sociocultural systems arise and develop to reduce the vulnerability of human populations to conditions that can weaken or destroy that population, or subject it to deprivations that are defined by their sociocultural systems as being undesirable. Through such systems, human populations provide themselves with food, shelter, water, energy, and other vital goods and services. They also seek to satisfy less vital aspirations that arise out of their socially generated values and ideologies. The point is that, in a very real sense, sociocultural systems arise to prevent or control disasters and crises that threaten the survival and well being of human populations. They are the instruments through which the human species protects itself against environmental conditions that have the potential to produce severe negative impacts for human communities.

Disasters occur when one or more of the sociocultural systems that a population depends on fail to provide an adaptation to the environmental conditions which surround it, or when one of these systems produces, from within its own technological order, an event that threatens the population. The problem of understanding disasters then amounts to understanding the relationships between particular types of human systems and the environmental conditions to which they are related as adaptive devices.

To achieve such understanding, however, it is necessary to be able to conceptualize the vulnerability of particular human systems to disruption by environmental events or forces or to internal collapse, because disasters occur when the limits of vulnerability of a system are exceeded. Because human systems are interdependent and linked within chains of relationships that form the structure of societal and intersocietal networks, the challenges associated with solving the environmental problems allied with disasters are terribly complex. It is easier, therefore, to identify vulnerability rather than to reduce it because of this characteristic of structural complexity. For example, one can imagine solving vulnerability to earthquakes, floods, and hurricanes by moving populations away from exposure to these threats, but this would require an almost total reorganization of society and a redistribution of resources that would present vast economic, political; and social problems which, in themselves, might prove to be disastrous.

To exploit the perspective of general ecology, it is necessary to create a cognitive apparatus that conceptualizes the nature of so-called human sociocultural systems in terms that are compatible with a more traditional ecological perspective. By doing so, it might be possible to understand the social constraints on environmental policy, and, therefore, on our capacity to reduce vulnerability to disaster through the use of public policy. In the following few pages one such perspective is offered and discussed.

An Ecological Perspective Toward the Structure of Sociocultural Systems

We take the view that the culture of any society is comprised of a population of separate, bounded systems, each of which provides a set of cultural patterns that "solve" a particular adaptation problem or set of related adaptation problems (Bates 1991; Pelanda 1990). Each such cultural system may be thought of as a shared cognitive apparatus that stores a set of behavior patterns in the form of an encoded information system. These systems should be thought of as being bounded and evolving separately as systems as they are employed to generate "real" behavior on the part of a set of human actors in relationship to each other and in relationship to an environment. Thus, within a given culture there are many separate cultural systems that have evolved to solve distinct adaptation problems such as the problems of shelter, food production and consumption, transportation, biological reproduction, family life, industrial production, trade, and, indeed, every type of adaptation problem faced by a human population.

The overall culture of a society should not be viewed as a single integrated system with a pervasive overall logic that controls all aspects of human behavior within one integrated system. Instead, cultures consist, at least in complex societies, of thousands of separate bounded sociocultural systems, each with its own adaptive agenda, and its own internal organization or order. Parallel to this is the idea that society contains a population of bounded social or behavioral systems that are created and operated on the basis of bounded cultural systems, stored as shared cognitive systems in the minds of the members of society. These separate social systems are generated and maintained as these distinct cultural systems are used to pattern social behavior. This idea yields the view that society, itself, should not be thought of as a single, unified system but as a structural network consisting of the relationships among many systems that are themselves separate and linked only by contingent forms of interaction and exchange.

When we say a society consists of a population of bounded social systems, we are referring to entities such as particular families or particular economic organizations, for example, factories, banks, stores, and so forth. Such social units are the separate systems making up the network of systems comprising the structure of a society. Each of these separate systems exists within an ecological field within which it adapts to a particular environment (Gallino 1988). Each system operates separately to solve its own adaptive problems. As these separate systems operate, they interact with particular other social systems in their environment, being impacted on by them and impacting on them as they do, and as a consequence, evolving or changing in relation to their changing environment. Together, however, these many separate systems do not form a single integrated larger system. In this image of social systems the individual is viewed as an actor who participates in many separate social systems. Thus, the actor can be represented as a "position set" in the Mertonian sense, or as a station structure in the Batesian sense (Merton 1957; Bates and Harvey 1975).

Instead of thinking of society as a single social system, we can think of the population of separate social systems comprising the elements of a society as forming an *ecological network* within an *ecological field.* Because these two terms are essential to the argument being presented, it is important that they be clarified in more detail before returning to the discussion of disasters.

Ecological Network

An ecological network consists of a population of differentiated and specialized interacting systems that occupy "niches" in a "division of labor." Because the systems are specialized and differentiated within a "division of labor," they are interdependent in that they draw resources from each other or impact on each other in some other fashion. This interdependence arises out of the specialization of the systems in the outputs they produce and in the resulting specialization of their need for resources. Because of their interdependence, to survive or to achieve their goals, each system in the network must interact with some of the other systems in the same network so they can obtain needed inputs.

Ecological networks are so constructed that the various individual systems making up the network are in *categorical* rather than *unit specific* interdependence with each other. This is a crucial point because it is this characteristic that justified the claim that, by interacting in a set of exchange relationships, the various systems comprising an ecological network *do not* form a higher order or larger system, but remain autonomous systems in their own right.

To understand the term *categorical interdependence*, it is necessary to examine its opposite, *unit specific interdependence*, which prevails among the components inside the boundaries of a system. There are two forms of relationship that may be posited among the units of a network. One form involves a tie between exact particular units that persists through the entire interval of time during which a network exists as a structured entity. Such a relationship makes it structurally mandatory that the two units interact and, in the case of exchange relationships, that they engage in input-output relations. Such a case can be illustrated by the example of two work groups or departments within the boundaries of a single organization. For example, there is a unit specific link between the sociology department at X University and the office of the Dean of Arts and Sciences at that university. This relationship joins two particular work groups that are parts of one particular organization to each other within a single system. The sociology department is not left the option of dealing with other dean's offices in other colleges at other universities as the source of administrative decisions affecting its operation. Within the boundaries of systems the various units of structure are *always* in unit specific dependency relationships with each other.

In contrast, *categorical interdependence* exists when a given structural unit must depend on other structural units in its environment for needed inputs, but where there is no structural specification as to which particular other unit in a class or category of similar units with which it must relate. An example would be the case illustrated by families and their relationships to grocery stores. Because of the division of labor in our society, each particular family requires groceries to perform its internal functions, but it is structurally free to associate with *any* grocery store or supplier of food to obtain groceries. In this case the category families is dependent on the category grocery stores, but particular association patterns are not fixed by any controlling structure. Instead, relations between specific units are contingent on numerous factors that change constantly as events occur within and between units in an ecological network.

This form of contingent network, involving categorical interdependence, is the basis of ecological networks. In such networks bounded, anonymous systems interact on a contingent basis to obtain inputs and to dispose of outputs, and as they do they impact on each other. In the process resources are transferred from one part of the network to another. The best example of such a network in the case of social phenomena is the "market" in a capitalistic economy. In a market, buyers and sellers representing separate autonomous systems interact to achieve ends set within the boundaries of separate systems.

In ecological networks, systems with the same resource requirements stand in a competitive relationship to each other because they draw their resources from a common pool. This competition becomes most apparent when resources are scarce and when the success of one system in obtaining resources means the failure of others. This is like saying that families, a type of social unit, are in competition with each other for food supply, income, housing, and other desired resources. When famine strikes, families are in severe competition for food. When there is a housing shortage, they are in competition for housing.

In the case of categories of units that are interdependent, yet autonomous as systems, the relationship is better characterized as conflict because a unit in Category 1 attempts to benefit by drawing resources from a unit in Category 2 from which it obtains inputs. When it does, it attempts to obtain resources on terms favorable to itself, but not necessarily favorable to its supply source. For example, a clothing manufacturer's interests are in obtaining textile products of its own choosing at the lowest possible cost and under the most favorable conditions to itself. But the textile manufacturer's interest is the opposite. It seeks to sell its product for the highest price "the market will bear" and to deliver it on terms favorable to itself. Success on the part of the clothing manufacturer in reaching its objective results in a lowering of the level of success for the textile manufacturer.

In ecological networks comprised of organisms, these statements are more easily grasped. In this case let us suppose we are dealing with animals as the units of the network. The network consists of a population of individual animal organisms of various species, each species occupying a niche in a biological "division of labor." Each individual animal may be thought of as a separate biosystem acting on its own (or sometimes in organized packs, herds, or flocks) to obtain resources from its environment or to defend itself from threats that could injure or destroy it. Individual members of the same species who depend on the same environmental resources are in competition with each other. For example, let us over-simplify by saying that lions live by eating zebras. Lions are therefore in competition for zebra meat even though, at times, they may cooperate in making a kill. On the other hand, lions and zebras are in conflict as species and as individual members of species. When a lion wins the game, a zebra loses. This example uses a "zero sum" case to illustrate the point that conflict characterizes the relationship among categories of systems. It should not be overgeneralized because, in some cases, the winning and losing is relative rather than absolute, but it does illustrate the idea of conflict between categories of independent systems.

The point of this discussion is that ecological networks are structures that are *not controlled* by a *central systematic mechanism* that structurally forces the various units to work together to preserve the survival of the entire network, as if the network were a system in which the parts respond to an overarching control mechanism. Instead, the survival of the network and of species and individuals within the network depends on the outcome of multidimensional conflict and competition occurring on a small scale in thousands (or millions) of separate episodes or encounters. This means that units in the network may be destroyed, as can whole species of units, whereas other parts or elements in the same network may prosper and develop without being destroyed or injured. Which systems are destroyed and which are negatively impacted depends on their *relative vulnerability* to other systems. Vulnerability, in turn, depends on the capacity of individual systems to defend themselves from environmental threat. Disasters occur at both the individual and "species" level when the defenses of individual systems and classes of systems are vulnerable to events that expose them to conditions beyond their adaptive capacity.

Ecological Field

Before this discussion is complete, it is necessary to consider the idea of an ecological field. This concept goes beyond the idea of an ecological network and gives us the capacity to deal with the relationships among networks and between networks and their collective physical environments.

An ecological field can be thought of as a spatially defined territory containing a set of ecological networks and the substratum of resources and environmental conditions that surround them. The field is a populated space filled with physical objects and forces in which ecological networks form and interact, and from which they draw resources. Within this field, species of systems evolve as they are subjected to conditions and events occurring in the field. For general ecology, an ecological field contains the following interrelated elements:

1. A population of human sociocultural systems, operating as contingent ecological networks.
2. A population of biotic systems including animals, plants, and micro organisms operating as "biotic" networks.
3. A store of physical resources arranged in a geographic pattern, including minerals, water and air, and so forth. (The nonliving physical environment comprised of matter.)
4. A set of physical conditions including climate, weather conditions, sunlight, seasonal variation, and associated energy phenomena.

These various types of elements interact, forming the ecological field within which the fates of individual systems, as well as those of species of systems, are determined by an extremely complex set of evolving contingencies. At the same time, physical resources and environmental conditions undergo transformation, feeding back on the biotic and sociocultural networks, as all elements within the field change in relation to each other.

The crucial characteristic of an ecological field is the fact that the actions or responses of various elements in the field are not determined by a set of factors operating as the mechanism of a mega-system, "super organism," or "machine," but by uncontrolled or uncoordinated random or "chance" processes controlled by complex sets of probabilities or contingencies that, themselves, arise out of the conflict and competition among the many separate systems populating the field.

The interaction among systems comprising ecological networks, as they compete or come into conflict, is characterized by behavioral "maneuvering" or "adaptive structural change" aimed toward satisfying the needs of individual systems, or classes of systems, or by attempting to avoid negative impacts. This "maneuvering" and morphogenic adaptation on the part of all of the systems in the field sets or produces the particular contingencies to which individual systems are forced to adapt.

These adaptive attempts are not governed by a master control mechanism that operates according to some principle directed toward the preservation of the field, its component networks, or the individual systems comprising such networks; all evolve together in the direction of positive adaptation and, therefore, survival. Instead, the future of all elements in the ecological field is determined by the complex set of contingencies emerging out of the autonomous operation of a large number of independent systems. In this process, the individual systems comprising networks are vulnerable to destruction, as are the species of systems vulnerable to extinction.

Nevertheless, within the field, systems respond, and in the long run, evolve in the direction that increases the probability of survival. If they do not evolve, they disappear from the ecological field. If, on the other hand, they survive, they do so by reducing the random behavior of their own subelements within their own boundaries by being able to control their relationships, as autonomous systems, to other elements within and without the ecological field; therefore, they are not exposed to seemingly threatening random impacts from their environment.

The reduction of surprises and random effects for one system, however, may introduce surprises and seemingly random effects for other systems in the same field. Thus, reducing the vulnerability of one system or species of system may increase the vulnerability for others. Because of the uncontrolled, uncoordinated nature of the interaction among separate autonomous systems, which are governed independently by their own separate systematic mechanisms, the accumulated effects of the actions taken by large numbers of separate systems produce "epiphenomena," which can injure or destroy the environmental conditions on which many individual systems depend. This is like saying that human systems, in their attempts to improve their short-range adaptation to the environment, may use up the very resources they depend on for survival, or destroy the environmental conditions that are necessary to their survival.

It is the accumulated effects of individual systems operating as autonomous entities adapting to complex short-range contingencies that result in long-range vulnerability to ecological crises and disasters. Human systems, for example, evolve internal orders that allow them to control their relationship to the set of environmental conditions which they face in the short run. This is mandatory because failure to do so could mean short-run failure and extinction. Systems failing in the short run are not around to evolve adaptations to the threat of long-range failure. They tend to adapt themselves to the set of high-probability contingencies that they face on a day-by-day basis and by so doing are prepared to survive as long as low-probability events do not occur. Furthermore, short-range adaptive behavior may actually, although yielding a high level of short-range success in meeting system needs and aspirations, lead to a level of long-range vulnerability that can destroy the system.

For example, there may be social, cultural, or economic advantage for a social unit to locate itself on a floodplain or in a coastal area exposed to occasional hurricanes, or perhaps in a highly seismic zone. These advantages grow out of contingencies related to competition and conflict within economic or sociocultural networks, which give a short-range competitive advantage to systems who so locate themselves. However, by achieving short-run competitive advantage, such systems are exposed to less frequent, lower probability events with deadlier impacts such as floods, hurricanes, or earthquakes. The location of human populations engaged in agriculture in regions with exposure to periodic droughts and water shortages may yield short-range adaptive benefits within a complex ecological field when rainfall is high for a time. But the same adaptive decisions that bring temporary success in the economic market expose the same population to disaster during droughts.

We can state two rules that characterize the short-range adaptive strategy, which characterizes systems within an ecological network and their long-range "evolution" as systems.

1. In the short run, human systems will choose to position themselves within an ecological field through developing an internal order that provides solutions to high-probability contingencies, with high-level payoffs. High probability-events always dominate the short-run adaptive process.
2. In the long run, those systems that use short-range adaptive strategies to high-probability vulnerabilities will expose themselves to long-range, low-probability events and will experience disasters. They may, as a result, be "selected out" of the ecological field when they succumb

to disasters. Those systems that escape disasters and negative selection will be those whose short-range adaptations also, more or less accidentally, prepare them to withstand low-probability contingencies.

In interpreting these statements, it is important to realize that the contingencies, and therefore the vulnerabilities, faced by any human system are complex and multidimensional, and that they vary widely in their importance to the survival of the system. For example, the human organism can survive as a living system longer without food than without air or water. Therefore, the systems that provide the organism with air, and which must deal with the contingencies associated with it, take precedence in short-range adaptive processes over those associated with food or water. A business organization's need for income in the short range may take precedence over its need to avoid a disaster produced by a flood or an earthquake, which has a low probability.

In the sense that disasters are the result of the vulnerability of a system to events in the ecological field, they are produced by attempts on the part of systems to adapt to a complex matrix of contingencies that are present in the whole ecological field in which they are located. It must be understood from this perspective that the form which any system takes in the short range adapts it to a set of vulnerabilities that, if ignored, could quickly destroy it because of the high probability of short-range threats. This means that systems cannot change "at will" to adapt to long-range, low-probability events without, at the same time, exposing themselves to being vulnerable to higher probability short-range threats.

For example, lower vulnerability to earthquakes on the part of families might be achieved in a developing country that uses adobe houses as an adaptation to economic as well as climatic conditions by building reinforced concrete structures. Such structures, however, cost a great deal more, and often are not as well adapted to climate. The higher cost might expose the population to economic threats that, in turn, produce negative impacts on their ability to pay for education or food. Malnutrition might arise, and homelessness might even occur, by reducing the capacity of people to spend money on food or housing! Because of this sort of trade-off, which is inherent in the very nature of ecological networks where complex contingencies prevail, public policy is not free to impose any long-range solutions it chooses to deal with low-probability events on units in the ecological network. There is a limit to disaster preparedness within any ecological field that is a function of the economic, social, and cultural costs of long-range solutions in terms of producing short-range vulnerability.

It is tempting to think, as radical ecologists do, that an instant and complete moratorium should be placed on the use of pesticides, chemical fertilizers, or large scale agriculture in developing countries. But there are consequences of such policies that can produce new vulnerabilities, and which can completely alter the capacity of the system dependent on these practices to deal with other complex contingencies within the ecological field.

Summary

So now we return to disasters. In a real sense, the potential for large and small disasters, defined broadly as events that overwhelm a human system and its capacity to function, is at the very root of sociocultural evolution, indeed, of biological evolution itself. Human sociocultural systems, as well as biological systems, evolve or experience morphogenic change as a means of reducing their vulnerability to disasters, large and small. To understand disasters, it is therefore necessary to understand the adaptive strategies of systems and their vulnerability to events occurring within the whole ecological field in which they operate as systems. The events we have been calling disasters on the basis of a "folk definition" are merely dramatic examples of a general problem, easily observed because of their magnitude and because of their relative rarity. Millions upon millions of small "disasters" occur each day for millions of separate systems, disaggregated in space, and therefore regarded as small crises. But they are made of the same "stuff" that the larger events are made of as far as understanding the underlying processes involved.

Disasters, as normally defined by disaster researchers, present a priceless opportunity for the social as well as the physical and biological sciences to understand the very root processes by which both biological and sociocultural systems evolve, thereby impacting on the nonliving environment. By extending our perspective past those dramatic collective events we currently call disasters to encompass research on the relationship of human systems to their environment within the framework of "general ecology," we might succeed in making a contribution to general sociology and to ecological theory, as well as to systems theory and thereby to all of the sciences.

To conclude, an ecological approach to disasters should focus on the attributes of social systems that generate social vulnerability. In this approach, vulnerability is regarded as an intrinsic property of all systems that are evolving in response to selective forces to survive contingently rather than permanently within an ecological field. Social systems are viewed as evolving by giving high priority to developing the cultural and, therefore, the cognitive capacity to deal with high-probability routine events, giving lower priority at the same time to the development of a capacity to deal with low-probability events. A better theoretical representation of the ecological etiology of disasters could provide a firmer basis for applied programs that might go beyond the autonomous, contingent, piecemeal adaptations which occur uncontrolled within an unmanaged ecological field, offering the opportunity for policy interventions that could reduce long-range vulnerability to low probability, but high-negative impact events. This is, at least, our moral hope.

Part Two
Collective Behavior and Social Movements

Introduction

Although the chapters included in this section deal with a range of topics in the field of collective behavior and social movements, they have several things in common. First, many of the chapters provide a badly needed, richly deserved critique of both common-sense myths about collective behavior and misguided scholarship in the field itself. Second, several chapters report the results of solid empirical research and advanced analytic methods used to study forms of collective behavior, such as panic, that are often overlooked by researchers or incorrectly dismissed as unresearchable. Third, considered as a group, the chapters make some significant contributions to theoretical reasoning in the collective behavior/social movements area. Particularly notable are the linkages made between collective behavior and social movements—phenomena that increasingly have come to be seen as separate and disparate as the field has evolved—and between the more enduring, organized features of the social order, on the one hand, and emergent social phenomena such as panics, crowds, rumor, and "hysterical contagion," on the other.

The chapters in this section are arranged in an order that reflects both the continuum between emergence and institutionalization and the micro-macro dimension. In the first chapter, Norris Johnson, William Feinberg, and Drue Johnston discuss behaviors that emerged in response to the Beverly Hills Supper Club fire. This 1977 event, in which 165 persons were killed, was characterized in the mass media as an incident that led to uncontrolled panic flight, and highly individualistic, competitive behavior. Through a careful reconstruction of the event, the authors show that, except in one extremely threatening location, the behavior of the patrons at the time of the fire was neither individualistic nor focused exclusively on self-preservation. Instead, preexisting social bonds were maintained, and a considerable amount of altruistic behavior occurred. Interestingly, these researchers found that concern for the group, rather than individual survival, was quite marked at the time of the fire, even in situations in which patrons believed they were facing an imminent threat of entrapment.

In contrast with the Beverly Hills case, which involved extreme threat, the "crush" at Hillsborough, discussed in the chapter by Jerry Lewis and Michael Kelsey, occurred in a context of extreme enthusiasm and excitement. Situations like the Hillsborough crush and the Who Concert stampede that occurred in Cincinnati a little more than a decade earlier are often mistakenly classified as "panics" and considered analogous, comparable, or parallel to episodes of collective flight. Rather than making that conceptual error or treating the Hillsborough episode as merely one more event in a bizarre tradition of soccer violence, the authors correctly view the crush as a type of craze or collective enthusiasm. When the characteristics of the "crush" as a social phenomenon are separated analytically from its tragic consequences, the reader becomes aware that the behavior in question has actually more in common with fad behavior and convergence behavior in disaster situations than it does with panics and hostile outbursts. Using Neil Smelser's value-added theory of collective behavior, the authors show that an appropriate explanation of the crowd action that was observed must go well beyond notions that emphasize subcultures of violence and "soccer hooliganism" and must look instead for the antecedents of craze behavior in the social structure of the event. Among the most important structural antecedents noted in the Hillsborough case were factors that greatly limited access to the perceived benefits of participation in the event as well as social control practices that increased pressure for access to the site rather than reducing it.

The chapter by Robert Stallings is highly innovative in its treatment of another underresearched form of collective behavior, "mass hysteria." The distinctive contribution of this chapter is that it takes as problematic what most other sociological treatments of the phenomenon assume: the existence of hysterical illness. Using examples that involve similar kinds of symptoms that could be labeled as "hysterical," exhibited by people with different social statuses in markedly different social settings, Stallings shows that the definition of a particular set of behaviors or symptoms as hysterical depends to a large extent on the amount of power and status possessed by the group involved in those behaviors, as well as on the "framing" of the episode by those who have the power to define behavior for others. Through his treatment of various cases of mass hysteria, Stallings shows that in our society children, women, and persons of low social status are more susceptible to having their physical complaints dismissed as hysterical than are members of high status groups who, by virtue of their status, can resist being so labeled. The Stallings chapter not only sheds new light on the phenomenon of collective hysteria, but it also serves as a very important warning to sociologists and others who too willingly and uncritically accept mass media, medical, and psychiatric definitions of the social behavior they study.

The chapter by Dennis Wenger and Thomas James concerns both disasters and collective behavior. Much of the discussion centers on establishing the nature and extent of volunteer behavior in the 1985 Mexico City earthquake. The authors find that volunteer behavior was relatively widespread (although not as common as some would have predicted) and was related not only to residents' sociodemographic characteristics but also to predisaster roles. Through an analysis of the factors affecting residents' participation in the various kinds of helping behavior that occurred, the authors compare and contrast collective mobilization in "consensus" crises such as disasters and "dissensus" crises such as riots. Their findings with respect to the Mexico City case, like studies of mobilization in riot situations, disprove convergence explanations, which emphasize what participants in collective behavior have in common. At the same time, the authors show that mobilization for collective action in consensus crises like disasters differs in important ways from mobilization in conflict situations.

Like a number of other chapters in this section, Ralph Turner's chapter on disaster rumors and rumor control in the United States and China both corrects misunderstandings about collective behavior phenomena and illustrates the manner in which collective behavior is socially structured. Challenging what he terms the *folk concept* of rumor, which sees rumors as inherently outlandish, distorted, and nonfactual, Turner argues that rumor ought more appropriately to be viewed as one among a range of information-seeking strategies available to individuals and groups. He shows that earthquake rumors in California occurred in a context of other information-seeking activity, such as paying attention to mass media information, engaging in informal discussions, and consulting "expert" information sources. He notes that when accurate information is available, heightened rumor activity can have positive effects. Turner also shows that even though decision makers and disaster planners in China and the United States share similar scientific assumptions about the nature and origins of rumors, differences in societal structure lead to marked differences in rumor management strategies.

The chapters by Benigno Aguirre and Lewis Killian both represent attempts to bring about a much needed reexamination of the relationship between collective behavior—that is, emergent, less structured, and more ephemeral noninstitutionalized social behavior—and more highly organized social movement phenomena. Aguirre argues convincingly that those scholars who make sharp distinctions between collective behavior and social movements not only misunderstand the history of the field and the contributions of classical scholars, but also lose the theoretic and research opportunities that a more unified view could provide. He shows clearly why it is not useful to rely solely on perspectives that focus on organization and rationality in social movements to the exclusion of everything else. For Aguirre, an overemphasis on rational calculation models of movement activity, although once perhaps a necessary corrective, now brings diminishing returns.

Lewis Killian's chapter is also a reaction to the increased tendency in the field to separate collective behavior and social movements and to look on most work in the collective behavior school as irrationalist in emphasis and as lacking use for the study of social movements. He questions the conceptual accuracy of perspectives that seek to make sharp distinctions between the rational and irrational components of social behavior. He also defends the emergent norm perspective and related collective behavior approaches from those critics who would overemphasize the contribution of cognition, rational calculation, and structure to social movement activity, while downplaying feelings, symbols, interaction, and innovation.

Verta Taylor's chapter, written primarily from the perspective of resource mobilization theory, addresses the important question of how social movements survive during periods when movement activity declines. She develops the concept of an "elite-sustained" movement to explain how the positions advocated by social movements are preserved during periods of dormancy and non-mobilization. She shows that during the period when women's issues were not on the public agenda, there remained a core of women's rights activist groups that held a vision advocating social change and that provided a significant degree of social support for members. Moreover, the organizational structures and programmatic emphases developed by these relatively "elite" sets of organizations influenced the nature of the larger, more comprehensive women's movement that emerged in the 1960s. Thus, rather than conceiving of social movement careers as linear and social movement outcomes as bimodal or bipolar (e.g., success versus failure), it is necessary to think in terms of "active" and "abeyance" movement cycles that can be expected to vary in both intensity and duration. Similarly, rather than thinking of movements as involving either a mass base or elite sponsorship, it is necessary to acknowledge that the same social movement can take different forms, depending on the historical context.

The concluding chapter in this section, written by Gary Marx, is perhaps the most general and comprehensive chapter in the group, and it is in some ways the most ambitious. The topic considered is the relationship between societal diversity, social change, and societal cohesion. Marx discusses evidence from a variety of sources suggesting that U.S. society is becoming increasingly heterogeneous and diverse; that residential patterns and work arrangements appear to foster greater social isolation; and that rates of various types of social participation are on the decline. He then asks whether such patterns indicate that U.S. society is becoming increasingly disorganized and masslike and whether negative consequences will result. Fortunately, he counsels against such dire predictions, pointing out that declines in certain types of social activity (e.g., participation in religious and labor organizations) have been more than offset by the growth of other activities (e.g., membership in nonprofit organizations and attendance at cultural events). Marx makes a convincing case that what we are now observing and experiencing is not a loss of societal cohesion, but rather a change in the foci around which members of society choose to organize themselves.

The Marx chapter also contains an implicit critique of much of the work that is being done using social indicators. Too frequently, the manner in which social indicators are used make it difficult not to conclude that society is becoming increasingly fragmented and disorganized. What Marx shows convincingly is that signs of declining social attachment along some dimensions should be interpreted in light of other trends that signify the emergence of new forms of attachment and new types of social participation.

Microstructure and Panic: The Impact of Social Bonds on Individual Action in Collective Flight from the Beverly Hills Supper Club Fire

NORRIS R. JOHNSON, WILLIAM E. FEINBERG, and DRUE M. JOHNSTON

Quarantelli's discussions of panic over three decades (1957, 1981) provide the basis for presenting and evaluating three sets of assumptions about collective flight. Quarantelli's work recognizes the social nature of most collective flight and consistently emphasizes the importance of social ties to others in preventing panic; in defining panic he postulates that under the most severe threat, however, social bonds may break down with panic flight ensuing. His point is made most clearly in the assertion that panic flight (although rare) takes on a nonsocial character in disregard of usual social relationships and expectations, with "even the strongest primary group ties ... shattered and the most socially expected behavior patterns ignored" (Quarantelli 1957, 193).

We interpret this definition of panic as reflecting a *severe threat* model of collective flight that contrasts it with two other views of which it is a synthesis. One, the classic *organizational breakdown* model, focuses solely on individualistic responses to perceived threat, postulating that threatened individuals will ignore others and compete ruthlessly for limited escape routes. In its exaggerated form, this model leads to the inference that "crying 'fire' in a crowded theater" leads immediately to irrational, nonsocial behavior that reflects efforts for self-preservation at all costs.

The competing model, which was emphasized in Quarantelli's work and which we label simply a *social* model, assumes that the typical response to threat is "normal" social behavior in which normative expectations and role relations are maintained. Social organization does not break down, and any competition for exits is regulated by social ties and social norms. Disaster researchers have consistently argued that flight involving unregulated competition rarely occurs (Quarantelli 1957; Keating 1982), and some research reports show the impact of social bonds on response to threat (Sime 1983; Johnson 1988).

We noted in earlier papers (Johnson 1987b, 1988; Johnston and Johnson 1989) that social organization, including social bonds, was crucial in preventing unregulated competition in the flight from the Beverly Hills Supper Club fire, lending support to the social model. Much of the support for that conclusion came from qualitative analysis of data gleaned from interviews with survivors of the fire. This chapter subjects the conclusions to further test with a quantitative analysis of the same data, and uses the results of the tests to further evaluate contrasting views of collective flight.

Background

Sociologists' discussions of collective flight are usually framed within collective behavior theories, and "panic" is often characterized as one form of collective behavior. The explanations differ in their particulars, variously viewing the flight as an emotional response to an hysterical belief (Smelser 1962), the result of a norm emerging to legitimate individual, competitive behavior (Turner and Killian 1987), or a rational response to the reward structure of the situation (Kelley, Condry, Dahlke, and Hill 1965). In most instances, however, the aggregate response is postulated to be uninfluenced—or minimally influenced—by social bonds. In short, the prevailing view is that competition in panic episodes is not regulated by social ties and social norms.

Group Ties within Collective Flight

The problem poses two related research questions. The first is the question of the extent to which social bonds are present among those engaged in collective flight; the second question is the impact on flight of the bonds that do exist.

Aveni (1977) showed more than a decade ago that ties to others were important elements in crowd formation. A similar contention can be made for collective flight. Organizational breakdown models (e.g., Turner and Killian 1987) assume that social bonds are initially present within the aggregate that engages in collective flight, but that these bonds break down under threat. Other more purely individualistic models that have organizational breakdown as a consequence (Smelser 1962, Kelley et al. 1965) virtually ignore ties.

Although studies do not measure social ties in collective flight, some do investigate the impact of social organization on flight. Sime's (1983) work on the flight from a fire at a resort on the Isle of Man is the example most related to this work. He concluded that individuals attempt to escape with group members—particularly family members—to whom they have strong psychological ties. Similarly, Smith's (1976) examination of flight from an explosion in the Tower of London emphasized the crucial function of primary group interaction in the ensuing flight. Work on less related phenomena such as a bank run (Mann et al. 1976) and a crush in which concertgoers were killed while trying to enter an arena (Johnson 1987a) also showed that the selfish, disorganized behavior supposedly characteristic of such occurrences did not appear. Previous research on response to the Beverly Hills fire showed that occupational roles (Johnston and Johnson 1989) and other social roles continued to be performed (Johnson 1987b, 1988), although some evidence suggested the beginning of unregulated competition in one room in the club (Johnson 1988). Overall, the available research supports the conclusion by disaster researchers such as Quarantelli and Dynes (1972) and Keating (1982) that the unregulated competition implied by organizational breakdown models rarely occurs, and is most supportive of the social model of collective flight.

Research Procedures

Conceptual Model

Our intent in this research is twofold. First, we investigate whether or not there were group ties in this particular collective flight, and then we examine—in relation to the theoretical discussions earlier—the impact of the ties on the nature of the flight response. Our working hypothesis is that multiple bonds tied Beverly Hills patrons to others present in the building and that these ties had an important impact on flight from the fire. However, the discussions reviewed earlier permit competing inferences concerning the relationships among entrapment threat, social bonds, and the emergence of individualistic, nonsocial behavior. The classic organizational breakdown model can be interpreted as postulating that as threat increases, the breakdown of social order will be more complete, thereby increasing individualistic behavior. The social model (as well as Sime's more psychological, affiliative model) suggested that as entrapment threat increases (implying as well an increased threat for those to whom one is bonded), behavior reflecting concern for others to whom one was bonded would also increase (or at least remain constant). Quarantelli's description of panic requires consideration of a combination of the two models: the social model is appropriate up to some threshold of severe threat, beyond which increasingly individualistic behavior (or organizational breakdown) occurs. The hypothesized relationships might be diagramed as in Fig. 4.

Data that bear on the theoretical formulations are available from the fire

Figure 4
Models of Relationships Between Entrapment Threat and Social Response (Holding Social Bonds Constant)

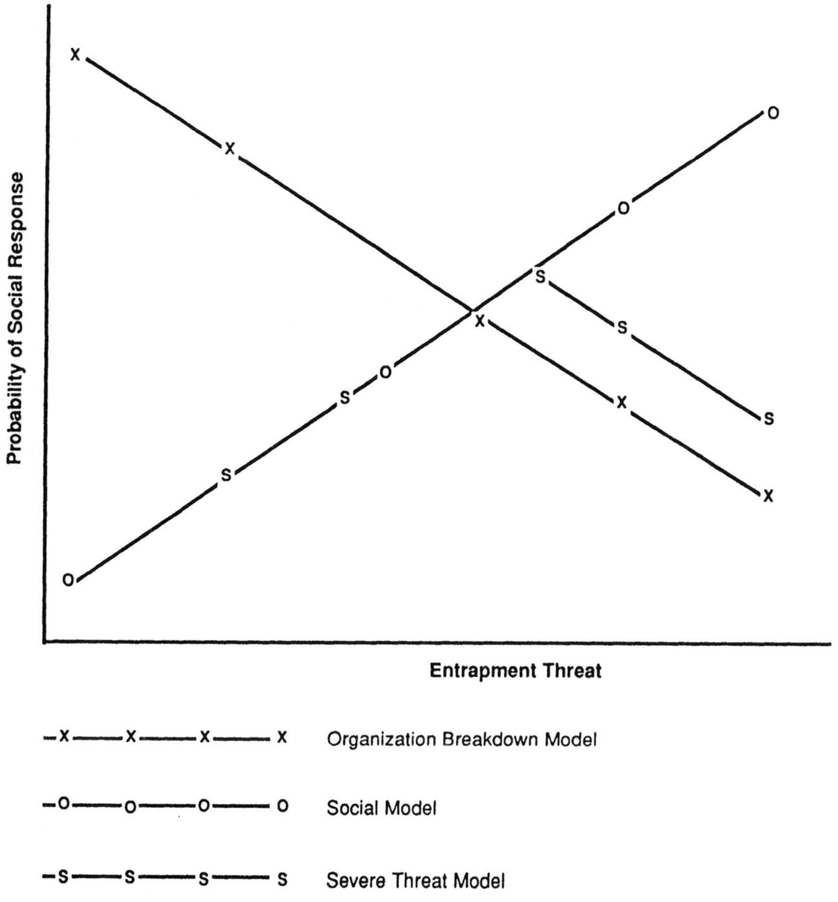

at the Beverly Hills Supper Club in Southgate, Kentucky, in which 165 people were killed and in which the prototypical conditions for organizational breakdown and ensuing panic flight seemed present.

The Setting: Beverly Hills Supper Club

Some two thousand five hundred people were in the Beverly Hills Supper Club, located in a suburb of Cincinnati, Ohio, on the night of the fire in May 1977. They were dispersed through the several rooms of the club, which served various functions (see Fig. 5). All who were killed in the fire died of smoke inhalation and were found within thirty feet of an exit.

The fire began in the then vacant Zebra Room at the front of the building. Patrons in the distant Garden Rooms, from which there was easy access to the outside, exited with no awareness of the fire other than having been quietly told by waitresses to leave. The patrons in the front of the building—in the main dining room, the main bar, and the Viennese Room—exited with slightly more difficulty. Patrons in the large, centrally located Empire Room first moved toward the main entrance, where they found themselves blocked by the line of patrons fleeing from the front rooms; many fell in line at the end of the queue, but many others retreated through the kitchen to the rear exits. In the Crystal Rooms, directly above the fire, patrons became aware of the fire only after the only exit with which they were familiar—a spiral staircase that led toward the fire—was in flames, blocking the only escape route known to them. These patrons began searching for a way out before an employee appeared and led them, in the dark, down a narrow service stairway and out through the kitchen. Two Crystal Room patrons died.

All the other fatalities were in the Cabaret Room, the main showroom in the rear of the building, where approximately one half of the patrons present (one thousand two hundred to one thousand three hundred persons) had gathered to hear singer John Davidson. Because the showroom had its own staff and service bar, there was little communication with the front of the building after the show began; thus, the showroom patrons were the last to be warned. The room was filled to capacity, and the available exits did not lead directly outside. Movement toward those exits was at first orderly, but as the exits began to be blocked, the crowd surged toward them, and the competitive conditions of panic may have begun to develop (see Johnson 1988).

Data Source

The data are from depositions obtained by the Kentucky State Police (KSP) from survivors of the fire. In an inquiry into causes of the fire, investigators

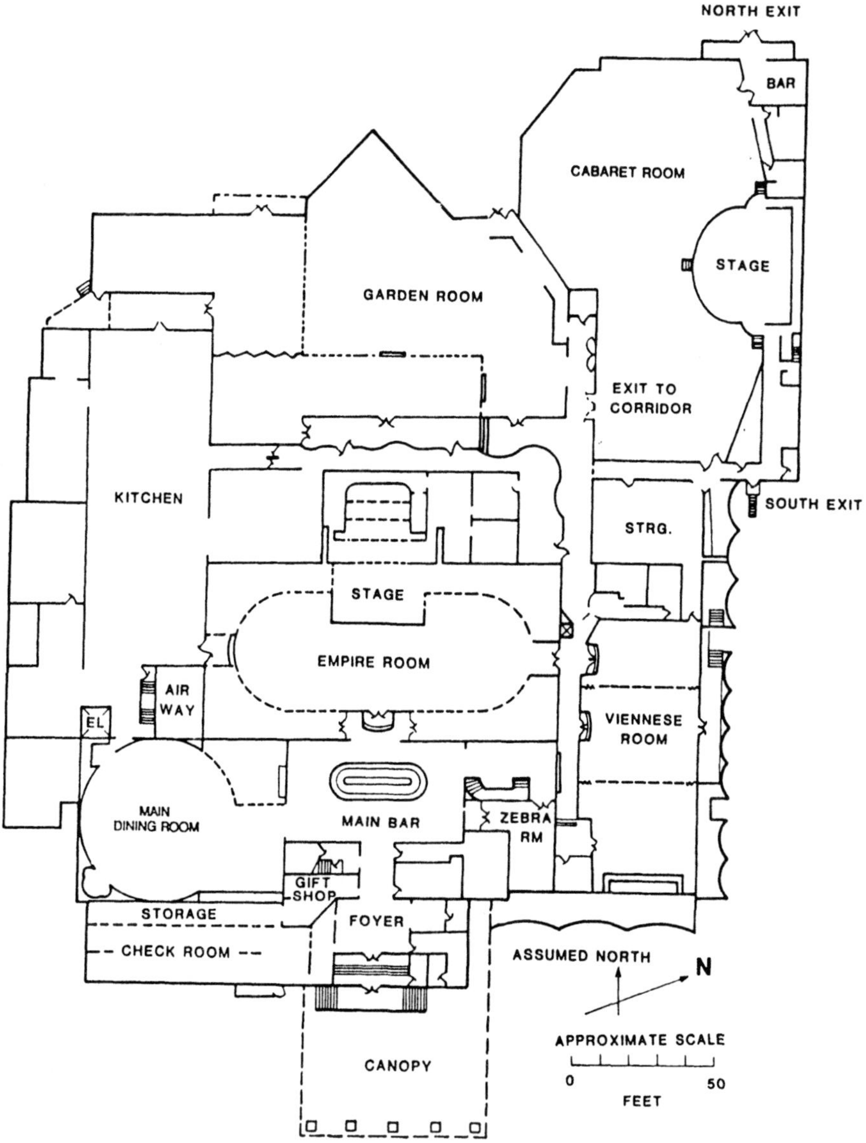

Figure 5
Diagram of First Floor, Beverly Hills Supper Club

from the KSP interviewed more than six hundred persons—including 465 patrons—present at the club on the night of the fire. Police officers from a small nearby city got an additional thirty-two depositions from patrons who were part of a group that traveled by bus to the club. Typed transcripts of 345 interviews with club patrons (313 KSP interviews plus the thirty-two depositions) were available for analysis. The 465 patrons interviewed did not constitute a probability sample of all patrons who survived the fire. Also, the 345 transcripts[1] available for analysis comprise only an availability sample of all completed interviews with surviving patrons and may not be representative of all of those interviewed. Thus, statistical inference must be made with caution.

To make use of the transcripts, theoretically relevant questions were devised and included within a questionnaire; the transcripts were then read for responses to the specific questions contained in the questionnaire. Responses were coded to make the data accessible to quantitative analysis.[2] The coded information does not comprise a set of responses to a series of specific questions but represents the patrons' reconstructed narrative accounts of their actions. The analysis must be interpreted with these limitations in mind as well.

Measurement of Variables

As part of the coding procedure, each respondent's transcript was examined for evidence of any social bonds to others within the club and for evidence of the impact of those bonds on response to the fire and exit from the club. Six types of social bonds were coded and used in this analysis:

1. Spouse
2. Dating partner (or opposite sex companion)
3. Family member other than spouse
4. Friends (other than dating partner)
5. Secondary group
6. "Nested" secondary group

The final category emerged when we discovered that some persons were in groups that were part of even larger groups.[3]

In the latter part of the analysis—focusing on the impact of social bonds on response to the threat—a distinction was made between primary ties and secondary ties; spouse (or dating partner), other family members, and friends are considered as primary group ties.

To code the impact of social bonds—a *social response* measure—we recorded each mention of an action that reflected a concern for other group members. The categories coded were:

1. Expression of concern for another group member's safety
2. Joint decision on response after learning of fire
3. Effort to stay together
4. Delay in exiting or returning to assemble group
5. Exiting together
6. Expression of concern over involuntary separation from group member
7. Assisting group member (directing, leading, lifting, carrying)

The items were summed to create a social response index, ranging from 0 (reported none of the activities) to 7 (reported all of the activities), with 0 being totally nonsocial and 7 being highly concerned for others in the group.[4] The seven categories were collapsed into four for some subsequent analyses.

Theories of panic flight emphasize the importance of a perception of the possibility of exits being blocked—an entrapment threat—as a crucial factor leading to the breakdown of ties to others. A measure of entrapment threat was developed based on comments of interviewed patrons concerning their difficulty in exiting the building. The measure was coded into four categories, from lowest to highest:

0. No threat. Patron walked directly out the exit, unimpeded by others or by smoke or flames.
1. Mild threat. Patron was forced to join a queue to file through an exit behind others.
2. Moderate threat. Patron was forced to choose an alternate exit after the first choice was blocked by people, smoke, or flames, or after being redirected by an employee.
3. Severe threat. Patron reported being unable to exit without crawling under smoke or over others who had fallen. (Category also includes all who were carried from the building by others, or who were hospitalized with injuries.)

Results

Results are reported first for the existence and frequency of social bonds among the 342 cases for which the relevant data are available. Then an analysis of the impact of the bonds on exit—that is, for the expression of a social response—is presented.

Table 6. Frequency of the Various Social Bonds Among Patrons of the Beverly Hills Supper Club on the Night of the Flight from the Fire

Bond Type	Number Patrons Reporting	% Patrons Reporting*
Husband/Wife	216	63.2
Dating Pair	33	9.6
Family Member (Other than Spouse)	75	21.9
Friend	141	41.2
Secondary Group	164	48.0
Nested Secondary Group	75	21.9

* Total exceeds 100% because of multiple memberships.

N=342

Existence of Social Bonds

Table 6 shows the frequency distribution for the various social bonds. The most frequent social bond (63 percent) was to a spouse, with an additional 10 percent to a dating partner (combined in subsequent analyses). More than one fifth (22 percent) of the respondents were with a family member other than a spouse, and just more than 40 percent were part of a friendship group.

In addition to having these primary group ties, nearly one half of the respondents (48 percent) were part of a secondary group (e.g., the large group attending a banquet in the Empire Room), and nearly half of that group were part of a nested secondary group as previously defined.

Table 7 shows patterns of bonds among the patrons, ordered to illustrate more clearly the possible patterns. With spouse and dating partner combined into a "couples" category, five separate bonds are possible. The upper panel of the table shows all patterns involving marital (and dating partner) bonds, almost three-fourths of the cases; the lower panel shows a similar ordering

Table 8. Most Frequent Patterns of Groups Found Among Patrons at the Beverly Hills Supper Club on the Night of the Flight from the Fire

	Couples*	Other Family	Friends	Secondary Group	Nested Sec. Gp.	N	%
1.	+	0	+	0	0	67	19.6
2.	+	0	0	0	0	41	12.0
3.	+	0	0	+	+	35	10.8
4.	+	0	0	+	0	29	8.5
5.	+	+	0	0	0	28	8.2
6.	0	0	0	+	+	24	7.0
7.	+	0	+	+	0	20	5.8
8.	0	0	0	+	0	16	4.7
9.	+	+	+	0	0	13	3.8
10.	0	0	+	0	0	13	3.8
11.	0	0	+	+	0	12	3.5
12.	+	+	0	+	+	8	2.3

N = 342

* Includes married couples and dating pairs.
+ = Bond reported.
0 = Bond not reported.

Table 7. Patterns of Social Bonds Found Among Patrons at the Beverly Hills Supper Club on the Night of the Flight from the Fire

Couples*	Other Family	Friends	Secondary Group	Nested Sec. Gp.	N	%
+	+	+	+	+	2	.6
+	+	+	+	0	1	.3
+	+	+	0	0	13	3.8
+	+	0	+	+	8	2.3
+	+	0	+	0	3	.9
+	+	0	0	0	28	8.2
+	0	+	+	+	2	.6
+	0	+	+	0	20	5.8
+	0	+	0	0	67	19.6
+	0	0	+	+	35	10.2
+	0	0	+	0	29	8.5
+	0	0	0	0	41	12.0
0	+	+	+	+	1	.3
0	+	+	+	0	2	.6
0	+	+	0	0	5	1.5
0	+	0	+	0	6	1.7
0	+	0	0	0	6	1.7
0	0	+	+	+	3	.9
0	0	+	+	0	12	3.5
0	0	+	0	0	13	3.8
0	0	0	+	+	24	7.0
0	0	0	+	0	16	4.7
0	0	0	0	0	5	1.5

* Includes married couples and dating pairs.
+ = Bond reported.
0 = Bond not reported.

N = 342

for those without the marriage/dating pair bond. These patterns are presented simply for illustration and are not further analyzed, but they clearly illustrate the potential density of social bonds.

Table 8 arranges the patterns in order of frequency, including only those patterns with eight or more cases. The most frequent pattern is a grouping that includes a married couple with friends, most often another married pair. The next most frequent patterns are a couple alone and a couple that is within a larger secondary grouping, in most cases a couple that attended the savings and loan association banquet (Empire Room) along with fellow employees. That most of these patterns involve the marital (or dating pair) bond illustrates the strength in addition to the density of the bonds present within the club.[5]

The crucial fact for this report is simply how much social bonds were present in the club. At least one primary bond was reported by 297 patrons (86.8 percent), although forty-five patrons (13 percent) reported only a secondary tie of either or both types; only five individuals (1.5 percent) reported no ties at all. It is clear, then, that almost everyone at the Beverly Hills was bound by social ties to others copresent, and many were embedded in dense networks with multiple bonds.

Impact of Social Bonds on Patrons' Response

The next step is to determine whether the presence of social bonds led to a social rather than an individualistic response to the fear of entrapment. Specifically, does the existence of ties to others prevent a nonsocial response to threat, that is, panic flight? A social model could suggest that the nature of the response is determined primarily by social bonds. In fact, following Sime (1983), a greater threat might lead not to a more individualistic but to a more social response, as the greater threat produced greater concern for the welfare of those to whom one was tied.

On the other hand, following the organizational breakdown model, we might expect that the primary factor determining the response would be the severity of the threat. Thus, here we could expect the nature of the response (social response) to be associated with entrapment threat. As threat from the fire and fear of entrapment increased, the response would be increasingly individualistic. Therefore, under this model social response should be negatively related to entrapment threat.

Finally, from Quarantelli's (1957) discussion and definition of panic, we might expect that if the perceived threat is mild, bonds will be maintained and the response will be social rather than individualistic; however, if the threat of entrapment is severe, social order may break down and individualistic responses will prevail. The crucial test is whether or not the measure of social response changes as the perceived severity of the threat changes. If the social response measure declines once some threshold of entrapment threat is reached, the results would be compatible with Quarantelli's definition of panic. If, on the other hand, acts reflecting a social response continue to increase or remain constant, support for the impact of social bonds—the second model—can be seen.

The central question of this research, then, concerns whether or not (and how) entrapment threat influences the way social ties affect individual behavior in collective flight; the discussions we reviewed lead to different expectations about the nexus of entrapment threat-social bonds-collective flight. To assess the possible links among these concepts, we chose initially to examine the relationship between social bonds and individual behavior in collective flight.

We thus considered how the social response score varied with patterns of social ties, ignoring entrapment threat. Table 9 shows the most frequent patterns of social ties arranged in descending order of average social response score; after omitting cases of infrequent patterns, 301 of the total 345 patrons are left to consider here. In general, those patterns that include a couple bond (either spouse or dating partner) produced, on average, the highest social response. The only noncouple pattern to produce a high-average social response was the relatively infrequent one involving only friends and secondary ties. The patterns that included only secondary ties yielded the lowest social response scores. Overall, there is considerable variation both in social response scores and in the social bonds that existed among the patrons whose police transcripts were coded.

The results in Table 9 seem to point to a significant connection between social bonds—particularly primary bonds—and the degree to which patrons reported engaging in a social response. To explore further this apparent connection, we pursued a series of regression analyses[6] of the data gathered from the transcripts. We further restricted this analysis to those patrons ($N = 253$) who had been in one of three locations: the Cabaret Room ($N = 143$), the Empire Room ($N = 81$), and the Crystal Rooms ($N = 29$), locations in which the threat was most severe.

We must repeat the earlier caveat about the lack of randomness in the samples and add two other cautions. First, membership in the sample was subject to selection bias because transcripts were obviously not available from those patrons who lost their lives in the fire; such bias is a particular problem for considering those in the Cabaret Room, in which all of the deaths except two occurred, and less of a problem when analyzing those in the Crystal Rooms, the location of the other two patron deaths.[7] We are using samples of *survivors* of the fire, and our attempts to infer from their behavior about the behavior of all who were attempting collective flight from the fire must be considered cautiously. Second, these are retrospective data, with all the attendant problems of participant reconstruction and revision.[8]

Table 9. Mean Social Response Scores by Social Bond Pattern Found Among Patrons at the Beverly Hills Supper Club on the Night of the Flight from the Fire, Ranked by Social Response Score (Includes Only Most Frequent Patterns)

Couples*	Other Family	Friends	Secondary Group	Nested Sec. Gp.	Social Response Score	N
+	+	+	0	0	2.39	13
+	0	+	+	0	1.89	18
+	0	0	+	0	1.77	26
0	0	+	+	0	1.67	12
+	0	+	0	0	1.65	66
+	0	0	+	+	1.63	36
+	+	0	0	0	1.45	29
+	+	0	+	+	1.38	8
+	0	0	0	0	1.33	40
0	0	+	0	0	1.00	13
0	0	0	+	+	0.58	24
0	0	0	+	0	0.31	16

* Includes married couples and dating pairs.
+ = Bond reported.
0 = Bond not reported.

All 253 patrons in our more restricted sample were first included in a single analysis in which we sought the equation best predicting an individual's social response score from information regarding primary and secondary social bonds, how much the patron was subject to an entrapment threat, and the room in which the patron was located. Dummy variables were used to indicate the presence or absence of primary and secondary bonds, and two dummy variables were used to distinguish among the three rooms. The entrapment threat variable (threat) took values between 0 and 3. The characteristics of the best-fit prediction equation are shown in Table 10.

Each of the social bonds variables and the measure of entrapment threat have a significant positive effect on the social response score; however, the

Table 10. Multiple Regression Analysis (OLS) of Social Response Scores, Patrons (N = 253) from Three Rooms at the Beverly Hills Supper Club

Predictor	Parameter Estimate	Standard Error	Beta
Constant	0.042	0.216	----
Primary Tie (D)	1.317***	0.179	0.456
Secondary Tie (D)	0.279*	0.129	0.137
Threat	0.150**	0.056	0.159
Room 3 (D)* Threat* Primary Tie (D)	-0.282*	0.108	-0.161

R^2 = .198

Adjusted R^2 = .185

* p < .05 (2-tailed);

** p < .01 (2-tailed);

*** p < .001 (2-tailed).

(D) indicates a dummy variable.

Room 3 (D) is a dummy variable denoting the Crystal Rooms.

negative effect of the interaction of threat and primary tie within the Crystal Rooms (Room 3) is particularly noteworthy[9] Thus, within the Crystal Rooms, patrons with primary ties displayed decreasing social concern with increasing threat, when compared to patrons in the other rooms who were comparably bonded and threatened. The considerable magnitude of the effect of the primary ties variable, reflected by the size of its beta relative to the others, is consistent with the pattern shown in Table 9.

Table 11 shows the results of our search for the best equations for predicting social response of patrons within each of the three rooms. The equation for the Cabaret Room patrons indicates significant positive effects for primary ties, secondary ties, and entrapment threat; these results thus provide some support for the social model. The equation for the Empire Room patrons attributes statistical explanation only to the existence of a primary tie, whereas the entrapment threat variable provides no significant effect[10]

Table 11. Multiple Regression Analyses (OLS) of Social Response Scores, Within Each of Three Rooms at the Beverly Hills Supper Club

	Room 1	Room 2	Room 3
	Cabaret Room	Empire Room	Crystal Rooms
Predictor	Parameter (SE)	Parameter (SE)	Parameter (SE)
Constant	- 0.208 (.339)	0.143 (.317)	0.374 (.701)
Primary Tie (D)	1.558 (.314)	1.587 (.331)	0.582 (.353)
Secondary Tie (D)	0.505 (.237)	Omitted	Omitted
Threat	0.151 (.063)		1.368 (.669)
Threat^2			- 0.481 (.187)
N	143	81	29
R^2	.214	.225	.282
Adjusted R^2	.197	.215	.195

(D) indicates a dummy variable.

As we should expect from the results of Table 10, the Crystal Rooms equation is again quite different. A significant positive effect on social response is evident for the threat variable alone, with a significant negative effect for the square of the threat variable[11] The effect of threat on social response here is thus curvilinear, increasing as threat increases from zero to one and decreasing as threat increases beyond a score of one. Here then is some evidence to suggest the occurrence of panic as defined by Quarantelli.

Thus far the results provide tentative evidence for the social model (from the Cabaret Room results) and for the severe threat model (from the Crystal Rooms equation). None of the results yet are consistent with the organization breakdown model.

A better test might be made by examining regression equations for social response as threat condition is held constant within each room context, be-

Table 12. Multiple Regression Analyses (OLS) of Social Response Scores, for Each Threat Level (Parameter Estimates and Standard Errors) Three Rooms at Beverly Hills Supper Club Combined

Predictor	Threat = 0	Threat = 1	Threat = 2	Threat = 3
Constant	0.467 (.199)	0.333 (.361)	0.875 (.313)	0.286 (.419)
Primary Tie (D)	0.908 (.224)	1.399 (.373)	0.761 (.349)	1.714 (.457)
Room 3 (D)* Primary (D)				- 1.600 (.529)
N	71	92	41	49
R^2	.193	.135	.109	.308
Adjusted R^2	.181	.126	.086	.278

(D) indicates a dummy variable.

Room 3 (D) is a dummy variable denoting the Crystal Rooms.

cause we know that the room context affects the impact of the other variables on social response. The limited number of cases unfortunately prevents such a simultaneous control, so we are limited to holding either room or threat level constant. The results of holding threat level constant are shown in Table 12.

The parameter estimates shown in Table 12 offer no support for the organization breakdown model but do show evidence for both the social and severe threat models, depending on room context. For those with only secondary ties (primary tie = 0), the predicted social response scores (constant) are not monotonic in their variation as threat increases. Individuals with only secondary ties might be considered more likely to be susceptible to organizational breakdown in collective flight, but the lack of monotonic decline in social response here suggests otherwise[12]

Ignoring for the moment the separate effect attributable to patrons in the Crystal Rooms (Room 3), we can add together the parameters for primary tie with the respective constants under each of the threat conditions to get the predicted social response scores. Except for an apparently insignificant decline between threat conditions 1 and 2, the predicted scores increase as threat increases,[13] thereby lending support to the social model. That support must be considered restricted to the contexts of the Cabaret and Empire rooms, however.

On the other hand, the severe threat model gains support from considering the responses of those with primary ties in the Crystal Rooms. Inclusion of the interaction term under the most extreme threat shows an inverse U-shaped pattern for social response as threat increases, with average social response dropping below 0.5 under the most extreme threat. Again, this is compatible with Quarantelli's definition of panic; it also shows that room context was critical to the kinds of responses patrons made to the situation.

Discussion

On the basis of the data, we can conclude without equivocation that any model that does not assume the presence of social ties is inappropriate for this collective flight episode, and probably for others in which social ties exist. Social bonds abounded in the club. All respondents in this analysis of 253 patrons reported at least one primary or secondary group bond, and most reported bonds of multiple types.

It is also clear that the immediate flight from the building was not individualistic, which permits us to reject (as has been so often done) models that predict immediate unregulated, nonsocial flight. Values on the social concern measure indicate that patrons fled as members of groups, often hesitating in their flight to ensure that others to whom they were bonded were exiting as well. Most exited along with group members, and others expressed concern while exiting over being separated from those with whom they had relationships. To the extent that these post facto reports of behavior are not distorted, we can conclude that threatened patrons did not immediately behave in a selfish, individualistic manner.[14]

The fact that entrapment threat also was generally positively associated with social response also leads to rejection of a simple breakdown model and lends further support to a social model. Generally, as the threat of entrapment increased, greater—rather than lesser—concern for group members was expressed. The Empire Room was an exception, but the minimal variation in entrapment threat scores from that room probably account for the anomaly.

Which model best depicts social response under entrapment threat depends on the room of the club in which the respondents were located, with one model appearing to fit the data from the Cabaret Room, and a very different model fitting the data from the Crystal Rooms. The effect of both primary and secondary ties and of entrapment threat was positive in the Cabaret Room, providing support for the social model. In the Crystal Room, however, the interaction effect of threat and primary tie is negative. The results indicating that concern with primary groups members decreased as threat increased are what would be expected if panic as defined by Quarantelli occurred. After a certain threshold of threat was reached, concern for others began to decrease rather than continue to increase. This response would represent panic, in which "even the strongest primary ties may break down." Thus, we might conclude that panic did occur within the Crystal Rooms, although not in the other rooms of the building.

This room effect could be due to the particular panic-producing characteristics of the situation within the Crystal Rooms, or might be simply an artifact of having data only from survivors in the Cabaret Room.

The patrons of the Crystal Rooms confronted certain conditions that might have been more conducive to panic than those conditions initially faced by patrons in the Cabaret Room. The Crystal Rooms, located directly above the main dining room, were the only rooms on the second level of the club and were reached by a carpeted (and flammable) spiral stairway (indicated by A in Fig. 6). By the time Crystal Room patrons confirmed the existence of the fire and learned that they should leave immediately, the stairway was in flames. While they searched for an alternate exit, the lights failed and the area was almost totally dark. One potential escape route was located—a locked service door that led to air conditioning equipment on the roof—but the patrons could not force open the door. Finally, an employee escorted the patrons along a very narrow corridor down narrow, wooden service stairs (B in Fig. 6) into the kitchen and eventually out a rear exit.[15] In contrast, the announcement of the fire came in a more "calming" manner in both the Empire and Cabaret Rooms, perhaps leading to a less threatening definition of the situation. An employee took the microphone from entertainers on the stage of the Cabaret Room and directed patrons toward two available exits. In the Empire Room, the announcement was made by an official of the organization sponsoring the banquet and to which the patrons belonged. The organizational hierarchy appeared to have some impact on the orderly exit (see Johnson 1987b; Johnston and Johnson 1989).

The recognition by Crystal Rooms patrons that the exit had closed, the initially unsuccessful search for an alternate exit, and the additional threat produced by the failure of the lights might have led to the breakdown of ties. The narrow corridor and stairway may also have simply prevented certain actions that were later coded as exhibiting concern for group members. For instance, returning or hesitating to locate group members might not have been possible within the crowded, narrow corridor and stairway.

Figure 6
Diagram of Second Floor, Beverly Hills Supper Club, Showing Exit Routes

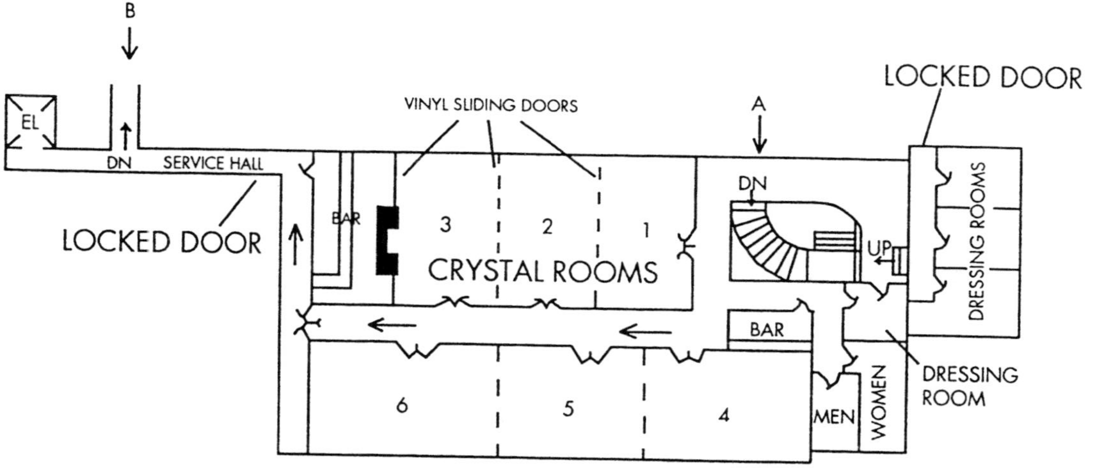

A. Main stairway to first floor entrance

B. Service stairway to kitchen

→ Exit route for most Crystal Rooms patrons

An alternative explanation is that the same behaviors occurred in the Cabaret Room, but those who were most severely threatened and who ignored primary group members did not survive. As a consequence, the available data might have altruistic responses overrepresented. Because it is inevitable that studies of this sort obtain data only from survivors, such alternatives can never be ruled out.

Notes

1. Not all the 345 transcripts were used for all the quantitative analyses. Some analyses considered only patrons who were in the Cabaret Room, Empire Room, or Crystal Rooms, for reasons later explained.

2. The interest of the KSP was not in testing social theory; consequently, the investigators asked many questions not related to our interest, and when they asked questions directly related to sociological interest, they often did not ask them of all respondents.

3. For example, one room was occupied by a banquet sponsored by an organization of savings and loan associations. Within that larger group were nested groups from specific institutions; for instance, employees of a single organization or branch sat at a table together. Thus, someone could be with his or her spouse and another couple (friends), seated with one spouse's immediate co-workers within the larger group banquet.

4. Note that the "strength" of ties is not considered; that is, family ties and secondary group ties are equally weighted in the index. Also, multiple bonds of a single type are not considered. The number of family members or the number of friends in a group may have important implications for response, but here we focus only on the existence rather than the size of the groups.

5. The distribution of patterns varied by room within the club, reflecting the variety of functions the cavernous club could accommodate simultaneously. Within the Cabaret Room couples with friends were most common. A wedding in one Garden Room brought together family and friends, and secondary groups do not appear, but within the Empire Room the most common pattern was a couple in a nested secondary group, that is, a husband and wife, one of whom was an employee of a savings and loan association. This relationship between room and number and types of ties is potentially important in determining the response and is explored later.

6. We used ordinary least-squares (OLS) analyses throughout.

7. None of the patrons in the Empire Room lost their lives in the collective flight.

8. One cannot imagine realistically having any other kind of data for this kind of actual collective flight situation.

9. Parameter estimates are more than twice their respective standard errors for all these terms. Multicollinearity is not problematic for these parameter estimates.

10. We must point out that 86 percent of the Empire Room patrons reported the same level of entrapment threat, making it difficult statistically for threat to show a significant effect. As well, so many of these patrons—and those in the Crystal Rooms—had secondary ties with others also present that the secondary tie variable had to be omitted from the prediction equations for these two rooms.

11. Although the primary tie variable shows no significant effect—the parameter estimate is less than twice its standard error—exclusion of this variable reduces the

amount of statistical explanation by the other two terms to insignificance. We thus consider the primary tie variable as providing a necessary control for uncovering the curvilinear effect of the threat variable. It is also noteworthy that the effect attributed to primary ties in the Crystal Room is approximately one whole unit less on the social response measure than in either of the other two rooms.

12. Of course, the predicted social response scores under increasing threat conditions for those with only secondary ties also do not support the social and severe threat models.

13. If we add together only those estimates of regression constants that are significant (twice their standard errors) with their respective parameters, the increase in response score is monotonic as threat condition increases. The range of increase is not as great, however.

14. The potential distortion from socially desirable responses must be of concern, but examination of the transcripts of those interviewed shows the mention of many more unselfish acts of others than of selfish, individualistic behavior. This informal cross-validation provides some assurance that our conclusion is not distorted by socially desirable responses to questions about the patrons' own behavior.

15. The point at which the stairway enters the kitchen is between the Airway and Elevator behind the main dining room, as seen in Fig. 5.

The Crowd Crush at Hillsborough: The Collective Behavior of an Entertainment Crush

JERRY M. LEWIS and MICHAEL L. KELSEY

On 15 April 1989 an entertainment crowd crush occurred at a Cup Tie semifinal soccer match between Liverpool and Nottingham Forest at Hillsborough Stadium in Sheffield, England. The crush, which caused the death of 95 soccer fans and injured another 170, was Britain's worst sporting disaster.

This chapter uses Neil J. Smelser's (1962) "value-added" model of collective behavior to analyze data from a number of primary and secondary sources describing the Hillsborough tragedy. In particular, this analysis focuses on the period from 2:30 P.M. until 4:00 P.M.

Theoretical Considerations

Most crowd soccer violence has been treated as instances of what Smelser terms a *hostile outburst*. The Hillsborough event was not a hostile outburst or an example of panic that Smelser defined as "collective flight based on a hysterical belief" (1962, 131). Instead, in this chapter, we examine the events of Hillsborough as an entertainment crowd crush drawing on Smelser's views of crowd actions as craze behavior.

Smelser defined a craze as "mobilization for action based on a positive wish-fulfillment belief" (1962, 171). At first glance, it would not seem plausible to explain a crowd crush using a theoretical model based on a positive set of feelings. However, using the notion of the irony of social structure, a case can be made that is exactly what happened at Hillsborough. Moreover, it is likely that other crowd crushes in Britain such as the Bolton disaster and

This chapter is a revised version originally presented at the annual meeting of the North Central Sociological Association in Louisville, Kentucky in March 1991. We want to thank Kathleen Tierney for her helpful editorial and substantive comments.

the Ibroux stadium tragedy[1] are also examples of behavior based on wish fulfillment. We call this behavior an entertainment craze leading to a crowd crush.

Smelser Model

Smelser's model is divided into five determinants, each with a set of subdeterminants. The determinants are: (1) structural conduciveness, (2) structural strain, (3) growth of a generalized belief, (4) mobilization for action, and (5) social control. The precipitating event that some versions of the Smelser model use is treated under the growth of generalized belief (Lewis 1972). Our discussion begins with structural conduciveness.

Structural Conduciveness

This determinant refers to conditions of social structure that describe structural possibilities for an incident of collective behavior to happen. Structural conduciveness sets the parameters and constraints for the other components of the model, particularly structural strain in the context of conduciveness. The model (Smelser 1962, 175–76), related conduciveness to several subdeterminants: a differentiated setting, a reward-cost relationship that generates action, a possibility for the manipulation of resources, and a generalized medium of exchange.

An economically differentiated market develops independence from other social institutions such as the polity, the family, or religion. In a differentiated economic system the analyst could expect to observe competition for scarce and desired commodities based on the value of the commodities per se. In the case of a highly desired sporting event, there is a differentiated market of fans who generate competition for access to the event through the seeking of tickets.

A differentiated market develops when a finite number of people seek tickets and a situation of differentiation develops between the haves and the have nots in terms of tickets. A further differentiation occurs when the quality of the location of the seat is considered. For example, seats on the fifty yard line at an American football game are more desirable than those on the end zone.

A defined gain/loss calculation for an entertainment craze is the weighing of gaining access to the performance against costs including money, time, individual discomfort, and personal energy.

In the manipulation of resources, the person involved in an entertainment craze can opt to continue or stop his or her participation through the expenditure of money, time, comfort, and personal energy. Last, the ticket becomes a crucial medium of exchange and an entertainment craze develops. In summary it is argued that England has a differentiated economy that is conducive for the development of entertainment craze behavior.

Structural Strain

Structural strain occurs within the parameters of conduciveness. The researcher looks for strain in terms of ambiguity that is defined as anxiety generated by uncertainty. Strain increases as uncertainty increases about the allocation of resources as well as the profit and loss from alternative actions. For example, the occurrence of an important entertainment event automatically creates strain through ambiguity about whether or not access to the event will be possible. Hence, the uncertainty or lack of predictable outcome increases anxiety. Thus, alternative allocations begin to be considered such as live television or reading about the event. The actor must decide if these choices are equal to actually participating in the entertainment event. Last, in regard to facilities, what conditions encourage the craze to happen? The facilities are the opportunities to pursue alternative line of actions in reference to a highly desired entertainment event.

Growth of a Generalized Belief

Smelser's model instructs the analyst to look for beliefs that guide action. Wish fulfillment beliefs are "rooted in ambiguity stemming from situations of strain" (Smelser 1962, 94). But, in contrast to hysteria, wish fulfillment beliefs are positive. Persons guided by wish fulfillment beliefs are seeking something positive and have become convinced that what they seek is in short supply. Thus, in an entertainment craze, the belief is that an experience that is highly desired is in short supply. Ironically, these positive feelings can lead to tragedy.

We treat the precipitating event as a subdeterminant of the generalized belief. It serves to focus the belief on specific events. In a sense the precipitating event combines with the belief to propose a solution for the ambiguity found in the strain conditions.

Mobilization for Action

Here we look at the subdeterminants of an entertainment craze that include leadership, the initial and derived phases of the craze, and the *turning point*.

People and/or events provide leadership or keynoting to shape a craze. Most incidents of collective behavior change during the flow of events. Smelser refers to this as the process of moving from the initial and derived phases. In an entertainment craze we examine the initial response to the entertainment event and how it changes into the derived phase as speculation begins. Last, the turning point refers to the slowing of activity in the speculation phase.

Social Control

The Smelser model proposes that the investigator look at several dimensions of social control that shape a craze and prevent it from getting out of hand. The first of these, labeled "normative regulations" (Smelser 1962, 218), influence structural conduciveness. Normative regulations in an economic craze reduce or eliminate conditions of structural conduciveness in the market (Smelser 1962, 218) by restricting the possibility of speculations in commodities. In an entertainment craze, both ticket producers and consumers engage in speculation. A producer speculates that people will want to buy a certain amount of tickets to an event that could generate revenue which exceeds the costs. Consumers speculate in that they try to determine whether or not a particular entertainment event will be so valuable that a ticket and location should be sought regardless of the time, money, and effort involved.

Social control also occurs through leadership; leaders can discourage interest in speculation. In the case of an entertainment craze, leaders or peer pressure, can increase or decrease demand for an event through the level of interest they generate. Thus, leader communication influences interest, need, and speculation.

Last, information serves as a form of societal control by providing accurate facts on the status of the commodity. In an entertainment craze, the important information is an accurate account of how much seating capacity is available to customers.

Narrative

The Cup Tie was played at a neutral site, Hillsborough Stadium, in Sheffield, England, about sixty miles from Liverpool. As with most Cup Tie matches, the semifinal between Liverpool and Nottingham Forest on Saturday, 15 April 1989 was eagerly anticipated.

Hillsborough Stadium, opened in 1899, had been a regular FA Cup Tie site since 1912 and holds 54,181 spectators (Inglis 1989, 3). The crowd crush occurred in the terrace section of the Leppings Lane stand (west end) that has a capacity of about 10,100. The Liverpool fans had been assigned 24,256 tickets for the entire Leppings Lane end, and the Nottingham Forest supporters were allocated 29,800 tickets in the Spion Kop section of the stadium (Taylor Interim, 5). Rival supporters were located at opposite ends of the stadium. This segregation was meant to prevent violence in the stadium and to keep the two groups from meeting in the streets as they were coming and going to the match.

Although the kick-off was thirty minutes away, the coveted middle sections behind Liverpool's goal (Leppings Lane terrace) were near capacity by 2:30 P.M. At approximately 2:40 P.M., Roger Marshall, a police superintendent, made an urgent request for the start of the match to be delayed fifteen minutes so the crowd could get into the grounds safely (Brough et al. 1989, A13).

By 2:45 P.M., several thousand Liverpool fans, some without tickets, were jammed outside the Leppings Lane gate. These supporters were jostling for positions to force their way into the grounds. The behavior of some showed evidence of drunkenness and rowdiness that contributed to an atmosphere of confusion. The Taylor Report (Interim, 11) noted:

> The mounted officers were surrounded by the dense mass of people and became ineffective. Superintendent Marshall was in the midst of the turmoil. He extricated himself and stood on a parapet of the bridge to get a clear view. A drunken fan tried to push him off: a beer can was thrown at a mounted officer. But these were isolated acts by individuals; the menace came from the massive numbers single-mindedly determined to be in for the kick-off with time running out. At the back of the crowd fans were frustrated by the lack of progress as 3 o'clock approached. Some, mostly young men who had been drinking, tried to push and force their way forward. At the front, people were jammed together and against the turnstile walls. Some panicked as the pressure intensified. Some youngsters and women were fainting and in distress.

The turnstile computer counts indicated a couple thousand available places in the Leppings Lane stands but made no distinction between openings in the main stands and the crowded fenced sections within them. There was extra room in the Leppings Lane stands, but this room was only in the outer "pens" and not the middle sections. With the congestion building up outside to dangerous levels, at 2:52 P.M., the decision was made by Police Superintendent Marshall to open a gate into the stadium.

After opening the gate, the intense pressure from outside was dramatically transferred inside with the sudden surge of humanity. The crowd crammed into the narrow confines of the tunnel that was the quickest way to the terrace. The Taylor Report (Interim, 13) continued:

> The initial influx through gate C, augmented by entrants via the turnstiles, came through the tunnel with great momentum. Fans spoke of being swept through, feet off the ground. The 1 in 6 gradient accelerated their progress. Upon reaching

> the crowded pens, some wished to go back but were unable to do so. The new arrivals found themselves pushed forward and the pressure became intense. At 2:54 P.M., shortly after this influx began, the teams came onto the pitch. As usual, this was greeted by a surge forward. Many were now acutely uncomfortable and some were in distress. The mass seethed about but voluntary movement by any individual was difficult; hands down could not be raised and some fans found it hard to breathe. Still the flow continued through the tunnel causing further surges forward. There were shouts for help and for the gates to be opened to the pitch.

The match started on time at 3:00 P.M., and the cheers and shouts from the stadium accentuated an urgency already present for those who did not want to miss any action. The wave of Liverpool supporters burst out onto a terrace already jammed with fans. Opening the gate to the right of the turnstiles determined the crowd's focal points and the direction of its movements. The confined tunnel, the most direct way to the terrace, became the object of attention or the focal point. This tunnel was really a secondary focal point because it directed movement to the primary focal point (terrace), the already bulging middle pens.

Many of the victims were being crushed even before the gate was opened, and some were trampled in the tunnel on the way to the terrace. Even after the surge, many of those present (including police) were not aware of the calamity happening on the Liverpool terrace. The extreme pressure at the back of the crowd pinned the people at the front against the steel-wire fence like caged animals. Some of the victims trapped were women and children who thought being in front against the "safety" fence would provide a good view of the contest. Many of those trapped tried to climb the barriers that separated the spectators from the pitch. Some of the fortunate ones were pulled to safety above by fellow supporters in the upper Leppings Lane seats.

At 3:05 P.M., the match was stopped with indications that injury, and possibly death, had occurred. The players left the pitch at 3:06 P.M.

At first, the police thought the Liverpool fans who were trying to escape the crush by climbing the fence were mounting a "pitch invasion."[2] Nottingham Forest fans, not realizing Liverpool supporters were dying, were chanting and jeering as the crush continued. At 3:12 P.M., the police had to form a line to separate the rival supporters who were rushing to fight each other.

Around 3:15 P.M., a perspective on what was happening began to surface. Bodies were pulled from the terrace, and CPR was administered to some victims. Because of inadequate health facilities, some frantic fans used advertising signboards for makeshift stretchers. At approximately 3:25 P.M., a public address announcement finally provided some substantiation of the severity of the emergency situation at Hillsborough.

By 4:00 P.M., reports and assessments of the tragedy were becoming clearer. Ninety-four people either suffocated or were trampled to death (one more later died in a hospital), and another 170 more injured in the Hillsborough entertainment crowd crush. Eighty-eight men and seven women were killed. Their ages ranged from ten to sixty-five with the mean age of about twenty-three years (*The Guardian* 1989, 2). Most of the dead supported Liverpool.

In the other semifinal, Everton (Liverpool's rival neighbor) reached the final by beating Norwich 1–0. There would be no celebrating in Everton or anywhere else on that day as a sports spectacle turned into a social disaster.

Analysis of the Hillsborough Stadium Crowd Crush

This analysis uses Smelser's model, both determinants and subdeterminants, to order and explain the data about the Hillsborough entertainment crowd crush. The analysis begins with structural conduciveness.

Structural Conduciveness

What conditions were generated by the social structure that facilitated the crowd crush? The social elements present that create possibilities for collective behavior to occur need to be analyzed as an entertainment craze. In the Hillsborough tragedy the analyst looks at four subdeterminants: (1) differential setting, (2) reward/costs, (3) general medium of exchange, and (4) manipulation of resources.

Differential Setting. The differential setting sets the condition that supporters make decisions about the game independently of other factors such as family, work, religion, or other recreational activities. Participating in the game involves a variety of choices and decisions that may lead to an entertainment craze. Fig. 7 illustrates the decision flow of the differentiated setting. Whether or not one decides to be involved is the first decision and this pertains to the importance or interest in the event. The Cup Tie was an eagerly awaited and emotion-filled sports spectacle that can be compared to America's National Football League (NFL) conference championship or a major league baseball championship series. If no involvement is the choice, there are other decisions, but wanting to be a participant leads to other levels of decisions.

The match had powerful appeal as represented by the thought of a mourner, Albert Cardus (*The Plain Dealer* 1989, 9A):

> They took our industries away and they took our jobs away. But they can't take our football away from us, surely.

The article continued (*The Plain Dealer* 1989, 9A):

Figure 7
Decision Flow Diagram of Differential Setting

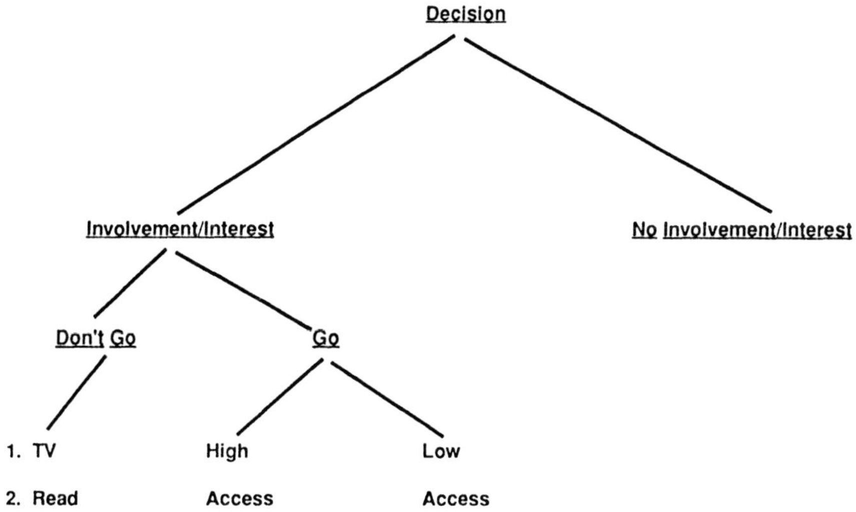

> This once-mighty city on the River Mersey has two teams, Liverpool and its cross-town rival, Everton. As in years past, both are in the final stages of competition for the Football Association Cup, the Super Bowl of British soccer. The saying here is that the city will occasionally "lend the cup out"—that is to say, Liverpool's teams do lose—but not for long.

A leading researcher John Williams, an English sociologist, noted (Condie 1989, Sport 4):

> We must recognize that soccer causes immensely strong emotions among supporters. It is a powerful force which can be both positive and negative in its effect. You can't really blame young men in Liverpool for travelling miles to support their team—even if they do not have a ticket. For many of them, football is all they have going for them and the chance of getting into the ground is enough.

Wanting to be involved, the next step is to decide whether or not to go to the match. Going to the match creates a condition of ambiguity that is predicated on access. Deciding not to go enables the participant to choose alternative ways of being involved such as watching the match on TV or reading about it. Hillsborough stadium holds 54,181 spectators and the interest generated by Liverpool and Nottingham Forest supporters indicated a full capacity crowd would be on hand.

Ticket allocation had a direct bearing on the probability of access and became a sensitive issue (Taylor Interim, 1989, 5):

> The effect of the decision was that for this all-ticket Cup Tie, Liverpool was allotted only 24,256 places as against 29,800 for Nottingham Forest. This, although average attendance of supporters at home matches was substantially higher at Liverpool than at Nottingham. Moreover, with standing tickets at £6 and seats at £12, Nottingham Forest had 21,000 standing places compared with Liverpool's 10,100. So, Liverpool's allocation was more expensive as well as smaller. Understandably, Liverpool was aggrieved by the allocation of places and tickets. They sought with some support from the host club and the FA to have it changed in 1988, but the police were adamant. To switch ends would, in their opinion, have involved rival supporters crossing each other's paths when approaching the ground thereby frustrating attempts at segregation and creating a risk of disorder. In 1989, when the same plan was proposed, Liverpool again challenged it. The police, however, maintained their view, adding that those who had attended in 1988 would be familiar with the arrangements and that any change would lead to confusion.

Because they had been banned from European competition because of the Heysel riot (Lewis 1989), the Liverpool supporters needed a positive soccer experience. This factor accentuated the condition of high demand that already existed. If one decides to go to the match and access is available, the second subdeterminant should be considered.

Reward/cost ratio. Participating in a highly desired event, especially one perceived by fans of being a one-time event, is the reward or benefit of attending. The Liverpool-Nottingham Forest Cup Tie of 1989 was an important one-time event. The reward for participating is balanced against its costs.

Costs can vary depending on the individual, but the primary ones are (1) money—not only for tickets but also for all expenses related to the event such as transportation and food; (2) personal energy—effort expended to get to the match, during the match, and the return trip home; (3) risk taking—knowing there is a risk of injury or death is a cost but for some a reward because it adds excitement to the match for some supporters. After the Bradford fire and Heysel riot, this condition became more realistic; and (4) discomfort, which can be related to the energy expended, to the conditions of the game itself such as overcrowding, or standing in the terraces for several hours.

In regard to stadia and terraces, Lewis (1982, 419) wrote:

> The typical football park was built around the turn of the century and consequently does not have the amenities one sees in many American sport complexes. In England the parks are located, particularly in the first division, in large cities, often in run-down sections of the city. More important than park locations are the standing terraces. Terraces do not have seats; rather they are concrete steps with

barriers at waist height behind which fans, generally young fans, stand. Terraces are usually located behind the goals, but can be parallel to the field as well. The terraces are slanted, quite sharply in places, and can go as high as three or four stories.

General Medium of Exchange. A ticket not only serves as the medium of exchange for access but also is a variable of social control. The capacity of Hillsborough is 54,181. Thus, allowing people in with tickets should not result in a dangerous overcrowding condition provided that only 54,181 tickets were sold.

The unfair ticket distribution provided motivation for some fans to attempt to get into the stadium without possessing a ticket. At Hillsborough and other English soccer stadia, this is accomplished by people already in the grounds throwing their tickets back to others or by large groups of supporters rushing the gates. If police or stadium personnel tear the tickets, a measure of social control is present preventing people from throwing their tickets to others outside the stadium.

Having a ticket does not automatically guarantee that the supporters will be able to see the match. Unlike American stadia the terraces have no permanent seats, although this is beginning to change. A ticket for the terraces at an important English soccer match guarantees nothing. Further, the terraces behind the goal are highly valued and are an important part of the tradition of football. One letter writer (Lindsay, *Sunday Mirror* 23 April 1989, 6) commented:

> It is alright [*sic*] for the big-wigs with seats in directors boxes or VIP suites to say "do away with terraces." The terraces are the engine-rooms of soccer, where all the real spirit of the game gathers. Bulldoze them and the heart will go out of Saturday afternoon soccer.

Some fans even tried to bribe their way into the match. At an inquest held sometime after the crush a gateman reported that "ticketless fans waved 10 and 20 pound notes to try and gain entrance, and others threatened to smash the gate down unless it was opened" (*The Times* 1990, 6).

Manipulation of resources. Manipulation established possibilities for the collective behavior to occur by creating desire for the event that results in a high demand. Sportscasters, writers, and other members of the media contribute hype to an event already rich in interest. The teams, the league (FA), and the sporting establishment in general create the demand with the intention being to portray the game as a one-time event.

Manipulating the entertainment craze has socioeconomic benefits but can also be a primary contributor to overcrowded conditions and crowd crushes. Soccer, the national game embedded in English culture, establishes demand of a magnitude of desire rarely seen in American sport.[3]

In summary, all the conditions for the structural conduciveness of an entertainment craze were met: there were differential settings that shaped various decisions: rewards and costs associated with attending the event were evident; the general medium of exchange was a ticket and this was present; manipulation of the game created desire and demand that culminated in the Cup Tie being a highly desirable event.

Structural Strain

Smelser's model indicates that strain should be examined in relation to the concept of ambiguity as indicated by the presence of uncertainty. Structural conduciveness sets the parameters in which strain can be present. Structural strain is the ambiguity that exists in the social structure.

In considering the structural strain of an entertainment craze, most ambiguity could be related to the anxiety generated by the uncertainty of the event—strain at the level of facilities. The two major dimensions of uncertainty would be: (1) ticket availability for the match, and (2) individual access to the match.

Thus at the start of the match, there were four types of spectators as illustrated in Fig. 8.

A Type I person has both a ticket and a greater-than-average expectation of access to the terraces. Such persons have little or no uncertainty and, hence, no anxiety about being able to see the match. As noted earlier, a terrace ticket only grants access to the terrace in the form of a place to stand.

Type II people have no ticket but nevertheless have high expectations regarding access to the stadium, and in particular, to the terraces. These people would likely exhibit a high level of anxiety and would accentuate the strain condition by pushing or sneaking in without purchasing a ticket.

Type III fans have a ticket but see their chances of access as lower than average. Type III supporters thus also experience considerable anxiety because crowded conditions or their position in the crowd could restrict their access. People in this category are likely to push or shove out of frustration.

There is no anxiety in the Type IV condition (no tickets and low expectation of access) because these fans accept their situation and stay outside the stadium. Except for being near the excitement, their motivation for being at the match is not to see the match per se, but to simply be present at the event location.

Although we have no direct empirical evidence on the distribution of these theoretical types in the crowd, we can provide some insight into the process of how shifts from one type to another might happen.

As we noted in the narrative, Gate C was opened at about 2:52 P.M. thereby increasing the pressure, which continued as the teams came on the pitch at 2:54 P.M. The entry of the teams caused excitement. Many fans

Figure 8
Typology of Spectators

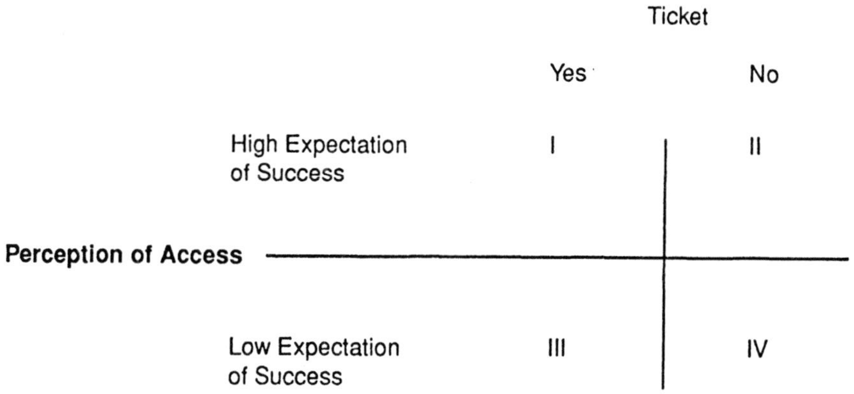

outside likely responded to the roar of the crowd, which signaled the match was about to begin. At this moment, many in the crowd who had tickets probably shifted from Type III's to Type I's because their desire shaped their perceptions. Eight minutes later we think this happened again when Peter Beardsley, a well-known Liverpool player, made a shot that hit the crossbar of the goal. As the Taylor Report (Interim 13) noted:

> There was a roar from the Liverpool fans and at the same time a powerful surge forwards in pen 3. The several surges which occurred after the influx from gate C carried the pressure down the pens towards the pitch. The force became such as to twist and break two spans of a crush barrier towards the front of pen 3. The evidence does not establish with certainty when this happened. Probably it was triggered by the surge at 3:04 P.M. But I am sure it occurred after the influx from gate C so greatly increased the pressure in the pen. When the barrier broke those whom it had supported were projected towards the perimeter fence. Many fell and the involuntary rush of those behind pressed them down. The crushing force was transmitted and dispersed so that all along the front of pen 3 fans were pressed hard up against the low wall and the wire mesh of the fence above it.

In summary, the source of strain was the uncertainty and anxiety about availability of a ticket and accessibility to the match. Anxiety was most reflected in Type II or Type III categories of people. Type I spectators, having tickets and a perception of high probability of access, had a propensity for low anxiety and ironically, had an increasing probability of being the crush victims.

Growth of a Generalized Belief

The generalized belief directs the behavior of the fans by preparing them for action. In an entertainment craze, wish-fulfillment, that is, the conviction that action will help participants obtain desired results, is the belief. Beliefs create a "common culture" (Smelser 1962, 82) for participants in an episode of collective behavior. Generalized belief processes can be observed at both the individual and social level. At the individual level it is necessary to determine how feelings of ambiguity and anxiety emerge and to chart the development of beliefs. At the social level it is necessary to investigate how much these beliefs become shared. Using the Smelser value-added model can be difficult at this point because, in the absence of interview or direct observational data, it is not always easy to determine the individual feelings or the extent of generalization and sharing of those caught up in an episode of collective behavior.

We think that there were three types of beliefs present at Hillsborough. The first was that the Cup Tie was a worthy sports spectacle. Being able to be part of this one-time event was the motivation of the supporters. Thus, the fans were not just spectators, they were important participants in the spectacle itself. This also related to the second belief, that the match between Liverpool and Nottingham Forest would be close, exciting, and worth seeing.

The third belief was that there were certain places in the terraces that were more desirable than others, namely, in the center of the pitch, behind the goal, and above the security fence sight line (Lewis 1982). The Taylor Report made it clear that most of the fans who cleared the outer perimeter thought they had a chance to get into the terraces for a desired location. Taylor (Interim, 12) wrote:

> The largest entry, however, was through gate C. In the five minutes it was open about 2,000 fans passed through it steadily at a fast walk. Some may have had tickets for the stands. No doubt some had no tickets at all. The majority had tickets for the terraces. Of these, some found their way either right to pens 1 and 2 or left through the dividing wall to 6 and 7. But a large proportion headed straight for the tunnel in front of them.

All four types of spectators exhibited some or all of the ingredients of the wish-fulfillment beliefs. Type II and Type III people, exhibiting more anxiety, had greater demands to satisfy their wish fulfillment because they had to overcome greater uncertainty. Type I people had the generalized belief that they would see the match. Type IV spectators had neither the belief nor an expectation of observing the match. All of the fans had the overriding belief that they wanted to participate in the Cup Tie, in one form or another. The chairman of the Football Supporters Association commented (Condie 1989, Sport 4):

> There is a tremendous feeling about being part of the crowd; a feeling of greatness. Football is the only food in that department. It's why people travel so far to watch it.

Mobilization for Action

Leadership. Leadership for the entertainment craze comes from three sources: the media, the Football Association (FA), and peers of the football supporters. The media wrote a great deal on the Cup Tie particularly because the match involved major teams—Liverpool and Nottingham Forest. The media hype, both print and electronic, was similar to that found at an AFC or NFC championship, with the winners going to the Superbowl. In the case of English soccer, the winners of the Cup Tie would go to Wembley.

The second source of leadership, the FA, did not make correct decisions about the distribution of the tickets. Liverpool supporters felt they had been badly treated both in 1988 and again in 1989. Thus, the actions of the FA resulted in increasing the speculative pressures on tickets, access to the terraces and favorable location.

Peers also provided leadership, particularly in reference to speculation about whether or not to attend the match. For example, the Taylor Report described the patterns of travel to the match (Taylor Interim, 5):

> Most supporters were expected to come by road although rail transport was available. Nottingham supporters travelling by train would arrive at the main line Sheffield Midland Station whence they would be directed and escorted northwards by the police. They would arrive at the ground along Penistone Road. Nottingham supporters travelling by coach, mini-bus or car would approach Sheffield from the south mainly on the MI. Again their arrival at the ground would predominantly be via Penistone Road or Herries Road.

Initial and Derived Phases. The second aspect of mobilization in the value-added model involves the initial versus the derived phase of the collective behavior episode. Smelser argued that in all types of crazes ranging from economic to religious, the analyst should be able to locate two distinct phases, the initial and the derived.

The initial phase at Hillsborough was the high interest in the match and the passionate quest for tickets. The derived phase (speculative) began when the Liverpool supporters got to Hillsborough and attempted to get into the match. This was particularly true for Category II fans who had no tickets and believed they had a high probability of getting in to see the match. Thus, the speculative phase began when supporters put physical pressure on other fans to get into the stadium.

The Turning Point. The value-added model predicts that every craze has a turning point. This means that there is a moment when activity slows. It may be sharp or gradual. In the case of Hillsborough, it was quite sharp. Indeed, the speculative phase of the craze probably lasted a little more than one hour from the time the crowd began to build around 2:00 P.M. to the time the match was called off at 3:06 P.M. In the next section of the chapter we deal with the actions of the social control forces.

Social Control

In the value-added model, social control of a craze is a counterdeterminant. An entertainment craze involves lapses in social control in three areas: (1) normative regulation, (2) leadership, and (3) information (Smelser 1962, 218–19). All three mechanisms need to be analyzed in terms of initial and derived phases of mobilization.

As previously demonstrated in the decision flow of the differential setting (Fig. 7), a person could simply decide not to go. Opting out of the match is part of the initial phase of mobilization and could be related to having no interest in the match, or to the second level of decision making, which is having an interest or involvement in the match. The high demand and not being able to get tickets are examples of social control norms that relate to the second level of decision flow. In this situation, interested supporters could watch the match on TV or read about it. Having peers who are not interested and opt out would be an example of how norms could play a role in the decision to not go to the match.

The initial phase concerns the events leading up to the match at Hillsborough and not the social control situation at the stadium. Police and peers, examples of leadership agents, could have contributed to the decision not to go to the match. Having peers not interested in the match could influence the decision, as could the perceived image of a strong, regulating police force determined to prevent people without tickets from gaining access to the stadium grounds. All this could have lessened the speculative pressure on access to the terraces.

The media illustrate how information, or a lack of information, contributed to the initial phase of mobilization. Promoting the match contributed to the tremendous demand, but later after a sell out was assured, the media could have disseminated information warning the fans without tickets of the futility of going to the match. This did not seem to happen, according to our evidence.

The three mechanisms of control also appeared to operate the derived or speculative phase of social control. We have argued that the derived phase begins at the stadium. Proper social control in the derived phase would have been to eliminate speculation, converting Type II spectators to Type IV. Increasing the amount of people who are passive and accept their condition results in reduced strain. All three mechanisms of control would operate to convert Type II spectators to Type IVs as reflected in the Typology of Spectators (Fig. 8). Type II people, having no ticket and a perception of high-probability access to the terraces, used improper social control actions such as ineffective gate control by police to gain access to the match.

The police, as leadership agents, opened the gate to relieve a dangerous overcrowded condition. The Taylor Report noted (Interim, 11):

> Superintendent Marshall realized the crowd had become unmanageable. Although loathe to do so, since it was contrary to basic police strategy, he decided to request the exit gates be opened to relieve the pressure. Otherwise, he feared fatalities would occur. Other senior officers outside the ground agreed. At 2:47 P.M. he radioed control to permit the gates to be opened. At 2:48 P.M., whilst Mr. Duckenfield was considering the request, gate C opened to eject a youth who had climbed in with no ticket. Immediately, fans outside took advantage and about 150 managed to get in before a mounted officer enabled the gate to be closed again.

Police leadership will be debated throughout the lifetime of the Hillsborough controversy. Police actions fueled the fantasy of Type II people, which was to gain access to the match even without a ticket. By transforming these fans to Type IV people (those who accept their situation of no access and stay outside the stadium), the mechanisms of social control could have reduced speculative pressure on the terraces. Exactly the opposite occurred.

A second leadership issue involved the Liverpool supporters. Our knowledge of English soccer crowds suggests that there were probably leaders among the Liverpool fans urging their mates (particularly Type IIs) to rush the gates to reach the tunnel and eventually the terraces. This encouragement probably contributed to the crush by reinforcing the attitudes of Type IIs and preventing them from becoming Type IVs.

A lack of information, coordination, and communication by the police at the stadium site intensified the derived phase of mobilization. Opening the wrong gate directed the focus of crowd movement to a narrow tunnel that led to the already bulging pens behind the Liverpool goal. Information on the dangerous overcrowding of the Leppings Lane terrace, especially the middle pens, was not communicated. The lack of stabilizing norms, leadership, and information was the reason social control forces failed at Hillsborough. Opening a gate to reduce crowd pressure encouraged the wish-fulfillment belief and the entertainment crowd crush.

Conclusion

The entertainment craze that occurred at Hillsborough contained the elements of Smelser's model of collective behavior. All five determinants: structural conduciveness, structural strain, growth of a generalized belief, mobilization for action, and social control were necessary for the entertainment crowd crush to occur.

Soccer is embedded as the national sport of England. The Cup Tie was sold out but attracted hopeful spectators who did not have tickets. Our Typology of Spectators reflected how ambiguity related to ticket uncertainty about availability and access to the match produced four types of spectators. The supporters were not content to simply follow the match through the media; rather, they wanted to be participants in the spectacle itself.

Several factors contributed to the overcrowded condition. The mass movement of people to a desired location in the terraces contributed as did an unequal ticket allocation carried out by the FA and supported by the police. Inadequate police coordination and communication resulted in a lack of social control. Many reasons can be mentioned all pointing to the situation in which there were too many people in a restricted space, the middle pens of Liverpool's terrace at Hillsborough.

Notes

1. The previous disasters happened on these grounds: Burnden Park (Bolton in 1946)—thirty-three dead; Ibroux (Glasgow, Scotland, in 1971)—sixty-six dead. (*The Times* 1989, 4)

2. A pitch invasion occurs when football fans rush on the field to stop the match when an opponent scores or an official makes a bad call.

3. For a discussion of the culture and passion of British soccer, see Desmond Morris (1981), *The Soccer Tribe*.

Collective Behavior Theory and the Study of Mass Hysteria

ROBERT A. STALLINGS

What is collective behavior? Who decides when interaction represents collective behavior rather than conventional social behavior? By what criteria are these matters decided? I argue in this chapter that studies of mass hysteria in the social sciences are seriously flawed and that uncritical use of the results of this research is detrimental to the development of a theory of collective behavior. Careful examination of the assumptions underlying research on mass hysteria leads directly to consideration of basic questions of what constitutes the essence of phenomena considered collective behavior and to the fundamental criteria by which collective behavior may be distinguished from other forms of social behavior.

I argue that mass hysteria is a specific interpretation or frame given to events rather than an inherent or objective property of those events. It is a social construction. The existence of mass hysteria depends on a specific answer to a basic question of social life, "What has happened here?" Why one particular answer to this question rather than another occurs—that is, why an event is defined as mass hysteria rather than something else—has as much to do with those who "own" the events as with the type of interaction displayed by the "victims." An adequate sociological account of an episode of mass hysteria must include *all* those who participate in it. When this is done, some serious inadequacies of collective behavior theory as a sociological frame for studying episodes of mass hysteria become apparent.

I undertake three different analyses before returning to an evaluation of core theoretical issues in collective behavior. These analyses involve examinations of how three separate institutions frame an episode of mass hysteria. By framing I mean how it is identified, described, and explained by people drawing on their cultural repertoires (Swidler 1986). My use of the term *frame* generally but somewhat loosely follows that of Goffman ([1974] 1986). The three specific frames that I examine are those of medicine (especially

I want to thank David Altheide, Joel Best, and Carol Warren for taking the time to read and comment on an earlier version of this chapter.

psychiatry), journalism, and sociology. Thus, the first three sections of this chapter deal with medical frames, media frames, and sociological frames, respectively. In the fourth section I use the logic of the negative case (i.e., I try to identify and describe *noninstances* of mass hysteria) to discuss the limitations of previous research. The conclusion contains my assessment of a more adequate sociological approach to studying mass hysteria and of the appropriate use of these studies in the subfield of collective behavior.

The Psychiatric Frame

To understand how the psychiatric frame identifies, describes, and explains mass hysteria, I turn first to the *Diagnostic and Statistical Manual* of the American Psychiatric Association (1980). *DSM-III*, as it is commonly known, is the standard medical classifying scheme for mental disorders. Unfortunately, mass hysteria is nowhere listed. It is treated as a residual category and confined, by implication, to the following category of mental disorder:

> Finally, Atypical Somatoform Disorder is the term applied to physical symptoms without an organic basis that do not fit the criteria for any specific Somatoform Disorder. (American Psychiatric Association 1980, 241)

Because the *DSM-III* offers little guidance, my description of the psychiatric frame is based on the most recent article available on the subject in the medical literature, a study published in the *Annals of Emergency Medicine* of "fifteen adolescent female students" treated in the emergency department at Humana Hospital in Anchorage, Alaska in 1987 (Selden 1989). These fifteen students had been triaged from a total of seven hundred students and faculty experiencing "a possible mass casualty, toxic inhalation exposure" at a junior high school. All fifteen complained of one or more symptoms including nausea, dizziness, headache, abdominal pain, and weakness or faintness. All but one of the patients reported that the symptoms went away within an hour of arrival at the hospital. Laboratory tests performed on the one patient who seemed to have the most serious symptoms disclosed nothing abnormal. Analysis of the air in the junior high school by personnel from the Anchorage fire department and from the Environmental Protection Agency "revealed no detectable levels of toxic gases" (Selden 1989, 893). All fifteen patients were subsequently released.

The longest section of the paper, "Discussion" (Selden 1989, 893–95), identified the evidence confirming that the Anchorage Junior High School incident was, in fact, an episode of mass hysteria. The author described selected features of the case and compared them with the "eight generally accepted criteria" that "differentiate epidemic hysteria from true organic illness." These criteria are from an article previous published in the *New England Journal of Medicine* (Small and Borus 1983). The following listing of the eight criteria is not taken from the 1989 Selden article but from the original 1983 publication, although the wording in each article is virtually identical. According to the earlier source, mass hysteria is recognizable by:

> the absence of laboratory results and physical findings confirming a specific organic cause; a preponderance of illness in girls or women; the apparent transmission of illness by sight or sound, or both; the presence of hyperventilation or syncope; a preponderance of illness in adolescents or preadolescents; benign morbidity, often with rapid spread followed by rapid remission of symptoms; relapses of illness in the setting of the original outbreak; and evidence of unusual physical or psychological stress. (Small and Borus 1983, 632)

The author marches through these tell-tale traits of mass hysteria in systematic fashion. "The smell of sewer gas served as the known precipitating stressor" (p. 893). "Also demonstrated is propagation of illness by sight and sound (only those with such direct contact were affected)" (p. 893). "The time course in this incident was also appropriate to explosive epidemic hysteria, with a total time of less than five hours from the time EMS [emergency medical service] personnel responded to the scene to the time that all patients had remission of symptoms" (p. 893). The "relapse of illness" criterion was satisfied when two of the victims who had been left alone at home (but not "in the setting of the original outbreak") by parents who went out for the evening became ill, one of whom summoned an ambulance to take her back to the hospital. Because he described the following features of the Anchorage Junior High School outbreak in the "Case Report" section of the article, the author does not explicitly mention the absence of laboratory and physical findings or the fact that all fifteen victims observed in the hospital emergency department were women who were "adolescents or preadolescents."

The author uses a table of the thirteen symptoms exhibited by the women taken to the emergency department to verify the presence of symptomology associated with mass hysteria (i.e., "hyperventilation or syncope"). Although only four of the victims experienced shortness of breath and only half (eight) reported feeling weak or faint, the author stated: "The symptoms exhibited by the patients in this report are similar to other reported incidents of explosive anxious epidemic hysteria" (Selden 1989, 894).

This research article has most, if not all, of the features Gusfield (1976, 1981, 86–101) associated with the "window pane" metaphor he used to examine the presentation of research in another medical specialty. That is, the author reconstructs the case so that the reader sees what has happened as an outcome of natural processes rather than his (and others') actions. Descriptions of the results of standard environmental, toxicological, and medical tests rather than decisions and judgments by human agents convey the facticity of this episode and imply a closure about its nature and its origins. What is surprising about the research report is how much it, in fact, does provide a glimpse of human beings making assumptions, decisions, and judgments. In so doing the article suggests that the degree of closure with which this episode was discussed is deceptive. One of the window pane's cracks is evident in the following quote: "Telephone consultation with a medical toxicologist at a regional poison center reinforced that further diagnostic evaluation or treatment was unnecessary" (Selden 1989, 893). The author concluded the discussion section with the sentence, "Diagnostic workup was also kept to a minimum" (Selden 1989, 895).

In other words, the diagnosis "explosive anxious epidemic hysteria" and the central criterion "absence of laboratory results and physical findings confirming a specific organic cause" are not completely independent. Examining physicians began to suspect mass hysteria when other specialists found no immediate organic cause for the illness; however, once mass hysteria was suspected, the momentum for further searching for physical causes quickly dissolved. In fact, the author explicitly recommended that his colleagues shorten the search process as much as possible during future episodes:

> The ease and success of treating an epidemic hysteria event will be directly related to the speed with which the incident is recognized.... Incident scene investigations and elaborate searches for suspected etiologic agents, like later ED [Emergency Department] patient evaluations, should be tempered by the fact that they may enhance and prolong the event. (Selden 1989, 894)

One wonders whether or not Legionnaires' Disease would have been "discovered" if this recommendation had been followed.

As this example shows, far from being a situation in which medical science identifies objective features in a straightforward manner, the framing of an incident as episodic mass hysteria results in part from negotiations between experts. Proper treatment of victims requires not only early diagnosis—what I would call successful framing—but also gaining control of the setting to exclude other competing but "incorrect" frames. The principal competing frame at the scene of these outbreaks is that the illnesses experienced are "real" in the sense of having identifiable physical causes. Initial adherents to this competing frame were not only fire department personnel and environmental specialists, but also by the implications of their actions, news reporters and camera operators, police, ambulance drivers, and other emergency responders. The following quotes from the same article reveal the "turf battles" over which frame will prevail, a contest analogous to successful social problems claims-making activities (Spector and Kitsuse 1973, 1977):

> the media often are implicated as an escalating factor in epidemic hysteria, both by their presence at the scene with vehicles, large cameras, and bright lights and then later by verifying the incident in the public eye by faithfully reporting it as a "real" event.... The EMS and police responses, in this report, played a similar role to that of the media described above, with the arrival of multiple emergency vehicles with lights and sirens, carrying personnel in full protective gear with respirators, gas analysis equipment, and large, loud smoke ejector ventilation fans.... Most authors believe that to perform an involved search for [an "undetected environmental toxin"] only serves to prolong the episode. (Selden 1989, 894)

Support for the author's position comes from the earlier summary article by Small and Borus (1983):

> We conclude with other investigators [three other works are cited] that intervention should include dispersion of the group and discontinuation of prolonged environmental and laboratory investigations once initial studies have proved nonproductive and the characteristic pattern of mass hysteria has become clear. (635)

These two authors imply that the medical frame does not prevail easily when they discuss a previous outbreak of mass hysteria in Massachusetts investigated by the senior author (Small):

> The eagerness among citizens and the media to attribute the Templeton outbreak [students at a junior high school had become ill during both the rehearsal and performance of the school's band] to chemical rather than psychological variables speaks not only to their heightened awareness of the former but also to their reluctance to accept the latter. (Small and Borus 1983, 634)

Another paper from the medical literature suggests that victims and their representatives constitute another source of indeterminacy in efforts to frame events as mass hysteria episodes. Brodsky (1988), in the latest review of "mass psychogenic illness" (MPI) in the workplace, showed how contentious the process of framing can be in the context of labor-management relations. He described how physicians and lawyers can prolong workers' symptoms "in the absence of an identifiable abnormality long after the outbreak has subsided" (Brodsky 1988, 658). Sometimes this is the result of otherwise good intentions on the part of the physician:

> physicians are not always honest and straightforward with their patients when they believe that the truth, in this instance that the individual does not have a physical disease, will cause patients to reject them or will otherwise work to the patient's detriment. (Brodsky 1988, 659)

Brodsky recounted other instances where physicians incorrectly asserted that patients who were "really" victims of mass hysteria were instead victims of organic disease:

> Based on one major episode of mass psychogenic illness in the work-place, in which the author was involved, it would seem that one of the more important factors in fixing patients' symptoms, in their belief that they have a physical disease, is the presence of a physician who has taken and maintained a position that the disease, mysterious and undiagnosed as it is, is the result of a physical illness, and that the physician can find signs of disease that other physicians do not recognize. The belief of such a physician convinces his patients that, in fact, they are ill.... Lawyers are experienced in finding physicians who will support their view that the patient has been damaged or is disabled, and such physicians further reinforce the patients' view that they do have physical disorders.... Public support in the form of the news media and legislative representatives reinforces the patients' belief that there is a conspiracy to suppress the fact that they have been exposed to some harmful agent. (Brodsky 1988, 659–60)

Later he concluded:

> Some reports [he cites only one, Faust and Brilliant (1981)] suggest that the diagnosis of MPI is really an excuse for not conducting an adequate evaluation of low-level environmental contaminants. (Brodsky 1988, 660)

One can see how employers could have greater enthusiasm for a diagnosis of mass hysteria than for commissioning further, possibly costly, testing of the workplace and presumably incurring liability if some agent were eventually found.

The conclusions I draw from this examination of the medical literature on mass hysteria are these. First, the origin of physicians' diagnoses of mass hysteria is a nexus of three things: a negative or inconclusive outcome of standard measurements by other experts (environmental engineers, laboratory technicians, etc.); the nature of the behavior exhibited by the victims (i.e., symptoms); and the nature of the victims themselves, specifically their age and gender. This third characteristic comes as no surprise. The gender base of hysteria was firmly entrenched in the cultural repertoire of psychiatry by the beginning of the twentieth century (Showalter 1985). The specific configuration of age (adolescent), gender (female) and hysteria in fact was so tightly drawn that it had its own name, "the daughters' disease," and (male) physicians used it, for example, to explain the "aberrant" behavior of early feminists (Showalter 1985, 145–47).

Second, insofar as the psychiatric frame is concerned, the cause of the episode is "out there," independent of the status, power, and authority of the physician, the diagnostic process itself, and the social standing of medicine. The successful diagnosis provides closure as to the meaning and genesis of the episode, with etiology established exclusively in the *interaction among victims*. Those doing the diagnosing do not see themselves as individual physicians or the institution of medicine as a necessary, indispensable part of the creation of mass hysteria.

Third, the psychiatric frame, nevertheless, requires some "selling" because there are other parties to the episode proposing competing explanations for the behavior. A recent dramatic example of this competition was the denial of Israeli physicians' diagnosis of mass hysteria among Arab school girls in the Israeli-occupied West Bank of Jordan by Arab physicians who contended that the girls had been subjected to poison gas by the Israelis in an effort to promote sterility (Hefez 1985). However, this selling process encompasses not only the actors involved in the immediate events but also the historical processes through which occupations gained ownership (Gusfield 1981, 10–13) of these matters (see the discussion of "rule creators" in Pfuhl 1980, 121–59; also Conrad and Schneider 1980, 17–30). An episode of mass hysteria is therefore a successful accomplishment in which one frame "wins" and others "lose." I turn next to an examination of a second set of actors, the news media, and to the media frame of mass hysteria to see in what ways news accounts bolster or weaken the claims of those promoting the psychiatric frame.

The Media Frame

News reports of mass hysteria present a structured account of the "facts" of the episode. Both form and content communicate the journalist's frame or format (Altheide 1985; Fritz and Altheide 1987), categorizing and giving meaning to the event for news consumers. As an example of the media frame I use an account of a recent incident in Santa Monica, California treated by the *Los Angeles Times* as an episode of mass hysteria. First I briefly describe the episode, then look closely at some formal properties of the *Times*'s account of the incident (e.g., headlines, photos and captions, article structure, and news sources) and at two aspects of the content of the account, portrayals of the victims and of the cause of the episode. At the end of this section I summarize what I think are some important features of the media frame regarding mass hysteria that reinforce the facticity of the psychiatric interpretation.

On Thursday evening 13 April 1989 the Santa Monica Unified School District opened its fortieth annual student talent show in the Santa Monica Civic Auditorium. The event, called "Stairway to the Stars," is a showcase for the musical talents of one thousand one hundred students in grades four through twelve. At 9:40 P.M., approximately three fourths of the way through the show, Santa Monica police evacuated more than two thousand five hundred spectators and performers from the auditorium because of the reported presence of toxic fumes. Eighteen young people, all participants in the music festival, were taken by ambulances to three area hospitals. All eighteen were examined, treated for their symptoms, and eventually released; none were hospitalized.

A search of the auditorium by the Santa Monica fire department using a toxic fume detector produced no evidence of the presence of fumes or any other toxic substances. Public health and fire officials subsequently labeled the incident an episode of mass hysteria. A second performance scheduled for the following (i.e., Friday) evening took place without another outbreak.

Although there were brief reports of the episode in most of the Los Angeles area news media, I only discuss two articles that appeared in the *Los Angeles Times*. The *Times* carried a one hundred-word article on an inside page in its "Metro," or local news section on the morning after the incident (14 April 1989 2: 2).[1] Written with essentially a "command post" point of view (Quarantelli 1975), the article reported the evacuation of one thousand five hundred people following a toxic release. Eight ambulances transported victims to area hospitals. Two victims were in critical condition. The article contained one identified news source, a Santa Monica police sergeant.

The second article appeared on Saturday morning also in the Metro Section (15 April 1989 2: 3, 10). This one thousand-word article reported that two thousand five hundred people had been evacuated on Thursday and that eighteen young performers had been treated and released from various hospitals. The absence of any toxic substances or fumes was noted, and the bulk of the article reported in depth about the incident as an episode of mass hysteria. Ten sources were cited including a victim and her mother, various fire department and school district officials, and two physicians.

The news headlines established mass hysteria as the official—and correct—explanation for what happened. The primary headline was: "Mystery Illness That Hit 18 Student Singers Is Diagnosed as Hysteria." The continuation headline (on page 10) was: "HYSTERIA: Mystery Illness Identified." The choice of verbs is especially revealing. "Diagnosed" suggests human agency (someone must have done the diagnosing), a medical context (physicians diagnose illnesses), and a certain amount of tentativeness (physicians do sometimes make incorrect diagnoses). "Identified" closes the door on any lingering uncertainty. With finality, the mystery has been solved, the perpetrator unmasked; case closed.

Photos and photo captions created a more balanced impression between the mass hysteria diagnosis and the belief that "something else" was the cause of the illnesses. Two black-and-white, 2″ × 3-5/8″ photos accompanied the article. One was a photo of the Santa Monica fire chief placed at the beginning of the article next to the headline. Beneath it was the caption: "It's tough to make it look like you're not ridiculing the kids, but the mind can really play tricks." The second picture was of the mother of one of the victims. This picture appeared on the continuation page (page 10) above the caption: "I feel that there was something, maybe in the earlier part of the day, that was present in the air that the children breathed. I don't believe that it was hysteria." Although not of equal weight (the fire chief's picture was arguably more prominent than that of the victim's mother), the photos and captions display more balance between the two competing explanations than do other properties of the articles. Whatever considerations might lie behind the editorial decision to use two photographs (rather than none, one, or three, for instance), the inclusion of the mother's photo rather than one of a county health expert who is quoted in the article, a woman frequently in the news whose photo is undoubtedly in the newspaper's files, is noteworthy.

The structure of the article, that is, the manner in which its paragraphs are organized to present the account to the reader, established mass hysteria as the more reasonable of the two explanations for the incident. The first ten (of twenty-six) paragraphs quoted experts and emergency service personnel who explained the reasons for the mass hysteria diagnosis. The middle ten paragraphs introduced—and disposed of—alternative explanations for the episode. Paragraphs 21, 22, and 23 dealt with the second (i.e., Friday) performance and concluded with the following quote from a young (male) performer: "'I think it's like they said on the news,' the teenager said. 'I think it was mass hysteria.'" The concluding three paragraphs recapitulated two other outbreaks of mass hysteria, one in Los Angeles County in 1982 and the other in Israel in 1988. In short, the thesis is laid out, an opposing point of view is presented and refuted, and the episode takes its place alongside others in recent medical history.

Three clusters of news sources appeared in the article. One cluster included two physicians, one from the Los Angeles County Health Department and the other an assistant professor of psychiatry at the University of California-Los Angeles (UCLA). Interestingly, the UCLA medical school professor, Small, is the senior author of the paper setting out the eight criteria for identifying episodes of mass hysteria quoted earlier (Small and Borus 1983). The second cluster consisted of several persons connected either with the evacuation or with the performance itself. They were the Santa Monica fire chief, a police sergeant, an official from the management staff of the auditorium, a representative of the school district, and an official with the city's cultural and recreation department. The third cluster included one of the victims, her mother, and another youngster who took part in the performance.

Journalists paraphrase and quote from their sources so that the facts of the story seemingly speak directly to news consumers. This maintains the illusion that journalists are merely neutral conduits between events of the recent past and the newspaper reader. However, the choice of news sources also reveals assumptions journalists make about the nature of events and about who is competent to explain them. In this case, the choice of news sources uses and perpetuates a psychiatric frame of the incident. Thus, journalists' use of news sources facilitates the continuity of cultural patterns by both drawing on a world taken for granted and inviting news consumers to view new events in light of that taken-for-granted world (Gamson and Modigliani 1989, 2–3).

Virtually all the news sources used in the account stressed two qualities of the "victims" of the mass hysteria: They were young (mostly teenagers), and they were female. The specific terms used included "student singers," "stricken youngsters," "the kids," "these gals," "these young ladies," "youngsters," and "the children." Such labels established social and psychological identities common to the sufferers consistent with the psychiatric frame.

In explaining the causes of the outbreak, the *Times*'s article failed to separate two questions that sociologists would keep distinct. One question is why the outbreak occurred when and where it did. The other is why some individuals but not others fell victim to the mysterious illness.

Quoting the psychiatrist from UCLA, the *Times* identified stress rather than toxic fumes as the cause of the episode:

> "I observe a few cases a year, most of them around this time when there are more assemblies and graduation ceremonies. Assemblies can be places where there is not only physical stress because of the crowded room but also psychological stress. The children performing are excited about being in front of the audience, but they are also anxious that their performances will fall short."

The quote implies that an "interaction effect" between stress and the makeup of people experiencing it is the reason outbreaks of mass hysteria are infrequent even though the supply of stressful settings is limitless. In other words, there may be daily occurrences of stress in other settings, but most people (e.g., adult males) have coping mechanisms that prevent the "'[m]ental anxiety or unconscious psychological conflicts'" from being "'converted into a physical symptom.'"

The article emphasized that the absence of any other measurable causal agent is central to experts' diagnoses of mass hysteria: "fire officials were unable to find evidence of fumes or toxic substances that could have caused the *phantom* outbreak Thursday night" (emphasis added). The physician from the county health department noted: "'But the key is that there is no pathological or objective sign of illness.'" The proposal put forth by the mother of one of the victims and her daughter, that something had been present earlier in the day that was not detectable when the fire department arrived at the scene, was dismissed in the article with the following statement: "Auditorium officials acknowledged that guard rails had been touched up with white water-based enamel paint in the morning, which could have left a lingering but harmless smell in the air."

A second set of "facts" about this episode requires another causal explanation. If "transmission of illness by sight or sound, or both" explains symptoms experienced by "late adopters," how are the symptoms displayed by the *first* person to fall victim to the illness to be accounted for? The answer provided in the news account is that at least one of the women who became ill *wasn't* a victim of mass hysteria. The first person to experience the symptoms was "really" sick:

> "All it takes is one person with a real illness and sympathetic people around who are tensed up for some reason, and it spreads like wildfire," said [the] associate deputy director of disease programs for the Los Angeles County Health Department.

Was such a person present in the auditorium? According to the article, this was the case.

> "In fact, one of these young ladies had a medical problem that was conducive to having her faint," said [the] director of cultural and recreational services for the City of Santa Monica. "Other youngsters followed the leader. It was a domino effect."

However, it turns out that there were actually two outbreaks of mass hysteria rather than one. The first occurred before noon during a rehearsal, some ten hours before the second (and, by implication throughout the article, only) episode.

> But anxiety among the performers apparently increased before noon when two girls fainted and a third complained of nausea during rehearsals at the auditorium. School officials said all three were friends and were seated near each other at the time.

The article did not mention whether the young woman with the medical problem conducive to fainting was one of the three who became ill during the morning rehearsal. Nor did it raise the question of whether there had been two women who were really ill, one during the morning rehearsal and one during the larger evening outbreak.

Why was there not a second (third?) outbreak during the Friday evening performance?[2] The absence of cameras was the explanation offered for the absence of hysteria during the second performance. Everything else presumably was the same (i.e., stress, the fear of "falling short" of expectations, etc.).

> Because authorities were concerned about a replay of Thursday night's mass reaction, cameras were prohibited from the auditorium [at Friday evening's performance]. "Part of the problem [Thursday night] might have been nervousness" from such attention, said Santa Monica Police Sgt. [. . .]

In summary, the media frame for this news account of a recent outbreak of mass hysteria could be summarized as the story of a mystery solved. The answer to the question, "What has happened here?" is, according to the account in the *Los Angeles Times*: "Some kids got sick, but there was nothing really wrong; it was a case of mass hysteria." Other aspects of what Altheide (1985, 13–21) calls the media "format" also shape this news account. For example, assumptions about time impose a temporal boundary about events, including the length of time required for a cause to produce an effect, the length of time for specialists to eliminate noncauses (e.g., toxins), and the length of time for the correct diagnosis to be identified. Assumptions about space embedded in the news format are also evident and impose a spatial boundary around the setting within which the relevant interactions took place, the spatial distances among the "victims," the spatial boundaries between victims and nonvictims, and so forth. The assumptions about time and space in the media format align nicely with those in the psychiatric frame, lending further, more subtle credibility to it. The news account is also a "window pane," passing along and thereby reinforcing medical versions of events and their etiology.

Both form and content in the news account establish the following "facts" about the events in Santa Monica: what had happened involved the medical domain; experts were certain about both the cause and the noncauses of the outbreak; such outbreaks are actually not uncommon; and they usually involve females in their teens.

The Sociological Frame: Mass Hysteria as Collective Behavior

It is now appropriate to examine the relationship between collective behavior theory and the study of mass hysteria in the social sciences, specifically the nature of the sociological frame provided by collective behavior for these outbreaks. There are a only handful of empirical studies of mass hysteria in this research literature. These include Schuler and Parenton (1943), Johnson (1945), Tumin and Feldman (1955), Medalia and Larsen (1958), Jacobs (1965), and Stahl and Lebedun (1974). Here I focus on the best known of these researches, a study of female textile workers in North Carolina who fell victim to the "June bug," by Kerckhoff and Back (1968; see also Kerckhoff, Back, and Miller 1965; Kerckhoff 1970). This is the most significant study of mass hysteria for the field of collective behavior for several reasons. It is often included in collected readings (e.g., Evans 1969), and it is widely cited in textbooks as a study of elemental "building-block" processes typical of all forms of collective behavior. The Kerckhoff-Back study is frequently used as an illustration of the operation of noncritical interaction, a concept closely related to "circular reaction" (Park and Burgess [1921] 1970, 382; Blumer [1939] 1951, 170–177) in particular and to social contagion more generally. The medical literature also contains numerous references to Kerckhoff-Back as "a classic example" of mass psychogenic illness (e.g., Brodsky 1988, 657), and medical researchers (e.g., Small and Borus 1983) frequently cite the monograph in their own papers.

Kerckhoff and his collaborators discovered the outbreak in North Carolina through the news media. First local television news broadcasts, then local newspapers reported a mysterious illness affecting workers at the "Montana Mills" dressmaking division in "Strongsville" (both are the authors' pseudonyms). Descriptions of efforts by the U.S. Public Health Service and county health officials to isolate the insects believed to be responsible for the outbreak of stomach pains and dizziness dominated the local news for nearly a week. News reports also included interviews with some of the victims. The news accounts, in short, identified for the researchers not only the beginning and end of the outbreak but also many of its key participants as well.

Kerckhoff and associates were initially troubled about whether or not the insect bites were the real cause of the outbreak. They refer to "the undeniably impressive symptoms the women had" and "the strength of the women's belief in the reality of 'the bug.'" However, they side with the medical experts in concluding that the outbreak was psychogenic with only a hint of having any second thoughts ("we seem to have no alternative in a case so well studied as this one"). Thereafter, they proceed unhesitatingly to establish through sociometric techniques an empirically based sociological explanation for what had happened under the assumption that the bites were not real.

Kerckhoff et al. ground their analysis in both the literature in social psychology, especially that linking psychological states to physiological symptoms, and in the literature in collective behavior, in particular Smelser's (1962) value-added model. The authors apparently found Smelser attractive because his model links structural strain (e.g., in the workplace, such as large numbers of overtime hours, production deadlines, etc.) to the growth and spread of a generalized belief (in this case, the bug bite and its symptoms). They treat the first woman to become ill (a person who had fainted five times during the preceding year) as the "precipitating event" necessary for the spread of subsequent illnesses. On the other hand, they make no use of Smelser's "agents of social control" in reference to the actions of people other than victims who can influence the career of the mass hysteria episode. In short, Kerckhoff and his colleagues selectively translate Smelser's theory of collective behavior from the macro to the micro level to fit some—but not all—of the circumstances of the dressmaking room episode.

The quasi-experimental research design used by Kerckhoff and Back included all the women treated for symptoms of the alleged bug bites (N = 58) in the sample. An additional 127 women were randomly selected as a control group. The authors created a second control group by separating out 21 women who reported symptoms similar to those of the victims but who themselves had not sought treatment. Two professionally trained female interviewers from the National Opinion Research Center (NORC) conducted standardized interviews with all research subjects. The authors present their findings in the form of percentages of the various groups in several tables, but they performed no statistical tests on their quantitative data. They summarize the data on sociometric ties with a diagram (Fig. 10) depicting relations among the victims (Kerckhoff and Back 1968, 117).

> Our general view of such cases . . . may be summarized briefly as follows: there are a number of people exposed to common sources of strain from which there are available but not easily accepted avenues of escape. The combination of the original strain and the added stress experienced because of available but forbidden solutions brings about a general state of tension whose intensity is probably increased by interstimulation among the people so situated. The state of unresolved tension leads to the experience of physiological symptoms. These symptoms become associated in the minds of the distressed people with some credible (but factually incorrect) "cause" of the situation, the connection deriving from what are probably random events that become interpreted in the context of the experienced discomfort. (Kerckhoff and Back 1968, 162).

The Kerckhoff-Back study illustrates three typical features of the sociological frame. First, they employ a social psychological conceptualization that is highly compatible with the basic elements of the psychiatric frame of mass hysteria, which emphasizes the presence of stressors, interstimulation among victims who are in close proximity, and the gender of victims. Second, the research and the report based on it are dependent on—and would not have been possible without—both the prior imposition of the psychiatric frame identifying the event as one involving psychogenic etiology and the news accounts chronicling the emergence of this frame. Third, by assuming the correctness of the psychiatric frame, these researchers, like others publishing research on mass hysteria in the social science literature, locate the causal mechanism for the outbreak exclusively within the work group (i.e., in the interaction among victims) and in its immediate social context (i.e., strain in the work environment). The role of other "agents" (Smelser's term) in framing the incident and in legitimating that frame is never taken to be an integral part of the research task.

Discussion

My examination of these three frames and the relationships among them leads me to conclude that these studies of mass hysteria are sociologically inadequate and of dubious value for students of collective behavior. This is true regardless of whether the collective behavior theory employed is contagion theory (which assumes hysteria victims fall under the uncritical influence of a socially transmitted suggestion), convergence theory (which assumes the copresence of individuals with similar psychological states), or emergent norm theory (which, as conventionally used, assumes the adjustment of separate lines of behavior based on victims' shared understandings of the meaning of the situation and of appropriate responses to that situation).[3]

The principal reason for these shortcomings is that all three theories of collective behavior draw a boundary around the situation that excludes the very actors who define a behavioral episode as mass hysteria. Focusing only on those who fall victim to the forces of contagion, coincidentally share the same state of mind, or create a definition of the situation under which the role of sufferer is appropriate ignores the success certain individuals have in imposing a framework of meaning on the behaviors after they have occurred, particularly the causes of these behaviors. The sociological reality of mass hysteria is not the behavior itself, but rather the successful imposition of the label *mass hysteria*.

This point has an instructive parallel in the research on urban ghetto riots by social scientists during the 1960s. A spate of sociological studies attempted to explain the outbreak of rioting as a function of the variable characteristics of urban places. Among the best known are studies by Lieberson and Silverman (1965), who focused on fifty-one years of race riots prior to 1964, and by Spilerman (1970), who concentrated on disturbances between 1961 and 1968. Two key features of the methods employed in these studies were the use of cross-sectional designs comparing samples of riot and nonriot cities and the use of newspaper indexes such an the *New York Times Index* to identify a population of riot cities. Unfortunately, Danzger (1975) demonstrated that the principal difference between riot and nonriot cities was the presence of wire service stringers in the former but not in the latter. In other words, the "existence" of riots for sociological purposes was as much a function of the actions of nonparticipants (e.g., employees of news organizations) as of the actions of the riot participants themselves.

The flaws in mass hysteria research stem from limitations in theory rather than in method, however. The research is fairly evenly divided between qualitative and quantitative approaches. Johnson (1945) employed a mixture of qualitative data from content analyses and descriptive quantitative data from the Census and newspaper sources in his study of a phantom anesthetist running loose in a small Illinois town. Tumin and Feldman (1955) used depth interviews, a keystone qualitative research technique (Lofland and Lofland 1984, 53–62), in their study of worshipers who experienced a religious vision in Puerto Rico. Medalia and Larsen (1958) employed survey research techniques in a study of pitted vehicle windshields in Seattle. Jacobs (1965) relied mainly on qualitative content analysis of newspapers in his examination of a slasher in Taipei in 1956. The study of female textile workers who fell victim to the June bug in North Carolina used sophisticated sociometric techniques (Kerckhoff and Back 1968).

The inadequacies to which I refer, therefore, do not seem to be the result of the methodological requirements of quantitative methods, as might have been the case in the study of ghetto riots. The problem resides in collective behavior theory itself, whether that theory is seen as involving contagion, convergence, or emergent norms. What each of these theories overlooks is that *power* is inevitably involved in the existence of mass hysteria, hence, ultimately in whether or not the episode represents an instance of collective behavior.

A long time ago, Howard Becker (1967) called attention to the fact that researchers unavoidably take sides in studying relationships between the powerless and the powerful. This is true in studies of mass hysteria as well. We routinely use news accounts to identify the existence of an outbreak, and we accept experts' assessments of its causes (as well as of its temporal and spatial boundaries), although not always comfortably (see Kerckhoff and Back 1968, 1–18). We have then gone on to concentrate exclusively on the victims as our research subjects without paying any attention to other actors who were indispensable in the social construction of mass hysteria.

The sharp difference in power between those exhibiting symptoms and those doing and reporting on the diagnosis of symptoms stands out very clearly in both the medical and media accounts. The following quotation (with emphasis added) from the *Los Angeles Times*'s article on the Santa Monica episode illustrates this difference:

> Health *experts* called the incident a classic case of mass hysteria, explaining that it usually afflicts *female adolescents* or *preadolescents*. (*Los Angeles Times*, 15 April 1989 2:3)

The experts—physicians, public health officials, medical college faculty members, and others—are superior to sufferers in terms of several dimensions of social status including education level, occupational prestige, public credibility, and access to channels of mass communication. Sufferers, on the other hand, lack the intellectual, social, and technical resources to prevent the event from being framed as one having psychogenic rather than organic etiology. Feeble protests such as the one by the mother of one of the victims in Santa Monica, who stated "I *feel* that there was something, *maybe* in the earlier part of the day" (emphasis added), have little chance of persuading either physicians or reporters. The Montana Mills women textile workers also strongly protested the presumption of mass hysteria—that is, the absence of any detectable insects—but were unsuccessful in convincing officials to define the situation differently (Kerckhoff and Back 1968, 7–12).

Warren (1987) made a similar point about the relationship between power and diagnosis in her recent study of women and schizophrenia. She reexamined the cases of seventeen housewives diagnosed during the 1950s as schizophrenic. These were women for whom the gender-based stresses of the male-dominated family of the 1950s took expression in forms of behavior that husbands and other authority figures found troubling. Defenseless against those who treated them as mad women, they also came to see themselves as ill. These cases had in common a powerlessness to counteract physicians and others who successfully labeled them as mad.

To publicly label someone as being ill, not from anything "real" but from a socially induced delusion, is to engage in a form of ridicule, rebuke, and reproach. Put differently, public accounts of mass hysteria are a type of public degradation ceremony (Garfinkel 1956). The Santa Monica fire chief recognized this (as did the editor at the *Los Angeles Times* who chose the captions for the article examined earlier) when he said "It's tough to make it look like you are ridiculing the kids, but. . . ." Successfully ridiculing and embarrassing someone through public discourse requires one to have greater resources of "cultural capital" (Collins 1987, 199–200) than the other. Those with fewer resources relative to their labelers will have difficulty countering the negative label.

What about the possibility that there are situations wherein would-be experts-as-labelers have fewer resources, less cultural capital, and so forth, than those they would publicly describe as victims of mass hysteria? Put the other way around, have potential "victims" ever mounted a successful counterattack to prevent their illness from being diagnosed as mass hysteria? Two examples come to mind, one an actual situation, the other a hypothetical but highly plausible one.

First, imagine a meeting of the president of the United States and the National Security Council. Now imagine that several of the participants, including the president, become mildly ill. Doctors are summoned, and the sufferers are taken to Bethesda Naval Hospital and kept overnight for observation, then given a clean bill of health and released the following morning. Meanwhile examinations of the ventilation system, the carpeting, the drapery, the coffee, the breakfast pastries, and so forth, turn up nothing that might explain the outbreak. At the hospital the attending physicians confer before stepping in front of a dozen microphones, dozens more cameras, hundreds of reporters, and, ultimately, interested observers in hostile as well as friendly governments around the world. What is the probability that the medical team is about to publicly degrade the "Leader of the Free World" and the top security experts of the United States by announcing that they have fallen victim to mass hysteria?

The second incident actually took place at a university in the southwest. The entire six-person staff of a development office became ill one day at work, although the severity of their symptoms varied a great deal. Headaches and nausea were the common complaint. The offices were closed, the staff sent home, and university safety and health officers summoned to search for the cause of the illness. None was found. A few days later some staff members again felt ill (one person subsequently took a one-month medical disability leave). More experts were summoned including outside consultants but with the same result: again, no cause found. One hypothesis suggested was that a fungus might be growing near the air intake that was being circulated by the ventilation system. Another round of symptoms, another inspection by environmental engineers, another inconclusive outcome. Meanwhile, no one in any of the adjacent offices (serviced by the same ventilation system) had ever experienced these symptoms, and derisive comments were made—always behind the backs of the development office staff—that the problem was "all in their heads." After yet another flare-up of illness, the development office was relocated to a suite on another floor, and "something" growing among the papers stored in boxes in the ill-fated quarters (mold was suspected) was offered as the latest version of the alleged cause of the problem.

Three things are of note in this sequence of events. First, the influence and assertiveness of the assistant dean who headed the development office was such that the laughter and derisive humor making the rounds were carefully concealed from the dean and the office staff. Second, no one in the university's administration blocked the repeated use of health and safety experts, including outside consultants, or the relocation of offices, even though both actions cost the university a considerable amount of money. Third, no one dared call in psychiatrists or other specialists with expertise in matters of hysterical contagion. At this writing the "real" cause of these repeated bouts of illness remains undiscovered.

The point of these two cases—one hypothetical, the other actual—is not that many instances of mass hysteria go undetected, unreported, or undiagnosed. Rather, the point is that whether an episode involving collective illness gets defined as hysteria or not depends on more than the combination of identifiable symptoms and unidentifiable causes. Mass hysteria is a function of power in two senses. First, there is the power of "ownership" in the sense of the lengthy historical process of gaining cultural and sometimes legal hegemony over the right to frame the situation. The term Conrad and Schneider (1980) used, *the medicalization of deviance*, applies to this process (see also Freidson 1970, 1986). Notice that in potential instances of mass hysteria both environmental and toxic specialists and psychologists and psychiatrists (but not sociologists) currently can claim the right of ownership.

Second, there is the power to label, that is, to frame a specific situation as mass hysteria. This involves more than a negotiated outcome between two parties, such as environmental and toxic experts on the one hand and medical or psychiatric experts on the other. It also involves in important ways the sufferers themselves, the so-called victims of mass hysteria. Where the cultural capital of labelers is matched or exceeded by those they would label, a nonepisode (i.e., a negative case) is more likely to exist. Where it is not, the label is more likely to be successfully applied, and another outbreak is added to the news, to medical history, and possibly to the sociological literature as well.

What course of action is open to those who would frame an instance of illness as mass hysteria when potential victims have sufficient resources to "cause trouble"? One is obvious. Experts can simply leave the mystery unsolved. The mysterious agents producing symptoms of illness could be said to have dissipated by the time officials arrived with measuring instruments. Another is to suggest that the instruments used were not sufficiently sensitive to measure the unknown agent that went undetected because it lingered only in minute concentrations. Or experts could argue that there is a totally new agent that existing instruments were unable to measure. Calls for a government-sponsored research effort like those heard after the first outbreaks of Legionnaires Disease could be made, for example.

What consequences do the psychiatric and media frames of mass hysteria have for collective behavior theory? The most serious consequence is that the sociological frame or conceptualization not only depends on but is determined by these nonsociological criteria. That is, if the outcome of interactions such as described here is to define the episode as mass hysteria, then collective behavior exists; if the outcome of these interactions results in some other definition—for example, that the illness resulted from a physical cause yet to be identified—then there is no collective behavior, hence, there is nothing to study.

There are parallels between current studies of mass hysteria and the treatment of such topics as crowds and riots early in the history of American sociology. That portion of the sociological frame dealing with collective behavior remains a cultural repertoire centered on forms of behavior that a century ago were deemed "abnormal." Rather than pathological, destructive, and antidemocratic forces as they were seen by conservative European writers, however, collective behavior in general and the crowd in particular were recast by Robert Park and other American writers in a more positive mold (Bramson 1961, 53–72). Yet, although the more overtly pejorative tone disappeared, the essence of the European conceptualization—such as the loss of critical reasoning implicit in the concept "circular reaction" (Park and Burgess [1921] 1970, 382; Blumer [1939] 1951, 170–71)—remained. For example,

> The crowd does not discuss and hence it does not reflect. It simply "mills." Out of this milling process a collective impulse is formed which dominates all members of the crowd. Crowds, when they act, do so impulsively. (Park and Burgess [1921] 1970, 385)

My argument is the same regarding mass hysteria. Although a pathological medical repertoire has been transformed into a nonpathological sociological one, there remains an exclusive focus on victim interstimulation and an ignoring of phenomena such as power and authority, both in terms of the historical acquisition of hegemony and in the situational use of claims to hegemony in negotiations of appropriate frames or interpretations of events.

There may be other reasons justifying the limited focus of these sociological approaches to mass hysteria. For example, the research by Kerckhoff and his associates (1965, 1968) strengthened the Durkheimian argument for social causation in such a seemingly nonsocial act as becoming ill (e.g., Durkheim 1951). Their quasi-experimental design, standardized interview schedule, and quantitative data provided strong evidence that sociometric patterns can distinguish between victims and nonvictims if insect bites or other causal agents do not.

Such total dependence on the psychiatric frame to identify the presence or absence of a sociologically significant phenomenon such as collective behavior is troublesome, however. For one thing, the certainty and closure of the psychiatric frame are deceptive and far from absolute. Sociologists also tend to see mass hysteria as an accomplishment, a fait accompli, rather than a process with more than a single possible outcome. This is due in part to the fact that sociologists typically learn of these events from news accounts in which the psychiatric frame is taken for granted.

News accounts of outbreaks of mass hysteria add their own closure to the closure already inherent in the medical story they are covering. This occurs through what Altheide (1985) called format considerations (e.g., newspaper articles have a beginning, a middle, and an ending; stories adopt one of the standard news themes; etc.). In fact, as I have suggested, the psychiatric frame masks a considerable degree of openness and uncertainty. I am referring to the difficulties in measuring low-level contaminants, an issue openly debated in the medical literature. More important, I mean the question of when to stop looking for physical causes. Continuing such a search can be costly. Public health and safety agencies may be reluctant to continue committing scarce resources for what medical experts would probably call a "wild goose chase." Such agencies would likely be more willing to continue the search if the president of the United States were a potential victim or when victims are powerful university development officers rather than teenage girls. The ability to use private resources to continue such a search is also probably highly correlated with power, as the review of mass psychogenic illness in the workplace suggested (Brodsky 1988). In other words, the extensiveness of the pursuit of possible organic causes itself is also related to the resources of potential mass hysteria victims.

In short, mass hysteria is a research topic for which collective behavior exists when others (using different frames) tell us it exists. We have no independent sociological criteria for identifying the phenomenon. Nonsociologists using nonsociological criteria determine for us what is institutionalized behavior and what is collective behavior. Sociologists borrow from the psychiatric frame its assumptions about both the timing of cause-effect connections and the locus of those causes. They then attempt to explain the outbreak exclusively in terms of the interactions among victims. This is true regardless of whether one uses contagion theory that assumes hysteria victims fall under the uncritical influence of a socially transmitted suggestion; convergence theory that assumes the copresence of individuals with similar psychological states; or conventional emergent norm theory that assumes the adjustment of separate lines of behavior under a common definition of the situation. We side with the authorities (Becker 1967), and our research implicitly reinforces one etiology (the psychogenic) rather than another (the organic).

Conclusion

This examination of mass hysteria suggests that we need a set of independent sociological criteria by which the phenomena of collective behavior may be identified. There remains too much unexplored territory in the no-man's-land between collective behavior and other forms of interaction studied sociologically. We continue to occupy trenches and fortifications laid down in this area nearly a century ago. Past studies of mass hysteria are relevant not because they provide data on elemental collective behavior processes but because mass hysteria, properly framed, confronts us with a fundamental question: What is collective behavior?

I do not have the answer to this fundamental question, but I do believe that an answer must come from careful consideration of the meaning of the terms *new* or the *novel* as applied to social phenomena. Most useful for this purpose are recent theoretical advances led by E. L. Quarantelli and his students. Their approach in essence identifies collective behavior as interaction using cultural repertoires that are new to settings or to participants in settings. The repertoires themselves need not be new or novel. It follows that collective behavior phenomena inevitably contain mixtures of institutionalized and noninstitutionalized elements (see Weller and Quarantelli 1973, 674–79; Quarantelli 1966; Dynes and Quarantelli 1968). Their work is grounded in a series of researches including an examination of black peacekeeping groups in ghetto riots (Quarantelli 1970), a study of women in a college dormitory whose lives were driven by a ouija board (Wenger and Quarantelli 1973), analyses of the emergence of intra- and interorganizational structures (Forrest 1972 and Ross 1976, respectively), an investigation of the sequences through which social organization emerges and evolves (Bosworth and Kreps 1986; Kreps 1985), and an examination of streaking on college campuses (Aguirre, Quarantelli, and Mendoza 1988), to name but a few.

In the variant of collective behavior theory being developed by Quarantelli and his associates there is much more to the study of an incident like an episode of mass hysteria than its causes (etiology). There are obvious questions of both the career and the consequences of such episodes. How are the routine relationships between performers and audience transformed into emergent relationships between those adopting the sick role and others responding to this role before the experts arrive? Do outbreaks of mass hysteria and their causes (e.g., toxic substances) cluster in time in the manner of fads? The point is that this variant of collective behavior theory can be applied in the study of mass hysteria in ways that do not depend on whether the adoption of the sick role is unconventional behavior resulting from psychogenic causes or conventional behavior produced by organic causes (see Gehlen 1977).

In summary, mass hysteria is not self-evident. It is a *social construction*, a label used under certain circumstances by specific human beings. If one is interested in an adequate sociological understanding of mass hysteria, then one must take into account the "political" context of these events, that is, the power relationships involved in the production of this label. Such an analysis requires, for example, an understanding of how professions acquire and maintain hegemony (e.g., Freidson 1970, 1986), how the interplay among actor, control agent, and setting leads to the use of one interpretation of events rather than another (e.g., Lemert 1951, 1972; Becker 1963; Schur 1971, 1984), and how cultural "tool kits" or repertoires (Swidler 1986) are used by physicians, reporters, and social scientists to construct a frame for these incidents (what Wuthnow and collaborators [1984] call cultural analysis). A reported outbreak of mass hysteria may or may not be a good opportunity to advance the development of theory in collective behavior, but it does challenge us to continue to develop the sociological criteria by which collective behavior may be distinguished from other forms of social behavior.

Notes

1. The Metro section contains local news of the day except for any major local stories placed on the front page. Called "Metro Digest," page 2 consists of brief descriptions of local news not reported on in detail elsewhere in the paper.

2. Evidence for the absence of further illnesses is circumstantial; there were no news reports that I am aware of on Saturday or Sunday noting that all went well on Friday.

3. This taxonomy of collective behavior theories is from Turner and Killian 1972, 12–25.

The Convergence of Volunteers in a Consensus Crisis: The Case of the 1985 Mexico City Earthquake

DENNIS E. WENGER and THOMAS F. JAMES

Within the field of collective behavior considerable attention has been focused on determining the social, demographic, and attitudinal characteristics of participants in nontraditional, emergent forms of collective action. Influenced by the theoretical assumptions of convergence theory (Turner and Killian 1987), this effort has focused on two primary concerns. First, numerous investigators have attempted to explain the behavior of emergent collectivities through reference to the characteristics and attitudes of the participants. Second, assuming the importance of the participants' prior characteristics and attitudes in explaining emergent behavior, considerable attention has been paid to examining factors that differentiate participants in various forms of collective action from nonparticipants. In effect, the second kind of research attempts to determine what "types of people" engage in collective, nontraditional action and what social and demographic characteristics facilitate their participation (Perry and Pugh 1978; Miller 1985).

Although theory and research in the convergence tradition can be traced to the early periods of the field of collective behavior (cf. Dollard et al.

This research was funded by the National Science Foundation under Grant # ECE-8610904. The assistance of Dr. William Anderson, the NSF liaison research officer, was most outstanding. Also, the authors wish to thank Dr. Jose Ruben Jara Elias, Alejandro Garnica Andradi, and the staff of the Instituto de Investigacion de la Comunicacion in Mexico City. They evidenced astute insight in conducting the initial research. We are indebted to them for providing these data to us. In addition, the authors are indebted to Dr. E. L. Quarantelli and Dr. Russell Dynes of the Disaster Research Center at the University of Delaware who were the Co-PIs on this project. This study was truly a collaborative effort, and their guidance and assistance in all phases of the research was valuable and most appreciated. In fact, these data are presented in the final NSF report on this research (Dynes, Quarantelli, and Wenger 1990). Finally, the authors would like to thank Dr. Benigno Aguirre of the Department of Sociology at Texas A&M University, who offered inciteful comments on an earlier draft of this chapter.

1939; Mintz 1951; Hoffner 1951; Smelser 1962), the examination of this issue became a small cottage industry during the 1960s and 1970s. One of the more comprehensively researched issues emerging from the studies of riots and civil disturbances during this historical period was the issue of "participation proneness." A number of studies attempted to examine the social and attitudinal characteristics of rioters and to compare these traits with those of individuals who did not participate in riots. Some of this research examined a selection of implicit, low-level hypotheses based on approaches such as riffraff theory (Perry and Pugh 1978, 177–83). Other research sought to simply develop a profile of the typical rioter (Caplan and Paige 1968; McPhail 1971).

McPhail's (1971) overview of this research found that the results were not impressive. Only 17 of 287 relationships discernible in ten studies were of moderate strength. The only independent variables to show a consistent relationship with riot participation were age, sex, ethnicity, and education (McPhail 1971, 1063–69). Young black males who had not graduated from high school were more likely to have participated in riots than other individuals. Consistent with his approach to collective behavior, McPhail argued that these characteristics were related to riot participation because these individuals were more likely to be "available" to participate due to more free time.

Therefore, with regard to riots or violent forms of emergent crowd behavior, there is a body of empirical data that indicates that, although participation may have been more extensive among some segments of the population, in general, riot participants were representative of the broader community and did not represent the riffraff of the system. For example, rioters were no more likely to be unassimilated migrants, unemployed, socially maladjusted individuals than nonrioters. Simply put, hypotheses emphasizing social-psychological or social structural predispositions to riot were discredited.

However, with regard to other forms of emergent crowd behavior, such as that associated with the collective, emergent, volunteer behavior that occurs during the immediate post impact period of natural and technological disasters, systematic studies are fewer in number. Although it has been widely observed that such behavior is massive and widespread during the emergency period of disasters (Barton 1970; Dynes 1970; Drabek 1986), there have been relatively few studies of the characteristics that separate participants from nonparticipants.

This brief chapter presents data related to these issues. Simply put, we consider whether or not factors can be identified that differentiate between those who participate in emergent volunteer activity during disasters and those who do not undertake such helping behavior. In effect, we once again consider the assumption underlying convergence theory that collective behavior results from the coming together of persons with similar sociodemographic characteristics and dispositions. Furthermore, we compare the results we have obtained for disasters with the previous findings regarding riot participation.

With regard to the latter concern, the working hypothesis of this examination is that differences will be observed between riot and disaster participants because of differences in the contextual nature of the social crises represented by these two phenomena. Simply put, riots represent instances of "dissensus crises," whereas disaster settings are inherently "consensus crises" (Quarantelli and Dynes 1976). To examine this issue, we present and analyze data from a survey of the population of Mexico City that focuses on the volunteer and collective action undertaken during the emergency period of the earthquake of 19 September 1985.

In effect, we have two major concerns. The first is to compare disaster and riot participants, and in undertaking our discussion, we first consider the nature of consensus and dissensus crises. Our second concern involves distinguishing between disaster volunteers and nonvolunteers. Thus we subsequently review some previous studies of volunteers in disaster. After presenting our data, we discuss the findings in light of issues relevant to collective behavior and disaster research.

Dissensus and Consensus Crises

Following the perspective of emergent norm theory, which assumes that all forms of noninstitutionalized, collective behavior, emerge from the crucible of social crisis (Turner 1964; Turner and Killian 1972, 1987), Quarantelli and Dynes proposed an important analytical distinction between two types of crises (1971, 1976). On the one hand, they noted that the crisis conditions that facilitate and characterize rioting crowd behavior are *dissensus* in nature. In a dissensus crisis there are basic disagreements and conflicts among the various participants regarding the courses and goals of action. In addition, a dissensus crisis is highlighted by the following characteristics: a heightening of social distinctions such as those based on race, ethnicity, or socioeconomic status; a psychological orientation of the participants to past and future, as opposed to present conditions; norms favoring vested interests; an internal threat; a disassociation with the community; and the presence of ambiguous problems.

On the other hand, Quarantelli and Dynes offered that natural disasters are examples of behavior emerging from *consensus* crises. In a consensus crisis there are basic agreements among the participants regarding the goals and objectives for action. In these settings there is a leveling of social distinctions, a present orientation, the emergence of altruistic norms, an external threat to the system, the expansion of the citizenship role and increased identification with the community, and the presence of immediate and non-ambiguous problems.

Those making previous comparisons of riots and natural disasters have observed differences in the patterns of looting associated with each type (Dynes and Quarantelli 1968). Given the diametrically opposed natures of these two types of crises, it is hypothesized that differences may also be observed in the characteristics of riot and disaster participants.

Previous Studies of Volunteerism in Disasters

As previously noted, compared to the rather voluminous research on riot participation, we have relatively fewer studies of participation in emergent, volunteer crowd behavior during disasters. Some of the earliest students of disaster did examine the problem. Fritz and Mathewson (1957), in their classic study of convergence during disaster, differentiated between five different types of people who converge on the scene of disaster. This distinction was based on their underlying psychological motivation and included the following types: the returnees, the anxious, the helpers, the curious, and the exploiters (1957, 29). These findings directly challenged the notion of convergence theory that individuals come to the sites of collective behavior incidents with similar, homogeneous motives and attributes (Wenger 1987). In addition, Form and Nosow (1958) examined the impact of family and gender roles and disaster-relevant skills on volunteering.

By the end of the 1960s, the great volume of volunteer and emergent group behavior in disasters had been well documented and even given a name, "mass assault," by Barton (1970). Based on the earlier NORC and NAS-NRC studies, Barton developed a theoretical model to explain the emergence of this mass assault. The magnitude of the mass assault has been subsequently reaffirmed in many studies, including those of Lechat (1976, 42), who observed that up to 75 percent of healthy survivors may engage in helping behavior, and Drabek et al. (1981, 97), who found that about 60 percent of the victims in Wichita Falls, Texas engaged in helping behavior. Most recently, Hull and Wenger (1992) found that approximately 95 percent of those interviewed in Charleston, South Carolina after Hurricane Hugo engaged in some sort of volunteer activity, whereas Mileti and O'Brien (1991) reported volunteer rates of 70 percent in Santa Cruz and 60 percent in San Francisco counties following the Loma Prieta earthquake.

Additionally, following the earlier work of Fritz and Mathewson (1957) many typologies and classifications of types of volunteers were developed. These efforts ranged from the general four-fold typology of disaster volunteers developed by Dynes and Quarantelli (1980) to the differentiation of four types of volunteers associated with local Red Cross operations by

Adams (1983). Holland (1989) discussed the issue of integrating volunteer and formal organizational activities in disaster response. Empirical efforts have varied from the qualitative case study of the formation and role structure of an emergent, disaster work group by Zurcher (1968) to the multivariate analysis of hospital volunteers undertaken by Stallings (1989), to the extensive, comparative study of search and rescue activity undertaken during the late 1970s by Drabek et al. (1981).

However, empirical studies that compare the social and demographic characteristics of volunteers and nonvolunteers are limited in number. For example, some of the earlier studies noted that males were more active in volunteering than females, particularly in rescue and relief work, and that younger persons volunteered at a higher level than older persons (Mileti, Drabek, and Haas 1975, 65). These findings, however, have been questioned by Perry et al. (1983) who found that men were not more likely to help unknown persons than women, and that younger people were not more likely to give aid beyond their immediate family. Nehnevajsa (1989), in a nationwide survey of 1595 respondents, found that women were slightly more willing to volunteer for community-disaster preparedness programs than men. Furthermore, stated willingness to volunteer was positively related to such measures as education and income, and negatively related to age. However, the differences were slight. In addition, the respondents were only asked about their willingness to volunteer in some future action. Their actual participation in postimpact, volunteer behavior was not examined.

Finally, recent studies provide conflicting evidence regarding the background characteristics of volunteers. In Charleston, "family resources" (a variable related to socioeconomic status) were strongly related to volunteering. In effect, those families with greater human and material resources were more likely to participate in postdisaster activities (Hull and Wenger 1992, 25–28). However, neither socioeconomic status nor gender influenced volunteering after the Loma Prieta earthquake. Those victims who experienced damage in either the Loma Prieta or pre-Loma Prieta earthquakes were more likely to volunteer (Mileti and O'Brien 1991, 5–6).

In summary, we have limited and somewhat contradictory evidence regarding the social and demographic characteristics of disaster volunteers. Furthermore, we have no studies that compare the characteristics of riot and disaster victims. This study addresses these two issues.

Research Methodology

Our data come from a survey of the residents of Mexico City in the aftermath of the 19 September 1985 earthquake. During the first week of October 1985 the Instituto de Investigacion de la Communicacion conducted a survey of 567 respondents from the general Mexico City area. The respondents were sixteen years of age or older and the sample was stratified with respect to household socioeconomic status and gender. In addition to gathering considerable attitudinal data, the survey queried the respondents concerning their own participation in earthquake-related voluntary activities subsequent to the impact as well as the volunteer activities of others in their households. Data were gathered on whether or not individuals had engaged in any earthquake related activities, the specific tasks undertaken, and the length of time that respondents devoted to those tasks. It must be emphasized that information was gathered on all forms of volunteering, not just participation in search and rescue activity. This information was obtained on 576 direct respondents and 2,357 other household members for a total sample of 2,924. The sample had a margin of error of 3 percent. To our knowledge it was the only survey undertaken of victim actions during this disaster.[1]

As part of a collaborative research effort supported by the National Science Foundation, the Disaster Research Center at the University of Delaware and the Instituto de Investigacion de la Communicacion analyzed this data. The findings related to volunteering have been presented elsewhere (Dynes, Quarantelli, and Wenger 1990, 84–91). At this time we discuss the data that pertain to the issues raised in this chapter.

The Research Findings

The Magnitude of Voluntary Activity

Previous studies of disaster indicate that the mass assault of volunteer, emergent collective action could be extensive. In fact, the popular wisdom held that it was immense. This impression was formed in part by the extensive mass media coverage of the postimpact period that portrayed massive, emergent search and rescue activity.

Our data indicate that the popular wisdom was correct, but only to a point. Only a relatively small percentage of the residents of Mexico City engaged in any volunteer behavior in response to the earthquake. As can be observed in Table 13, of the 2,922 individuals represented, only 285 or 9.8 percent engaged in at least some volunteer earthquake related activity during the three weeks following the disaster. If we only consider those respondents over the age of twelve, the percentage increases to 12.4 percent. In other words, about 90 percent of the Mexico City residents did not volunteer in any type of activity. However, if we assume the population of the city to be approximately twenty million, that percentage would suggest that somewhere between 1,740,000 and 2,180,000 individuals volunteered sometime during the three week period.

Table 13. Participation in Earthquake Related Volunteer Activities

	Volunteers N	Volunteers %	Nonvolunteers N	Nonvolunteers %	Chi Square	Probability
Total Sample	285	9.7	2637	90.2	N/A	N/A
Gender						
Men	192	13.2	1259	86.8		
Women	93	6.3	1378	93.7	38.7	< .001
Age						
< 18	32	2.9	1073	97.1		
18 - 29	136	17.3	650	82.7		
30 - 44	77	14.7	450	85.3		
45 plus	40	7.9	461	92.1	127.6	< .001
Residence						
Very Far Away	155	7.6	1883	92.4		
Away	84	14.6	490	85.4		
Near	34	15.1	193	84.9		
Close/Within	11	13.7	71	86.3	34.9	< .001
Socioeconomic Status						
High	57	25.7	165	74.3		
Middle	129	11.7	978	88.3		
Low	98	6.2	1494	93.8	92.1	< .001

Information was also gathered on the amount of time spent in volunteer tasks. On the average, volunteers worked for about five days and devoted almost ten hours per day to the effort. Using the middle of our estimated range of the number of volunteers, those figures translate into over ninety million volunteer hours. Therefore the response was truly massive by any standard!

Factors Relating to Voluntary Activity

What factors differentiated those who volunteered from those who did not? Did only certain types of people volunteer, or was the volunteer activity broad based? To answer these questions we considered the following social and demographic characteristics: gender, age, residence, socioeconomic status, education, and occupation.

Table 13 presents data comparing participation rates of men and women, different age and socioeconomic status groups, and groups based on residential location relative to damage areas. Socioeconomic status is positively associated with volunteerism. Within the highest SES category, 25.7 percent volunteered compared with 6.2 percent of those in the lowest. In addition, men were significantly more likely to volunteer (13.2 percent) than women (6.3 percent).

Age and location of residence relative to damaged areas were both related to volunteerism in a curvilinear manner. With regard to age the highest participation rates are found among those eighteen- to twenty-nine-years-old (17.3 percent) and those thirty- to forty-four-years-old (14.7 percent). As might be expected, fewer of those older or younger participated (7.9 and 2.9 percent respectively).

The data on residential location show that the highest rates of volunteerism were found among those who lived a moderate distance from the areas of greatest damage. Volunteerism was not as great from those residing within damaged areas or those from distant areas. In fact, the lowest partici-

Table 14. Comparisons of Volunteers and Nonvolunteers by Educational Levels

Educational Level	Volunteers % N = 93	Nonvolunteers % N = 434
No Schooling	4.1	95.5
Primary, incomplete	10.1	89.9
Primary school complete	5.2	94.8
Secondary, incomplete	10.6	89.4
Secondary school complete	17.7	82.3
Preparatory, incomplete	25.1	74.9
Technical	20.7	79.3
Professional, incomplete	33.8	66.2
Professional school complete	27.7	72.3
Post graduate schooling	65.1	34.9
	Total = 17.6	82.4

Chi square = 43.30088

pation rate (7.6 percent) was found among those who live the farthest away from the damaged areas.

Data regarding education are only available for 537 of the initially, randomly selected respondents, not for the entire sample of 2,922. These data are presented in Table 14. Education was positively related to volunteering (Chi Square = 43.3, p. = <.001). The highest rates were found for those with postgraduate schooling (65.1 percent), and professional school incomplete and complete (33.8 percent and 27.7 percent respectively). The lowest rates were found for those with no schooling (4.1 percent) or just a primary school education (5.2 percent).

The Characteristics of the Typical Volunteer

Because of the skewed distributions of the demographic variables and the disproportionate stratified sample that over represented higher socioeconomic levels, these rates of participation do not describe the mix of volunteers. A separate analysis (not shown) was undertaken of the demographic composition of the volunteer group. As opposed to examining factors that facilitated volunteerism, here the attempt was made to develop a profile of the typical volunteer.

This analysis indicates that the typical volunteer was a male (67.3 percent), eighteen to twenty-nine years of age (47.9 percent), from the middle class (45.4 percent), who resided outside the damaged areas of the city (84.0 percent). (All percentages are based on the total number of volunteers in the sample.) Furthermore, lower class volunteers constituted 34 percent of the total, whereas the upper class provided about 20 percent. Contrary to popular image, a majority were from the middle and upper classes, and those under the age of eighteen only constituted about one in ten of the volunteers.

Kinds of Volunteer Activities

Data were also gathered concerning the types of activities the volunteers undertook during the emergency. These tasks were many and varied. They ranged from search and rescue and debris clearance to collecting money and obtaining and transporting food and supplies. Some volunteers worked as translators, building inspectors, psychological counselors, medical providers, security personnel and blood donors. Some opened their homes as shelters.

In Table 15 these activities are grouped into seven basic categories, and it is evident that search and rescue and the provision of supplies were the dominant activities. Table 16 combines the smaller frequency activities into an "other" category and examines the behaviors of different volunteer

Table 15. Volunteer Activities

	N	%
Search/Rescue and Debris Clearance	117	41.9
Supplies	94	33.5
Medical and Psych. Aid	22	7.8
Sheltering	8	3.0
Transportation	10	3.7
Collection of Funds	5	1.8
Other	23	8.2
Total	279	100.0

groups. Clearly men (65.4 percent) were far more likely to assist in search and rescue and debris clearance than women (27.2 percent). Conversely, reflecting traditional gender role definitions, women were much more likely to engage in the provision of supplies (64.6 percent) than were men (32.8 percent).

Search and rescue and debris clearance was also strongly associated with socioeconomic status. Of those volunteers from the lowest SES category, 71.4 percent contributed to this part of the recovery effort compared to 47.2 percent of the middle-class volunteers and only 34.8 percent of those from the highest SES. Probably for reasons of health and physical strength, those eighteen to forty-four years of age were far more likely to lend their help to search and rescue than were those who were older or younger. Finally, it was those volunteers who resided the greatest distances from damage areas who were most likely to engage in search and rescue and debris clearance. Indeed, if we look at only low socioeconomic status males who are eighteen to forty-four years of age and reside a great distance from a damaged area of the city, we find that 86.7 percent worked at search and rescue and debris clearance, whereas the remaining 13.3 percent helped in activities related to provisions and supplies.

Regression analyses focusing both on volunteering and the type of volunteer activity were also undertaken and are presented in Tables 17 and 18.

Table 16. Volunteer Activities by Demographic Background Characteristics

	Search Rescue		Provisions		Other		Chi	
	N	%	N	%	N	%	Sqr.	Prob.
Socioeconomic Status								
High	14	34.8	23	58.4	3	6.7		
Middle	49	47.2	49	47.2	6	5.6		
Low	55	71.4	22	28.6	0	-	18.9	< .001
Gender								
Male	98	65.4	49	32.8	3	1.8		
Female	19	27.2	44	64.6	6	8.2	29.0	< .001
Age								
16 to 17	9	34.5	17	62.9	1	2.7		
18 to 29	69	62.1	38	34.1	4	3.8		
30 to 44	30	55.6	23	43.1	1	1.3		
45 & older	9	32.3	16	57.5	3	10.1	15.4	< .05
Residence								
Far	80	65.6	38	31.7	3	2.7		
Away	22	35.4	39	63.1	1	1.4		
Near	13	47.2	11	39.5	4	13.3		
Close	3	31.4	6	60.8	1	7.8	26.4	< .001

With regard to both analyses, distance from the area of devastation ceases to be a significant predictor of either volunteering in general or of type of volunteer activity. However, with regard to volunteering socioeconomic status, gender, and age are significantly related, with socioeconomic status, followed by age, being the most significant variable. (All these measures are significant at the .000 level, and the analysis produced a multiple r of .22 with an $F = 37.11$, which is significant at the .000 level.)

In considering the type of volunteer activity undertaken, gender, age, and socioeconomic status were also statistically significant predictors at the .01 level. However, with regard to the type of activity, gender was the most important variable, followed by age and socioeconomic status.

Table 17. Factors Associated with Volunteering

Variable	SE B	Beta	T	Sig T
Distance	.008138	-.017603	-.850	.3956
Sex	.010738	.120484	6.661	.0000
Age	.004868	-.078565	-4.330	.0000
SES	.009718	.054070	7.426	.0000
(Constant)	.038887		43.534	.0000

Multiple R .22016
R Square .04847
F = 37.10662 Signif F = .0000

Table 18. Factors Associated with Kinds of Activity

Variable	SE B	Beta	T	Sig T
Distance	.060006	.041202	.659	.5105
Age	.052413	.157869	2.808	.0053
Sex	.097085	.225184	3.982	.0001
SES	.069343	-.162704	-2.607	.0096
(Constant)	.309655		3.721	.0002

Multiple R .33984
R Square .11549
F = 9.13836 Signif F = .0000

Summary and Discussion

In this chapter we have presented data concerning the patterns of participation in emergent, collective behavior during a disaster. In this instance we examined factors related to volunteerism and participation in the mass assault and early recovery phases after the Mexico City earthquake of 1985. We observed that although the total participation was quite large in absolute numbers, this emergent activity by millions of residents of Mexico City was actually undertaken by a relatively small proportion of the population. Only about 10–12 percent of our sample engaged in any earthquake related volunteer activities; however, that amounts to millions of volunteers. Although the absolute number of volunteers was massive, the percentage of the population evidenced by this data is not consistent with previous observations that vast majorities of the population participate in this type of activity (c.f. Lechat 1976 and Drabek et al. 1981; Hull and Wenger 1992; Mileti and O'Brien 1991). The differences are probably due to the diffuse and localized nature of the impact in Mexico City, and the extensive geographical size and massive population base of the city.

We further considered some social and ecological factors that might differentiate those who participated from those who did not. Gender, age, socioeconomic status, education, and location of one's home relative to the damaged areas of the city all were related to the likelihood of volunteering. However, regression analysis eliminated the effect of location. Similarly, with regard to the type of volunteer activity that was undertaken, gender was found to be a significant predictor. Males were more likely to engage in search and rescue activity, whereas females participated in the provision of supplies and aid. Furthermore, search and rescue was more likely to be undertaken by those eighteen to forty-four years of age, whereas those older and younger provided supplies. Finally, search and rescue was most likely to be undertaken by those from the lowest socioeconomic status level, whereas those from the higher levels were more likely to undertake other tasks.

These findings have implications for both collective behavior and disaster research. With regard to the former, these data clearly support our working hypothesis. Different patterns of participation may be observed in disaster and riot settings. Although males appear to be more likely to participate in both types of events, the patterns of involvement for age, education, and socioeconomic status differ. In natural disasters, or consensus crises, participation seems to be higher for the upper and middle classes and those with higher educational credentials; in riots the reverse pattern is observed. Furthermore, it seems as though higher rates of participation can be found among older age groups in natural disasters.

These differences may indicate the influence of consensus, as opposed to dissensus crises. In the former, diverse factors such as the clear, imperative nature of the tasks associated with disasters, the emergence of norms favoring altruism in such settings, and the leveling of social distinctions during the emergency period are likely to facilitate participation by broader segments of the community, especially the middle and upper classes. In addition, the alteration of community priorities and values that results in a heightened emphasis on mutual support and social participation, but deemphasizing production, distribution, and consumption may free individuals from traditional, bureaucratic work roles and allow them to participate in nontraditional, emergent volunteer behavior during disasters. These dimensions are generally lacking during riots or dissensus crises (see Dynes and Quarantelli 1971; Wenger 1978).

In addition, the kinds of tasks that emerge during disasters appear to differ from those performed in riots, in that they are concrete and more embedded within traditional occupational roles. In contrast, in riots the roles of rioter, counterrioter, spectators, and the like, are somewhat vague; for most participants they are not strongly related to traditional social and occupational roles. This situation may facilitate the involvement of a wider range of participants from a greater variety of social backgrounds in the disaster situation.

Further, it appears that role acquisition during disasters is more likely to be based on normal, traditional roles than is the case in riots. In these data we have observed, for example, that upper-class volunteers were less likely to take on search and rescue and debris removal tasks, and were more likely to engage in managerial, professional, and support functions. The gender specialization we observed is also consistent with this proposition.

The findings from both this study and those of riot participation, however, raise serious questions about convergence theory and other views that see collective action as homogeneous and monolithic. Participation in helping behavior following the Mexico City earthquake was widespread and involved all segments of the population. Research on riots has suggested that there is no "riot prone personality," and the same appears to hold for participation in emergent disaster behavior. Differential rates of participation for various segments of the population have been observed in both types of crises, suggesting there is no single, homogeneous, dominant individual characteristic that determines participation. Rather, these patterns suggest there are differences in the social context and emergent role structures associated with these two different types of crises.

For disaster research, these findings provide valuable data on an under-researched topic, individual participation in the "mass assault." Few studies have been undertaken within the United States or developed countries; in developing nations, research has been nonexistent. Our findings support some observations culled from research within the United States. First, the data provide evidence that the "mass assault" can be found in different cultural settings. Second, the findings are consistent with earlier observations regarding the effect of gender and age on volunteering, and they are somewhat consistent (except for the opposite finding regarding gender) with the findings of Nehnevajsa (1989) regarding people's inclinations to volunteer in future disasters.

Third, the patterns of differential involvement in disaster-related activities support the "normalization principle," which argues that behavior during disasters is often routinized and patterned after normal, day-to-day behavior (Quarantelli and Dynes 1977). Fourth, the findings indicate that stereotypical views of disaster volunteers can be observed in other cultural settings. The stereotypical volunteer is a poor, young male. But the typical volunteer in the Mexico City earthquake did not fit that description. Perhaps the mythical image is a result of television coverage that focused on the exciting, emotional, and highly visible task of search and rescue; an activity in which younger males seemed to dominate. In contrast, other valuable helping activities, such as processing the massive amounts of clothing and material aid or providing food, shelter, or psychological aid, were comparatively invisible.

Finally, the findings indicate that volunteering must be viewed in a more complex manner. Specifically, different types of volunteer activity must be examined in future studies. It is not sufficient to simply ask, "Who volunteers?" We must also consider "Who volunteers for what types of activity?"

In conclusion, we want to call attention to limitations in these data. Critics could argue that we have compared apples with oranges and committed a cross-cultural fallacy, by in effect comparing findings on the participation in American riots with data from participation in a Mexican disaster. This justified concern is tempered somewhat by the consistency between our findings the limited earlier work and observations focusing on disaster volunteering in the United States. We hope that future empirical research will lend additional validity to our findings.

Note

1. The survey conducted by the Instituto de Investigacion de la Communication was based on a disproportionate stratified sample that overrepresented higher socioeconomic levels. Thus, all findings presented here are based on weighted calculations.

Rumor as Intensified Information Seeking: Earthquake Rumors in China and the United States

RALPH H. TURNER

The popular idea of rumor as an insidious tale, often malicious, that is accepted and spread by gullible people, has generally been replaced among serious scholars by more sophisticated conceptions—except possibly for Koenig (1984) in his analysis of marketplace rumor. Rumor is most often understood as a process of person-to-person communication (although often with the help of the mass media) through channels that are not institutionalized, about subject matter that cannot be satisfactorily verified by reference to the established culture or generally accepted authoritative sources. Theories of rumor have generally differed in the emphasis placed on answering different questions about rumor.

Theories of Rumor

The most widely accepted theory of the *incidence* of rumor is Allport and Postman's (1947), that rumor intensity is a function of importance of the topic multiplied by the ambiguity of available information. There is little disagreement with the importance principle, understanding it to mean importance to the participants in the rumor process. Although also widely accepted, the ambiguity principle is more debatable. Lind (1946) reported an immediate postwar rumor that lasted for several months in the Hawaiian Japanese community to the effect that Japan had won the war and that the American government was trying to conceal that fact from the American populace. It was widely believed, primarily among the Issei,[1] that the victorious Japanese army, headed by the emperor, would soon land in Hawaii and institute social and cultural changes that would elevate American culture

---

Translation of Chinese materials was made possible by a grant from the Dean of the Social Sciences, University of California, Los Angeles.

to the higher level that the Issei had left behind in Japan. In no literal sense could one say that the war's outcome was ambiguous after the emperor's public surrender and occupation of the Japanese mainland by American forces. The ambiguity was strictly subjective, involving a gross discrepancy between available evidence and one's deep beliefs about reality. A more adequate formulation might be a principle of ambiguity or the *impossibility of assimilating available evidence into preestablished worldviews and frames of reference*. This extension could help to account for many of the more fantastic rumors that abound in crisis situations.

At a more social structural level, Turner (1964) suggested four kinds of situations in which rumor is rife: when established accommodations bridging major social cleavages are threatened, when a rigid social structure impairs the circulation of information, when events threaten the normal understandings underlying daily action, and when there is a strong shared incentive to engage in normatively proscribed behavior. The first principle is illustrated in the rumors that abound during active or incipient interracial and interclass strife. The prevalence of rumor in military organizations and all deeply stratified organizations and communities and in totalitarian political regimes illustrates the second principle. The third principle explains the rumor in the Hawaiian Japanese community. The fourth principle was suggested by the anthropologist Raymond Firth's (1956) analysis of a recurring rumor among the Tikopia to the effect that a ship had been sighted. Although the rumor was nearly always disconfirmed by events, it supplied the justification for the people to abandon temporarily the dull, daily routine and gather in a festive mood to watch for the illusive ship. Shibutani (1966) did not fundamentally disagree with Allport and Postman on rumor incidence, but like Turner, explained it in more sociological terms. Conceiving rumor as a process of collective problem solving, he predicted rumor when information necessary for collective action was not available. Knopf (1975) criticized Allport and Postman for reflecting an individualistic bias, arguing that racial rumors are prevalent in a rigid social structure supported by racist ideology, with a hostile belief system, under conditions of stress.

Rumor theories also address *content and distortion* in rumor. Early assumptions that rumor necessarily leads to progressively more inaccurate accounts was refuted by Caplow's (1947) finding that the veracity of rumors in an army regiment was generally high, and by Bauer and Gleicher's (1953) observations on the Soviet "grapevine" as a source of information. To describe changes in content during the life of a rumor, Allport and Postman posited two processes, leveling and sharpening, guided by an underlying principle of assimilation. *Leveling* is a process of shortening and simplifying, which they contrasted to the popular belief that rumors "snowball." The evidence for leveling came principally from their laboratory experiments in the serial transmission of information, with subsequent support from some field experiments with "planted" information (DeFleur 1962). *Sharpening* refers to the selective retention of a limited number of details, so as to make the story more coherent. *Assimilation* identifies a psychological process characteristic of all perception to some degree, in which we tend to see things in a way that fits in with our preconceptions and biases.

There is general agreement with the assimilation principle and little challenge to sharpening. But the assumption of leveling has been generally discredited, as overlooking both the recirculation and problem-solving nature of rumor. In one of the first challenges, Peterson and Gist (1951) demonstrated the variety of detail that came and went and the imaginativeness of new themes added from time to time in the case of rumor in a midwestern city following the reported rape and murder of a fifteen-year-old girl. Buckner (1965) offered a theory to reconcile conflicting findings regarding leveling versus snowballing in rumor depending on the auditors' critical set, group involvement in the topic, and the close or diffuse nature of the group structure.

The all-inclusive concept of assimilation has been given more precise meaning by other scholars. Knopf speaks of the crystallization of hostile beliefs already held, linking those beliefs to a tangible event. Ambrosini (1983) emphasized common fantasies shared by a group of people and interprets rumor as a defense mechanism for individuals and for the group. A more situational meaning for assimilation comes from Prasad (1935), who argued on the basis of evidence from the great Indian earthquake of 1934 that rumor content takes the form necessary to justify the emotions people feel by presenting imagery consistent with those feelings. Writers such as Kapferer (1990) look beneath the explicit rumor content for a "hidden message" that expresses collectively shared but repressed fears and beliefs. In more general terms, Cantril (1941) spoke of the pursuit of meaning, and Festinger et al. (1948) identified the purpose of rumor as gaining cognitive clarity.

Other theoretical formulations have dealt with *differential participation* and with the *group formation* effects of rumor. Rosnow and Fine (1976) addressed both of these questions by proposing that rumor and gossip are exchanged like commodities in a marketplace for the benefits to be gained through the transaction. This analysis reflects the growing importance, during the last two decades, of rational decision models in explaining diverse forms of individual and collective behavior. Kapferer (1990) proposed that commodity value is greater for negative than positive information, which is one important reason why most rumors announce a mishap, catastrophe, peril, or treason.

Consideration of all these theories makes clear that distinctions must be made among different forms of rumor, particularly contrasting rumors—usually planted—that serve to create crisis situations, to rumors that emerge in preexisting crisis situations. Referring principally to the latter, we must clarify the general relationship of the rumor process to the crisis and what it is about the crisis that evokes rumor, if we are to have the basis for an adequate theory of rumor.

I find Shibutani's conception of rumor as a step in collective problem solving the most satisfactory general formulation of the nature of rumor. It is but one step further to conceive of rumor as *the upper end of a continuum of normal collective information seeking*. This requires a shift in analytic focus from attention to the tale and who started it to emphasis on the dynamics of the rumor collectivity.

Earthquake Rumors in California

In a study of unfolding events, media coverage, and public reaction over the three years following an earthquake near-prediction in southern California, it was possible to observe the continuum of information seeking (Turner, Nigg, and Paz 1986). Although most survey respondents gave television, newspapers, or radio as their chief source of information on the earthquake threat, a great deal of information was exchanged through informal family, friend, and workmate networks. Most of the time, public awareness and concern rose and fell in response to media attention to the earthquake threat. But when media coverage failed to address questions that were on the minds of many people, the relative reliance on these informal networks increased. At certain times people were unwilling to wait patiently for information to be made available to them and actively sought information from a variety of sources. Under certain circumstances this information seeking took the form of rumor mongering.

In seeking information, people in southern California first turned to what they understood as expert sources. A good indicator of the rise and fall of such information seeking is the record of requests received by the Office of Civil Defense in Los Angeles. Over two years, monthly totals for civilian information requests ranged from lows of five requests to highs of 103, 161, and 198. The initial announcement by the authoritative U.S. Geological Survey that the land surrounding a portion of the great San Andreas Fault nearest to Los Angeles was measurably uplifted—a frequently observed earthquake precursor—stimulated very few citizen requests. We attribute this passivity to two circumstances. First, it accords with the often documented tendency for people to define their situations as normal in spite of disaster warnings. Without social pressure to redefine the situation or undeniable physical evidence that a disaster was imminent, the "normalization" tendency prevailed. Second, because the media had no prior experience with earthquake predictions, they had very little to say about this warning. Without rich files to draw on and without a fund of personal experiences to report and mull over, the media could do little but report the straightforward and unsensational pronouncements of scientists.

The initial announcement had been carried on the front page of the *Los Angeles Times* on 13 February, but it was not until April that our indicators reflected a substantial rise in citizen information seeking. Four things happened in April. First, a researcher attached to the prestigious California Institute of Technology (Caltech) issued a very tentative "prediction" for an earthquake of about the same magnitude as the five-year-earlier destructive Sylmar–San Fernando quake, in the same general area, within one year, based on new and different evidence. Second, the California State Seismic Safety Commission issued a formal warning based on the earlier announcement of the uplift. Third, the new Caltech announcement became a point of political controversy as one city councilman urged the council to sue Caltech for damage to his constituents' property values. And fourth, the media by this time were ready to give featured attention to the earthquake threat and the political controversy. Citizen requests for information reached their highest peak during April, declined somewhat in May, but rebounded to their second highest peak in June on news of a devastating earthquake in northern Italy. An equally precipitous drop in July was followed by further declines in August and September before the next peaks in October and December.

During the April and June peaks, most of the inquiries were for more information or for verification regarding the announcements and events reported in the media. Often they reflected misinformation that was circulating through informal communication networks. But the misinformation was not pure invention; it consisted rather of misunderstandings that were quite understandable, taking into account the ambiguity of the announcements, the absence of technical background for interpreting such announcements, the justified arousal of anxiety, and some sensationalism in media treatment. Nor was there one single theme or plot that dominated the communications. This might be thought of as a period of quasi-rumors.

In October and November a different kind of communication became prevalent. There were two distinct waves of earthquake prediction rumors that could not be linked directly to any media-relayed predictions, whether of scientific, psychic, or other origin. During both waves we heard repeatedly that Caltech or its affiliate, Jet Propulsion Laboratory, had predicted a major earthquake in the immediate area at an imminent and usually specified date, but were keeping it secret for fear of disruptive panic in the community. The accounts were generally authenticated by identifying a source such as a secretary or cleaning person or wife of a professor. The wave of inquiries to Caltech and to civil defense authorities was similar to earlier waves in asking for clarification or verification. But the topic was different in that the inquiries were not directly stimulated by media attention and could not be interpreted as misunderstandings of authoritative public announcements. At one point an unknown man, identifying himself as a "Caltech scientist," called several hospitals and schools, urging them to be prepared for a severe earthquake about 21 October.

By 21 and 22 October, which had become the consensual dates for the earthquake to occur, people were reporting a variety of confirmatory evidence of the impending quake. But by 25 October, inquiries about this rumor had subsided. In mid-November a similar but smaller wave involving similar accounts occurred, coming to an end when the media discovered and featured a man who represented himself—falsely, as it was later revealed—as an experienced seismologist, and who advanced a specific prediction for an earthquake in Los Angeles on 20 December. At this point the earlier pattern of information seeking resumed, with a flood of inquiries seeking confirmation or clarification of the prediction or advice on how to cope. Delayed exposure of the man's true background and absence of credentials, first by the *Los Angeles Times* and then by other media, dampened the level of information seeking. But concern remained high until 20 December passed without a quake.

With this sequence as background, we may think of four levels in mass communication, which may become stages in a rumor process. At the first level there were a good many respondents who heard about earthquake warnings only from the media and participated in no informal discussions on the subject with family, friends, or workmates. Most respondents, most of the time, were at a second level, at which people sift, amplify, and clarify what they hear and read in the media through informal discussion with family, friends, and workmates. Already some of the elements of rumor process are present, as people seek consensus in clarifying ambiguities and choosing courses of action. In most circles, one or more people become identified as "local experts" because of relevant experience, reading, or training, and thus become local opinion leaders. Both correct understandings and systematic misunderstandings emerge and are crystallized in this process. This stage corresponds to the classic formulation of the "two step flow" of communication (Lazarsfeld, Berelson, and Gaudet 1948; Katz 1957).

At a third stage, large numbers of people become impatient and pursue information by calls to authoritative sources and by intensified efforts to locate and question local experts. Again both corrections and further misunderstandings result from this process. The information gleaned is enthusiastically or anxiously shared, and accounts and sources are compared. But there is as yet no outright invention beyond some speculative filling in of information gaps. At this stage information seeking has become an active individual and collective pursuit, replacing the normal passivity of the audience. Members of the community share a common concern and have become mutually sensitized in a way that fosters the rapid dissemination of any new information that enters the system. Much of what we call rumor is found at this stage of collective information seeking.

Finally, at a fourth stage, the demand for information can become so intense that people grasp at straws, imaginings become realities, and the need for action is so compelling that some people spread information and misinformation indiscriminately out of a sense of responsibility (Medalia 1959–60). At this stage, invention becomes rife as speculation and interpretation are quickly translated into fact.

Along this continuum from the second to the fourth stage, the changes are more matters of degree than of kind. The mixture of sharing and confirming information dominates the communication through informal networks at all three stages. Distortions arise and prevail at each stage. To call one stage rumoring as a qualitatively different pattern of communication is misleading. Instead, we distinguish the fourth stage by calling it "affirmative rumoring," because it is here that the actual invention of accounts as distinguished from ordinary distortion and misunderstanding take place.

It would be a serious mistake to describe this sequence in strictly cognitive terms. The sequence is marked by progressive intensification of the level and mutuality of affect. As shared emotional levels rise, Prasad's (1935) principle that people accept and then begin to invent accounts that serve to justify the affect that they feel becomes of paramount importance. If the regnant emotions are fear and anxiety, reassurances acquire less and less plausibility as the sequence progresses, and more and more events and pronouncements are seen as confirmation of the dreaded forecast. The same process can involve a developing emotion of elation or joyous anticipation, and the same principle of increasingly selective perception, acceptance, and invention can prevail.

We did not find Allport and Postman's main variables, *importance* and *ambiguity* helpful in explaining the rise and decline of information seeking and affirmative rumoring at different times during the year of the uplift, although they were underlying constants. Our evidence suggests that the threshold from information sifting to information seeking is crossed as comprehension of the threat messages *increases*, when the nature of the threat is specified more precisely and immediately, when the threat is authenticated, and when people think they know of an accessible source for further information. The earliest stages are marked by the greatest ambiguity, and ambiguous information is easily discounted as normalization prevails. It is, rather, the reduction of ambiguity that forces people to pay attention to unpalatable news and instigates intensified individual and collective information seeking. The threshold to affirmative rumoring is crossed when all but the last of these conditions exist and heightened expectations are frustrated by a decrease or dearth of information from the media and other authoritative sources. The key to affirmative rumoring is the conviction that something must be brewing but that authorities won't talk publicly. Survey evidence that nearly half the people believed that public officials and scientists were withholding some information from the public confirms this element in the process. The earlier deluge of media coverage raised people's expectations, and the one-year prediction (April to April) gave them a time dimension. With simultaneous early discussion of the advisability of releasing predictions and public ambivalence over their release, also documented by survey evidence, there was suspicion from the start that authorities would withhold information at a critical juncture to avoid mass panic. Thus, as media coverage of the earthquake threat waned, the suspicion that silence from the media and the authorities signaled intensifying threat grew until it reached crisis proportions. We might say, à la Prasad, that the affirmative rumoring supplied cognitive justification for the intensifying fear that people were feeling and sharing through their informal networks.

Rumor Studies in China

In 1979, a unit was established within the State Seismological Bureau in Beijing to study and deal with the social and economic effects of earthquakes, as a necessary complement to the already flourishing study of the physical aspects of earthquakes. The subject matter of the new division's research was identified as a new scientific discipline, and its name was translated into English as *seismosociology*. In fact, the small nucleus of researchers was not recruited from the field of sociology, but consisted of seismologists and persons trained in literature and languages. Rather than turning to Chinese sociologists for help, they looked primarily to Western and Japanese writings on the social and economic aspects of earthquakes. They drew heavily on the Fifth U.S.-Japan Symposium on Earthquake Prediction, held in Tokyo in 1977, and the UNESCO International Symposium on Earthquake Prediction, held in Paris in 1979, for their initial inspiration. More recently they have established liaison with scholars in some Chinese departments of sociology, and have become fully aware of the relationship of their concerns to the broader field of disaster research in the Western world.

From the start, understanding and controlling rumors of imminent earthquakes has been one of their principal assignments. In pursuing this task they have sent field-workers immediately to communities in which such rumors are reported. The field-workers quickly attempt to assess the nature, extent, source, and apparent consequences of the rumor, and institute steps to discredit the rumor, ameliorate its effects, and install programs to reduce community susceptibility to future rumors of this kind.

Rumor is defined by this group as being untrue, groundless, as not originating in seismological agencies, and as being spread through unofficial channels (Song 1987). Rumor is distinguished from *misdissemination*, which consists of exaggerated information that may have some factual basis (Li 1988). In practice, the two phenomena are treated together, and I disregard the distinction for present purposes. Like most recent and contemporary Chinese scholarship, seismosociology is more pragmatically than theoretically oriented. But from the published case studies and other analyses of earthquake rumors it should be possible to infer a theory that is implicit in the work.

Each of the rumor accounts reviewed projected a destructive earthquake, often of an impossibly high magnitude, with such cataclysmic aftereffects as land subsidence and flooding of whole counties. The accounts in some cases included reports of supernatural visions and magical prescriptions for protection from the earthquake. As is usual, rumors contained attributions to authoritative sources, frequently outside of China. Each rumor typically took several forms, with varying attributions, dates of occurrence, magnitudes, and other consequences. Frequently, changes during the life of the rumor could be documented.

Rumor sources could not always be established. But those that could included forecasts transmitted by foreign media, inferences drawn from unexplained official actions in earthquake preparedness or investigation, and leaks of disagreement or uncertainty among officials concerned with earthquake preparedness. Susceptibility to earthquake rumoring was enhanced by memories of past destructive earthquakes and the conditions associated with them, by communications from concerned relatives and friends abroad, by direct observation of such phenomena as anomalous animal behavior and sudden well water changes that people had been taught to look for, and by the conviction that government and party officials regularly withhold information from the masses. Behavioral effects of the rumors typically covered a range of actions from panicky and fatalistic, such as killing and eating livestock and spending savings, stockpiling food supplies, trying to leave the area, not going to work and postponing essential agricultural activities, to active information seeking by writing to local seismological groups demanding authentic information. In more traditional areas of China, ancient superstitions were revived as people practiced mystical rituals or ate sacred food to protect themselves, and fortune-tellers and witches were able to exploit the population to their own benefit. Community effects included exhaustion of food stocks in local markets, inflated prices for goods in great demand, serious declines in economic production and income from tourism, and loss of important agricultural crops.

The characteristics of these rumors do not differ greatly from those we have observed in California during the three years following public announcement of the "Palmdale Bulge" on the San Andreas Fault near Los

Angeles (Turner, Nigg, and Paz 1986) and since then. The suspicion that authorities know something they are unwilling to tell the public, the sensationalized nature of the forecasts, the authoritative attributions, confirmation through the selective use of folk signs such as anomalous animal behavior and exceptional weather conditions, and misinterpreted official actions are all found in both China and California. The main differences are in the degree of superstitious content, the extremity of public reaction, and either the origin or attribution to foreign sources. The first difference should not, however, be exaggerated. Seers of one sort and another had a heyday in California after scientists first issued their highly qualified warnings, and fully half the adult population said they would take seriously their own strong premonition that an earthquake was coming. A constant element in California rumors consists of reports of extreme reactions—highways jammed by cars leaving the area, precipitous drops in real estate property values, officials and scientists arranging to be out of California during the indicated days—but subsequent investigation consistently refutes these contentions. In California, the only significant incident originating outside of the United States was a warning from a Soviet seismologist, which, in spite of being perfunctorily reported in the media, did not become the subject of rumor.

The Chinese approach to rumor abatement gives the best indication of their implicit or explicit rumor theory. Uniformly stressed is the importance of getting to the scene and instituting an abatement program promptly. In one analysis it is suggested that rumor develops in four stages, namely emergence, dissemination, thinking (cognition), and action. The secret of effective control is to launch a program while rumor is still in the third or cognition stage (State Seismological Bureau 1985, 80–86), and before disruptive action has occurred on a broad scale. The standard abatement sequence begins with two information-gathering activities. First is assembling a detailed account of the rumor content and investigating the credibility of each of its elements. For example, if the warning is attributed to a particular scientist, governmental, or party unit, or external source, determine whether or not the attributed source did actually issue such a warning. If the rumor is buttressed by reference to signs that people have been taught to observe, such as well water changes and anomalous animal behavior, determine whether or not these phenomena are widespread or very localized and whether there is some other reasonable explanation for them. Rather than simply asserting that the rumor is untrue, which analysts say the people could not believe, the rumor can then be more credibly refuted point by point.

Second, scientists and personnel from local mass observation stations are assembled for an extended meeting in which all the potentially earthquake-precursory signs in the effected area are examined, so as to reach an independent conclusion concerning the probability of a damaging earthquake in the area in the near future. The conclusion to this meeting will normally be an unequivocal statement that a damaging earthquake cannot possibly occur in the vicinity in the near future, or that small earthquakes are expected but no damaging quake is possible. These two sets of information are then distributed to all the local governments and party leaders, and to the media, with instructions to disseminate them widely. In addition, through the network of mass observation units and the media, a program of general earthquake education is instituted. In nearly all the accounts, we are told that rumor abatement takes place quickly once these steps are taken. The exception was a case in which a small earthquake occurred just after public assurances had been given, resurrecting the rumor until the conclusion from a second meeting of experts was disseminated. For the longer term, mass earthquake education is emphasized, and a program of local monitoring for earthquake signs is maintained.

Again, Chinese and American approaches to rumor abatement are more similar than different. In both countries it is agreed that the answer to rumor is not less but more information, as well as coordination between scientists, government, and the media. But before noting the important differences, we must take note of the social structural differences that shape the respective contexts of earthquake abatement programs. First is the comprehensive and tight control of information and of the media in China. Although Chinese policy calls for mass earthquake education, state policy is not to release information about possible precursory earthquake signs. The people are to be told when the authorities believe it is time for them to take protective action (Haicheng Earthquake Study Delegation 1977, Turner 1978). Among the principal lessons said to have been learned from misdissemination rumors in Hebei Province in early 1984 were that potentially precursory anomalies should not be reported to government departments that must act on them until their implications have been made thoroughly clear, and that secrecy concerning the earthquake situation should be strengthened (State Seismological Bureau 1985, 94–96). Such tight information control means that any information leaks or supposed information from alternate sources is likely to be taken much more seriously than in California where ambiguous and even contradictory information is fairly continuously forthcoming. The Chinese's approach may be one reason why rumors seem to lead to more panicky and extreme reactions in China. On the other hand, information control facilitates the dissemination of unambiguous and uncontradicted counterrumor information. Furthermore, the American philosophy that looks askance on government attempts to control public belief leads, at least initially, to a laissez-faire approach to rumor so that decisive action tends to be delayed, and then tends to be taken more at the initiative of the media than of government.

A second important social structural difference is the more elaborate and fully developed infrastructure for dealing with earthquake predictions in China, established when Mao Tse-tung assigned top government priority to predicting earthquakes. At the top, scientific units such as the State Seismological Bureau in Beijing and provincial seismological bureaus do not separate research and application as American units generally do. The vast network of amateur seismological observation units is used both to gather earthquake information and to disseminate educational material and instructions (Turner 1978). The highly centralized Communist party organization, although under attack in recent years, permeates all levels of government and the media, serving as an effective vehicle for prompt transmission of authoritative instructions and information. Hence, it is probably easier for the Chinese to deal promptly and decisively with rumor, when initiative is taken at the highest levels, than it is in the United States.

A third social structural difference is the pattern of collective decision making in China. In California, if the governor were concerned about earthquake rumors, he or she would ask the director of the Office of Emergency Services to call a meeting of the state Earthquake Prediction Evaluation Council, made up of scientists. The council would then *advise*, but the ultimate decision and responsibility for any action taken would rest on the shoulders of the governor. In light of psychological *risky shift* (or *group shift*) theory, a group can often make a more unambiguous and, hence, risky decision than a lone individual (Haicheng Earthquake Study Delegation 1977, 263–66). No American scientist or public official would dare issue an unqualified declaration that no destructive earthquake can occur in the near future in a specific location, as the Chinese have done.[2]

A fourth social structural difference is the absence of a highly developed concept of liability in Chinese law. Indecisiveness and delay in acting by scientists and government officials in the United States is often occasioned by their concern over personal liability, in case events show their decision to have deleterious consequences.

These observations can be summarized by saying that there is a greater difference in the way earthquake rumors are handled in China, as compared to the United States, than in the respective scientific theories of rumor. Except for the Chinese fear that releasing ambiguous and incomplete information as it develops could spawn rumor flurries—a fear still shared by many American scientists and officials—the prevalent assumptions about rumor dynamics among American and Chinese scholars are similar.

If our conception of rumor as merely a highly intensified form of normal collective information seeking is correct, practice in both countries tends to set rumor apart too sharply from ordinary patterns of interpersonal communication. It is not so much the informal sharing of speculation and, often, misinformation, that distinguishes the folk concept of rumor, but the crystallization of accounts and the intensification of affect. Stated in another way, *diffuse rumoring characterized by subdued affect is a continuous and normal feature of society.* When it is provided with a more or less continuous flow of authoritative information, even when that information is often ambiguous, it can be educational and can often forestall the extreme reactions of *affirmative rumoring.*

Thinking of rumor as merely intensified information seeking also points attention to the normal rumor accompaniment to correct and authoritative information. I was not able to find Chinese studies of rumor accompanying accurate predictions. There is no published record, for example, of the nature of the informal communication that accompanied release of the highly accurate Haicheng earthquake prediction in 1975. If that communication had been carefully monitored, I question how different it would have sounded from the accounts of misdissemination and false rumor on other occasions, so carefully reported and analyzed.

One final point that seems to be missing from both American and Chinese rumor accounts and analyses is attention to the control aspects of rumoring. To what extent is affirmative rumoring also a demand that others accept the rumor account, and how much do rumor accounts include specific behavior demands? Recognition that rumoring is not only collective cognition but collective *behavior* may be the next important step in the development of rumor theory.

Notes

1. The Issei are the first generation Japanese Americans, born and raised in Japan, who migrated to the United States.

2. In light of the short span of earthquake history in the United States, the frequent discovery of previously unknown faults, and the suspicion that some earthquakes occur without unambiguous premonitary signs, American seismologists would seriously question whether such declarations can be scientifically justified.

Collective Behavior and Social Movement Theory

BENIGNO E. AGUIRRE

The purpose of this chapter is to examine the state of the specialty of collective behavior and social movements in the profession of sociology in the United States and to argue against the widespread view of collective behavior that exists nowadays. There is an increasing imbalance in the research attention given to social movements and collective behavior topics of investigation. Despite the continued importance of collective behavior processes, much greater attention is given today to the study of social movements than to the study of collective behavior, and this fact must have the effect, if it continues, of emasculating the study of collective behavior. I first speculate about the reasons for the imbalance in the attention given to collective behavior and social movements as topics of investigation in the specialty, and then offer some examples of the resulting misunderstandings and intellectual impoverishments. My conclusion is that the study of collective behavior should be of interest to social movement analysts.

Cultural Resonance

As Morris and Herring (1986) showed, the continued popularity of the study of social movements in the United States is very much linked to the emergence of the resource mobilization approach (RMA) in American sociology. Any comprehensive attempt to explain why the study of social movements is so popular today must also explain why the resource mobilization approach or theory is so popular in the United States and western Europe. The epigrammatic answer is that RMA is both fad and timely cultural product.

RMA made important and by now well known technical advances to the study of social movements (but see Zygmunt 1986). However, these advances

I want to thank J. Alston, G. Browning, R. Saenz, K. Tierney, J. D. McCarthy, J. Burk, M. Fossett, B. Finlay, D. Wenger, and J. Copp for their criticisms and suggestions to earlier drafts of this chapter. I am solely responsible for its contents.

do not provide sufficient reasons for its faddish popularity. Rather, an important although not very well explored set of reasons for its rise are the resonance, isomorphism, and plausibility of the approach vis-à-vis important cultural complexes in late twentieth-century U.S. society. Among these is the presumption that time, considered as a social and not just a chronological construct, is a scarce commodity. Moreover, both in the theory and the culture the emphasis is on pragmatism, rationality, manipulation of the individual by massive bureaucracies (and the resulting important, albeit limited, set of options available to the person to establish a sustainable style of life), the existence of a minimum set of constitutionally protected citizenship rights, and of a relatively decentralized political system that facilitates movement cooptation (Mayer 1990) as well as the activities of voluntary associations. The conceptual isomorphism between RMA's emphases and the popularity of models of rationality in contemporary sociology (e.g., Turner 1990) has also ensured a receptive audience for the approach.

The following comments on the dimension of time show some of the limits of this isomorphism. Cross-cultural differences in use and attitude toward time reflect differences in the material bases of social life and are very significant, although mostly unrecognized bases on which to construct a comparative framework for the specialty. In this reconceptualization, Staffan B. Linder's *The Harried Leisure Class* (1970) could offer guidance.

Linder's central insights are that time is a commodity, the identification of five sectors of use of time (working time, time devoted to personal work, consumption time, culture time, and idleness), the specification, in accordance to general economic theory, of the dynamic effects of economic development on the relative value of each type of time, and the identification of three types of cultures—time affluent, time surplus, and time scarcity cultures.

Linder's justly famous work is wrong in stating that culture time is inelastic. Many types of social movement action have the effect of increasing the productivity of culture time. For example, we write about antiabortion movements in the United States and their success in the block recruitment of Catholic practitioners without recognizing that participation in the antiabortion movement is a way of maximizing the profitability of culture time of practicing Catholics. For many Catholics, protesting the law through antiabortion organizational channels is part of their religious observance.

There are types of culture time activities that admit of value incrementation. In the previous example religious activities can be undertaken even as the person protests present abortion laws. The protest serves as testimony of the religiosity of the believer, making it unnecessary for the believer to allocate more time to the activity. The movement organization increases the value of culture time. In modern time-famine societies, the ease with which social movement organizations can mobilize people and obtain their goals is increasingly determined by their ability to establish these multiple and simultaneous time use action frames, so that participation in movement activities would occur as people participate in voluntary organizations of various sorts.

But this is not necessarily true in time affluent and time surplus societies, where time scarcity is not an issue, the cultural tradition of membership in voluntary associations is not as marked, and the risks attached to social movement participation may be much greater (e.g., Ekiert 1989). In some of these societies membership in a social movement is a master status and a life-transforming decision, and the resource most plentiful to social movement organizations is not money but people with time on their hands, with important, albeit so far largely unstudied, tactical and strategic consequences. To wit, in these societies the problem of social movement organizations (SMOs) may not be caused by free riding but rather by the phenomenon of convergence, or the organizational exigencies generated by too many people wanting something to do.

The exploration of these matters will have the effect of pointing out RMA's time famine sociocultural assumptions, and the need to consider its temporal, cultural, and social structural limitations as a conceptual framework. It is time to generalize social movement theory.

Faddish Organization

Aside from the cultural resonance of RMA, its rise in the specialty of collective behavior and social movements in American sociology exhibits many characteristics of faddish behavior (Aguirre, Quarantelli, and Mendoza 1988). As fad, RMA was an important cultural innovation to the sociological analysis of collective behavior and social movement; its emphasis on social movement organization and its use of complex organizational models of economic firms in the study of social movements was new and useful. It also reflected the social stratification of science in the profession. There were significant power differentials between the innovators and early adopters and the followers, for RMA spread from a few prestigious universities with large graduate programs, assuring it of an ample supply of talented graduate students. As fad, RMA also had the symbolic attractiveness, popularity, and large following of a few key and influential intellectual leaders in sociology, with power to monitor, through the editorial and review process, what was published in the major journals. Finally, as fad RMA operated in the absence of a critical mass of scholars that could dispute its claims and interpretations.

The faddish characteristics of RMA fits many of Lyn H. Lofland's (1990) perceptive criticisms of sociological practitioners' preference toward exaggerated manliness, imperialistic orientation, and intellectual isolationism in their work. Exaggerated manliness is the tendency to concentrate on "big topics, important issues, and ... hard techniques" (p. 316). An example of exaggerated manliness in the fad is the overemphasis on the study of political processes and outcomes to the exclusion of religious experiences, fads, fashions, collective celebrations, joyful occasions, and other presumably less politically important occurrences. Henig (1982, 286) makes reference to the presumably marginal and socially unimportant processes studied by collective behaviorists:

> Nor should it be surprising that some collective-behavior theorists have found the presumption of rationality to be unnecessary or even counter productive, when interpreting group mobilization as manifested in social crazes and fads. The stakes involved in buying a hula hoop, mailing a chain letter, or "getting into" punk rock may simply be too minimal to warrant the careful consideration that rationality implies. (Henig 1982, 226)

Elsewhere (Aguirre and Quarantelli 1983), we criticized this bias favoring explicit political analysis in the selection of topics of investigation.

The second preference is what Lofland refers to as the attempt to discredit other scholars' work as a precondition for the establishment of intellectual domination over the scholarly discourse in the specialty or profession and the framing of questions worth asking (p. 319). Examples of this imperialistic tendency in the fad is the widespread misrepresentation of the contributions of Robert E. Park, Herbert Blumer, and Orrin Klapp as well as the narrowing of the topics being investigated today (see below).

Finally, *intellectual isolationism*, Lofland's term for Pitirim Sorokin's "new Columbus syndrome" (p. 320–21), is also a characteristic of RMA as a fad. This is, in the words of Lofland, "the tendency of those who identify with a particular principality or small kingdom within sociology to be convinced that no one outside their own 'camp' ever had an idea that might be relevant to them." The syndrome in effect ignores the contribution of scholars past and present that are not users of the faddists' language and preconceptions.

Collective behaviorists today find themselves much like members of an endangered species: a strong external force narrows their habitat, ignores their differences, forces them to coexist in uneasy nearness to each other, and attributes the controverted weaknesses of each to all. Examples abound; Calhoun (1982, 210), for example, wrote that

> Berk and Weller and Quarantelli suggest that collective behavior can be characterized by concerted group activity when previous norms or social relationships fail to meet immediate needs. This is very close to Smelser's position as described above. (Calhoun 1982, 210)

Calhoun does not note the striking differences among Berk's rational view of crowd behavior, Weller and Quarantelli's modified emergent norm conception of collective behavior and their catholic understanding of the term as reflected in their famous typology, and Smelser's elaborate explanation of both the constitutive elements of collective behavior and the set of conditions required for different types of collective behavior to occur. Indeed, to equate Smelser's model of collective behavior, based as it is on a shortcircuiting of the normal relationships among the components of social action under the effects of the generalized belief, with a failure of institutionalized systems of norms, is to miss important differences between the two intellectual perspectives. Nevertheless, such grouping exercises are not uncommon nowadays. Of course, they are not a cabal's work, but the outcome of an ascendant, potent school of thought and of the distribution of academic resources (McCarthy 1991). It is generally true that, as McCarthy wrote (personal communication), "(e)mergent communities define themselves against images of the past. Scholarly communities function much like other kinds of communities. These images are always distortions of the past reality."

For quite some time now, the acknowledged wisdom in that part of the specialty oriented to the study of social movements is that there is such a thing as a unitary collective behavior position or approach (cf., Quarantelli and Hundley 1965), that this approach is heavily influenced by G. LeBon and the irrationalists, and that it trivializes politics, defends the status quo, and is politically conservative (for an evaluation of these claims, see Aguirre and Quarantelli 1983). For example, Gamson (1975, 132–33) wrote that the

> collective behavior paradigm, ... rests on a distinction between the politics of social movements and the politics of conventional groups and organizations.... Part of the appeal of the collective behavior paradigm is its serviceability as an intellectual weapon to discredit mass movements of which one is critical. (Gamson 1975, 132–33)

Characteristically, the critics use the term *collective behavior* to construct a distinct, homogeneous set of social science works, citing selectively what is, in fact, a heterogeneous and vast literature published over two centuries (e.g., Rule 1988) to provide contrast and counterpoint to their own, presumably more scientifically valid publications. Note, for example, how the term has been recently employed by six scholars.

> One view of recruitment and commitment in protest movements has been formulated within the crowd theory and collective behavior traditions.... Marginal, insecure, irrational people join these movements because they provide needed social direction unavailable from existing social institutions. (Hirsch 1990, 243–44)

> A spate of works ... has challenged the received wisdom (breakdown theory) that movement participants are a psychologically vulnerable lunatic fringe of malintegrated individuals. (Walsh 1978, 157)

> By the 1960s, collective behavior thinking, and a fortiori the notions of the irrationalists and mass society theorists, had lost their attraction as a lens for viewing civil violence. (Rule 1988, 116)

> The different variants of this (collective behavior) approach all share the basic assumption that individual deprivations, breakdowns of the social order, and homogenizing ideologies are important preconditions for the emergence of social movements. . . . Forming the basis of this noninstitutional politics are the backward, marginal, alienated elements of society. (Mayer 1990, 4)

> My analysis does not fit the thrust of classical collective behavior theory because I have found that rational planning, well-developed preexisting communication networks, and established leaders, institutions, and organizations, rather than psychic strain, spontaneity, and emotionalism, were the driving forces behind the movement. (Morris 1984, 278)

The advancement of this framework of interpretation in the specialty, to the effect that collective behaviorists are irrationalists and mass society theorists, marginalizes the contemporary study of collective behavior in the profession of sociology, for as received wisdom it ignores important differences in conceptual approaches to the sociological study of collective behavior. As a political tool, however, the strategy of claiming a unitary collective behavior approach is impeccable, exploiting as it does the ambivalence among collective behaviorists of different persuasions. An important reason for the relative inattention to the study of collective behavior is that its discreditation goes on unchecked. Collective behaviorists lack a unitary intellectual tradition and common training—what Quarantelli and Weller (1974) termed a *critical mass.* They are divided in their own intellectual presuppositions as to what collective behavior is and cannot mount a concerted and unified counterattack.

More than two decades ago, Carl Couch (1968), together with Ralph Turner and E. L. Quarantelli, among others, synthesized the emerging knowledge in the specialty. He most forcefully pointed out the myths surrounding crowd behavior, its presumed suggestibility, destructiveness, irrationality, emotionality, mental illness, lower-class participation, and spontaneity. For quite a long time now, many scholars who have trained in collective behavior have accepted Couch, Turner, and Quarantelli's brilliant rejection of the LeBonian tradition. We took this advance for granted, and it is particularly galling when some of us learn that after all these years we are LeBonians after all.

Thus, Zald wrote that it took social scientists' involvement in the social movements of the 1960s as planners and strategists to discover that "the irrationalist assumptions of the collective behavior approach seemed outmode" (1988, 8). More recently, James Coleman, in his influential *Foundations of Social Theory* (1990), advanced a similar claim:

> "(O)bservers of collective behavior have a wholly different conception (of persons). They are described as "excitable," "emotional," or "suggestive"; their behavior exhibits "contagion"; they are subject to "hypnotic effects of the crowd." That is, they are irrational, disorderly, unpredictable, and spontaneous. . . . (For) most of the theorists of collective behavior . . . their individual is a kernel of emotion, with no hint of rationality. (Coleman 1990, 197–98)

Hirsh (1990, 244), disregarding Turner's (1964) well-reasoned rejection of the riffraff explanation of collective behavior, added:

> They (Turner and Killian 1987, 32) stress that confused, impulsive people without institutional routines to direct them are the most likely recruits for collective behavior and argue that many individuals participate in collective behavior because they "feel insecure about themselves and their status in society" and receive psychic benefits such as feelings of power and righteousness that are "not directly related to the goals of the collectivity." (Hirsch 1990, 244)

Clearly, the collective behaviorists these scholars have in mind are not the same people whose works I have studied.

Misrepresentations

A manifestation of this attack has been a tendency to discredit scholars' entire life works because of reevaluations of some of their ideas. This is done without considering the complexities inherent in intellectual production or acknowledging that the very essence of scientific knowledge is to build on and progress beyond the work of our predecessors. As examples, I limit my comments to the treatment in the social movement literature of the work of Robert E. Park, Herbert Blumer, and Orrin Klapp.

The treatment of Robert E. Park in the contemporary, social movement literature is quite unsatisfactory. The tendency is to cast him simply as a person strongly influenced by LeBonian ideas (but see Oberschall 1973, chap. 1). Even Rule's careful treatment of Park's ideas continues the practice of juxtaposing rational and irrational explanations of collective behavior and social movement, placing Park and the "collective behavior school" in the irrationalist camp. Rule tells us that in contrast to the rationalists, Park offered explanations of collective behavior centered on emotional impulses and acts that were "consumatory rather than instrumental—not framed in reference to more enduring purposes and goals of the actor" (Rule 1989, 148). Here we see an interpretation of the past primarily as a function of present needs and concerns, telescoping all its complexity into a manageable text for the purposes at hand.

Rule is both right and wrong. He is right to remind us that Park's treatment of the crowd was strongly influenced by LeBon. He is incomplete in his treatment, however, for he does not mention many of the things that separate Park from LeBon. Park's abandonment of LeBon's racialist interpretive scheme, his treatment of the public (in his 1904 dissertation on the crowd and the public, where the latter is framed in terms of reasoned and rational discussion of issues) and sects, his attempt (later extended by Blumer) to systematically identify and order multiple forms of collective behavior, his emphasis on the adaptive functions of collective behavior, and his insistence in linking collective behavior and social movements with more general processes of social change, are very significant contributions that are unmentioned today. Likewise, the place that the study of collective behavior occupies in Park's catholic intellectual landscape goes unheeded, especially his central intellectual concern with the sociological study of human ecology and of competition and cooperation processes in urban settings.

Finally, the linking of Park and his assumed irrationality with the collective behavior school (as if the man and the specialty were the same) gives a false sense of intellectual stagnation. It is as if in describing Park's work one could also accurately portray the work of those who followed him, showing that nothing else of intellectual import had occurred in the so-called school since the founder's death. This, of course, is patently false, both in terms of the vigorous critical evaluation of Park's contributions by others in the specialty (Oberschall, 1973) and the controversies among different theoretical approaches to collective behavior, for example, between emergent norm, dramaturgical, symbolic behavioral, and value-added perspectives (cf., Turner 1964; Quarantelli and Hundley 1965; Weller and Quarantelli 1974; Wood and Jackson 1982; Smelser 1962; Brown and Goldin 1973; Wright 1978; Tierney 1980).

A similar misreading of Herbert Blumer's work is also quite the norm (cf., Zygmunt 1986). It is true that many of the attributes used by Herbert Blumer to conceptualize collective behavior are now rejected; in my view, by and large, this rejection is fully justified. For example, as we were taught by Professor Quarantelli many years ago, Blumer's proposal that circular interaction differentiated some elementary collective behavior from other forms of social organization in which symbolic interaction took place had the undesirable effect of marginalizing the specialty from the mainstream of the profession. However, contrary to what Morris and Herring (1986) wrote, it is not the case that all Chicago-oriented theorists argued for "a special sociology with its own theoretical principles to explain collective behavior." Not even Blumer was consistent on this point. For example, he never used the idea of circular interaction to account for publics, social movements, and fads and fashion. Of course, many scholars have justifiably found his definition of collective behavior contradicted by some of the types of collective behavior he identified, such as the mass and casual and conventional crowds. Others have criticized him for using the crowd as the prototype form of collective behavior (McPhail 1971), and for failing to adopt a social organizational approach to its study.

Even granting some of these criticisms, it is still the case that many of Blumer's other ideas remain worthy contributions. For example, where else in the literature is there a more perceptive and prescient complex and social organizational treatment of fashion (Blumer 1948) and of social problems (Blumer, 1971)? Blumer's conceptualization of social problems as collective behavior shows his full awareness of the impact of political dynamics in the setting of the public agenda. It also recognizes that public opinion takes place in many different institutional settings, from the neighborhood and the voluntary association, church, social club, political parties, and legislations, to the official bureaucracies in charge of enforcing the laws. Blumer made important contributions to our understanding of public opinion. For example, he repeatedly made the point that because of status and power inequalities in society, any survey or public opinion poll that assumed every opinion expressed was numerically equal to another opinion was sociologically incorrect.

Interestingly, there are at least two "Blumers": the early and the late Blumer of social problems as collective behavior, which constitutes one of the most important and powerful conceptualizations of social problems in use in the social sciences (Mauss 1975). It is inaccurate to gloss over these important contributions to say that Blumer is an irrationalist and a LeBonian. True, some of his early writings emphasized emotion and homogeneity of crowd participants. Nevertheless, later on his work was thoroughly infused with a symbolic interactionist perspective that recognized structural features of the society in determining the fate of social problem claims and the nature of industrial relations.

Even Blumer's early work on social movement has value in providing a point of comparison for subsequent scholarship. For example, as acknowledged by Zald and McCarthy (1989), it is easy to see a continuity between their concepts of social movement and social movement organization and Blumer's general and specific social movements (Blumer 1951). Although the later scholars' terms are expressed in terms of more contemporary ideas and reflect their unique and quite worthy theoretical contributions, it is still the case that the empirical references of the two pairs of concepts bear close scrutiny. Comparing the pairs helps us understand the development of knowledge in the specialty. The uniqueness and overlap implied by the pairs are useful for theory construction. The present-day doctrinaire and thoroughgoing rejection of Blumer's work renders our students ignorant of their intellectual past and ill-prepared to construct theories and thus to advance knowledge of these matters.

My third example is Orrin Klapp (1972), who many specialists either totally ignore or consider to be a mass society theorist. As with Park, Blumer, and other scholars similarly identified, in Klapp's case these labels are often carelessly applied and used to discredit their scholarship. Undoubtedly, Klapp's interest was in determining the effects of massive societal transformations on the individual personality and social relations occurring in the aftermath of World War II (Klapp 1978). In this context, he emphasized the ego deficits caused by the emergence and increasing dominance of one-way patterns of mass media communication, and the cybernetic effects of collective behavior and social movements, providing solutions to the strains and new points of equilibrium to social organizations.

Disagreements over his basic outline to collective behavior and social movement should not block an appreciation of his important contributions. Among these was his emphasis on the expressive and symbolic, as opposed to the instrumental side of collective behavior and social movement phenomena. Fundamental among Klapp's contributions was his insistence that we consider collective behavior and social movements from a dramaturgical and processual perspective (Klapp 1956, 1962). He claimed this paradigm would force us to appreciate the negotiated and half-articulated dimensions of roles characteristic of rapid social change. This insight is formally elaborated in Klapp's idea of the social type—the hero, villain, and underdog inhabitants of social worlds in flux (Klapp 1964).

Incorporating Klapp's idea of social type and its emerging role assumption, together with his detailed description of the origins and dynamics of the types, could advance our understanding of social movements. The present-day emphasis on a stable conception of roles, although matching the complex organizational stance common in the contemporary social movement literature in the United States, nevertheless restricts the range of behavior observed in most contemporary social movement research. It restricts our ability to link emergence with social movement action. The current emphasis on stable roles and well-defined complex organizations also blinds us to the importance of the unplanned and uncalculable effects of the right "timing" and "social circumstances" in the making and unmaking of public figures and social causes. Klapp understood the importance of these matters, and his playful awareness of the indeterminacy of historical events and social interactions—witness his skillful depiction of dramatic encounters—is one of the charming features of his work.

Payoffs

We need to move against present-day dogma to a renewed appreciation of the strengths and weaknesses of the contributions of scholars in the specialty area of collective behavior and social movements. We should not disregard the work of seminal thinkers merely because some aspects of their work are now considered unsound. In some important ways, recent continental scholarship in social movements bears some interesting resemblances to their ideas. For example, Touraine's (1985) emphasis on societal dynamics rather than statics and on the link between social movement and the creation of new social orders and struggles for the control of culture, as well as his insistence in couching his questions in terms of broad historical and national levels of analysis, are in some ways reminiscent of Park's own approaches and concerns.

There are important differences between Park and Touraine, such as Touraine's insistence that social movements are not a type of collective behavior. Nevertheless, the similarities between the two scholars are apparent, and should be studied at some depth to gain theoretical understanding. Indeed, Hannigan (1985) showed that important similarities exist between the traditional collective behavior school founded by Park and Alain Touraine and Manuel Castells' conceptualization of social movements, in terms of their assumptions regarding the historical context of social movements, their defining characteristics, the reasons for their formation, the basis for collective action, and emphasis on organization, sources of support, and other factors impacting on their success.

Or to give another example, Klapp's work should be useful in understanding and evaluating the work of Melucci (1985) and other scholars in what Cohen (1985) termed the identity-oriented school in the study of social movements, with its emphasis on the culture generated by the social movement and its value to participants. Klapp's emphasis on ego transformations and its collective behavior features would complement current scholarship. Here as elsewhere, understanding the history of ideas gives us a better perspective of the present and a better chance of improving our research and theorizing.

The concentration in modern RMA analyses on organizational features—on the social movement organization rather than on the social movement—comes at a price. Emphasis on structure rather than content narrows both the kinds of questions asked and the ways questions are asked. On the first point, witness, for example, the lack of research on the relative power of different collective memories, to use Shuman and Scott's term (1989; see also Saltz 1986), in mobilizing people for collective behavior and social movement participation. On the second, witness the article by Snow, Rochford, Worden, and Benford (1986) on frame alignment processes, an excellent paper that corrects important deficiencies in the literature. However, it reinforces the separation between people as potential collective actors and the social movement organizations attempting to mobilize them.

Frame alignment processes, it is argued, are put in place by social movement activists, leaders, or other officials or representatives. Alignments refer to the "linkage of individual and social movement organization interpretative orientations." In reality, however, the emphasis in the paper is from the organization to the individual, not the other way around. We are never told how the potential audience reacts to the marketing pitch of social movement organizations, and how this reaction affects the interpretative process. Analytically, it is a one-way approach very different from Snow's earlier use of Goffman's dramatistic perspective to make sense of victory celebrations (Snow et al. 1981).

To their credit, Snow and his colleagues recognized that people's willingness to participate changes over time and is not entirely explained by a rational calculus. They also conclude that future research must study "not only the nature and interpretive work and resources of (social movement organizations), but also . . . the degree of fit between the resultant framing or products of that work and the life situation and ideology of potential constituents." I would add that the collective behavior experiences of people are very important determinants of their receptiveness to movement frames.

Today there are many more studies of social movements concerned with strategies, recruitment, mobilization, selective incentives, and leaders rather than with the interface between social movement organization and social movement or with the feedback among episodes of street action, the occurrence of other instances of collective behavior, and social movement organizations. It is time to devote greater attention to the latter topics of investigation. Zald (1988) alluded to this research need when he acknowledged the lack of research on meanings and social movements, or the creation of collective symbols, metaphors, and concepts and their use by social movement entrepreneurs, and on "demonstration effects," or the fact that an incident often produces other incidents of its type. Of course, as Gary Marx (1979) showed, both of these topics of investigation are among the topics of central interest to collective behaviorists. Advancing our understanding of social movement dynamics is very much tied to the resurgence of the study of collective behavior. Future theoretical work in the specialty must not only become international and cross-cultural but also analytically catholic.

Similarities

The contemporary study of collective behavior is very closely tied to the study of social movement. Zald (1988) told us that the distinctive features of RMA are: the assumption that behavior entails costs and is thus rational "at some level;" that people may participate depending on their calculations of benefits and costs; that resources relevant to the task of mobilization may come from many sources, not just the aggrieved population; that sustained collective action requires organization; that costs are affected by powerful outsider groups and institutions; and that social movement success is problematic to determine and elusive to measure. In my view, roughly the same assumptions are made by most collective behaviorists working today.

Collective behavior research is couched at different levels of analysis than the level used by resource mobilization approach scholars in studying social movements. Thus, specific studies may find only some of the assumptions outlined by Zald relevant. Moreover, it is true that suboptimizing rationality, rather than a strong version of the rationality assumption, as well as a concern with individual values and perspectives, has characterized most contemporary collective behavior research and theorizing. Nevertheless, it is not difficult to show that similar assumptions about behavior are used by scholars in the two groups. For instance, Norris Johnson (1987) and his colleagues examined panic, the most extreme form of collective behavior, as rational collective behavior, in which existing ties of friendship, kinship, and occupation help explain the helping behavior of people at a supper club fire. The work of Jerry Lewis on sport violence (Lewis 1989 and in this volume) emphasizes the community and social class divisions operating among fans and the ecological structural features that encourage violence. Aguirre and his colleagues (1984, 1988) showed that collective behavior in Cuba is an elaborate social product, as well as that the occurrence, crystallization, and spread of the streaking fad across college and university campuses required planning, organization, as well as a number of structural preconditions. Most research analyzing the emergence of groups in disasters (Drabek, 1986) uses collective behavior frameworks and rational explanations of disaster-related behavior. McPhail and Miller's (1973) analysis of the assembling process emphasized the impact of communication processes and ecological features. Implicit in this model of the assembling process is the notion that people are rational actors engaged in a complex matrix of social interaction that produces their behavioral configuration and their propensity to participate in collective behavior events as well as the idea that people often act on the basis of incomplete information.

The similarity in assumptions between the resource mobilization approach and McPhail's work are particularly strong (for a review of the model see Miller 1985). His model of collective behavior makes no distinction between collective and noncollective forms of social behavior. Moreover, the model is grounded in symbolic and behavioral considerations, centered on the one hand on instructional stimuli that impact on people's participation in collective behavior and social movement action, and on the other, on an unusual attention to the actual patterns of behavior in common that occur in instances of collective behavior or action. Recently (1988), McPhail used Goffman's concept of the *gathering* as a bridge to systematically link much work at different levels and units of analysis, ranging from single gatherings to waves and cycles of gatherings.

John Lofland's explicit emphasis on collective emotion (1985), which he used to identify forms of collective behavior, does not link irrationality to emotion as an explanatory aspect of collective behavior. Instead, Lofland orders the vast amount of historical material available to him using the elementary forms of collective behavior formed by combining the dominant emotions of fear, hostility, and joy and the two organizational forms of the crowd and the mass.

Nor is the deemphasis on irrationality as an approach to research new among collective behaviorists. Tamotsu Shibutani's (1966) classic functional model of rumor is an attempt to explain rumor, a form of collective behavior, in sociological rather than psychological terms. Shibutani's rumor is not an emotional outlet or a defense mechanism used by troubled individuals. Instead, rumor is improvised news, providing information people need and are not getting from the mass media. Work on rumor has identified technical difficulties with Shibutani's model, such as its inability to account for the recurrent content of many types of rumor. Nevertheless, the model, which is not irrationally grounded, is an important contribution to the collective behavior literature and makes some of the same assumptions made by RMA scholars.

More than thirty years ago Gusfield (1955) traced the organizational dynamics of the Temperance Union and its movement from cultural center to periphery in the United States, explaining its decline not as an irrational collective act but rather as a matter of conscience and ideology; Turner and Surace (1956) studied the acting crowd during the zoot-suit riots in 1943 in Los Angeles. They did not assume the presence of an irrational mob, but instead posited the importance of preexisting culture and social organizational cleavages, hypothesizing that the "necessary condition for both uniform group action and unrestricted hostile (crowd) behavior is the presence of a symbol that arouses uniformly and exclusively unfavorable feelings in the group toward the object under attack"; Jackman (1957) related the frequency and form of protests in Japanese relocation camps to differences in the complex organizational features of the camps and in the emerging definitions of the situation in specific incidents. A few years later Wimberley and his associates (1975) described conversion during a religious crusade as a carefully planned and organizationally produced ritualized experience rather than a spontaneous event. Perhaps these cases are enough to provide support for my argument that LeBonian irrationality has been a small sideshow in contemporary collective behavior scholarship.

Conclusion

The separation of the study of collective behavior from social movements limits progress in our knowledge of social movements. This fad has not only devalued the study of collective behavior but also it has created important difficulties for social movement theory (for critical commentaries see Turner 1981; Jenkins 1983; Klandermans 1984). The repudiation of collective behavior has meant that the useful correctives inherent in collective behavior research have not had a full opportunity to influence social movement scholarship. In turn, this has brought about lamentable results: an impoverished sociopsychological foundation; a downplaying of the importance of public opinion, values, attitudes, grievances, spontaneity, ideology; an overemphasis on rationality and stable movement goals; an overreliance on unidirectional causal effects from the social movement leaders to the followers; a concentration on a limited scope of social movement effects. The resolution of these issues will depend in part on the eventual incorporation of insights generated by the collective behavior tradition.

The separation also unnecessarily devalues the intellectual currency of both areas of study. For example, Fireman and his colleagues' study of small groups' encounters with unjust authority (Fireman et al. 1979) is characteristically viewed as a study in the resource mobilization tradition. Nevertheless, it can be argued that their research is more relevant for understanding aspects of crowd behavior than for charting movement tactics. Although they give us many important insights on the process of small group mobilization, their experiment ends at the time that social movement organization becomes a possibility. Of course, Gamson's early book on organizational strategy is more convincingly couched at the social movement level, although here as well what he has to say about violence, organization, and group claims is of great interest to collective behaviorists. To insist in the separation of the two fields could stifle the intellectual growth of the specialty.

Fortunately, however, recent writings by Oliver (1989), Rosenthal and Schwartz (1989), and Rule (1988) recognized the empirical unity of social movement and collective behavior and the need for a holistic understanding of the phenomena. Rule's (1989) recent treatment of the rationality debate is an excellent example of this new writing. His conclusions that rationality is not enough to account for the central questions of mobilization and that "less rational" processes must be part of explanations and modeled in empirical research, are reminiscent of what Turner, Killian, and Quarantelli, among others in the "collective behavior camp," have been arguing for several years in their rejection of the presumed conflict between emotion and rationality and illustrate the emerging consensus. Unfortunately, however, present in the work of Rule and other social movement scholars is a persistent emphasis on political forms of social action or behavior (the two terms are substantively indistinct in my view) to the relative exclusion of the study of a whole array of social organizational forms traditionally considered central to the specialty. But the important point to remember is that theoretical work in the specialty is very quickly superseding the old divisions and categorizations.

New vocabularies and broader conceptual imagery are resolving the difficulties arising from treating collective behaviors and social movements as categorical units of analysis. Examples of this emergent theoretical emphasis is Lofland and Johnson's (1990) overarching concept of citizen surges, McPhail's (1988) gatherings, events, campaigns, waves, and trends, and Marwell and Oliver's (1984) collective campaigns. This emerging metaframework is still in flux, but augurs well for the future of the specialty, especially if its developing emphasis, as Bates and Peacock (1989) recommend for the general case, becomes the structural rather than the categorical modeling of social structure.

Are Social Movements Irrational or Are They Collective Behavior?

LEWIS M. KILLIAN

This venture into the sociology of sociology might be titled "The Strange and Perilous Career of Collective Behavior," for the history of this field of sociology has been marked by confusion, inconsistency, and self-doubt. LeBon, the man most often named as founder, was not a sociologist and never used the term *collective behavior*. Park (1921), the sociologist who introduced the concept and argued the need for the field, declared that the rubric was arbitrarily limited in his usage because it could properly include "all the facts of group life." The first book titled *Collective Behavior* did indeed deal with all forms of social interaction; the book and its author, Richard LaPiere (1938), are hardly ever mentioned in histories of the field. In recent years, this area has been threatened with extinction, perhaps oblivion. Some of its critics imply that it was never legitimate, but rather a misguided approach best forgotten. In particular, *collective behaviorists* (a term used here reluctantly) have been invited to keep their hot, emotional hands off the serious topic of social movements and to go study exotica such as skateboarding, streaking, and revelrous crowds of students. Some critics would amputate the field of social movements from collective behavior even though it has been conventionally an integral member since Park founded not only the specialty but also the "Chicago school" that, for many years, was almost coterminous with it. In their "critical mass" article, Quarantelli and Weller (1974) reported that a plurality of specialists who had done research in collective behavior or included it in texts they had written had either graduated from Chicago or studied under graduates.

Park's intellectual offspring was a frail child, however. Collective behaviorists remained a small community within the discipline of sociology for the next thirty-five years; their topic was peripheral to mainstream sociology, which focused on social stability and left the study of social change to cultural determinists such as Ogburn. Moreover, this community came to be characterized by self-criticism and lamentations over its lack of maturity. In 1947, Anselm Strauss, a student of Herbert Blumer, wrote, "Anyone who attempts to teach a course in collective behavior finds abundant descriptive and illustrative material for teacher purposes, but also finds a scarcity of generalizations dealing with each collective behavior phenomenon" (1947, 352). In 1957, Blumer himself declared that collective behavior "is clearly in need of systematic conceptual analysis" (1957, 128). In 1963, Neil Smelser prefaced his foray into the area by announcing that no coherent theory of collective behavior existed.

Yet the period between 1957 and 1970 was one of greatly heightened activity. In 1957, the first textbook since LaPiere's appeared: *Collective Behavior* by Turner and Killian. Four years later, two more members of the Chicago school, Kurt and Gladys Lang, published their *Collective Dynamics* (1961). Neil Smelser, a structural-functionalist with no roots in either the collective behaviorist or the symbolic interactionist traditions of Chicago sociology, brought forth *Theory of Collective Behavior* (1963). Although the two earlier texts were not judged to contain any unifying theoretical schemes, many readers felt that Smelser's *Theory* did. Yet as late as 1973, Roy Manning, reviewing the 1972 edition of Turner and Killian, said of collective behavior, "A dominant theoretical paradigm has not been achieved" (1973, 285).

The 1960s saw not only an increase in research and theoretical debate but also the beginnings of criticisms of collective behaviorists. In 1968, Carl Couch, an insider to the field, wrote his thoughtful critique, "Collective Behavior: An Examination of Some Stereotypes." In 1970 came the blast at the "irrationalist" bias of collective behaviorists by Currie and Skolnick, who obviously considered themselves outsiders—students of social protest, not of collective behavior (1970). By an amazing intellectual coup, they swept virtually all the previous theoretical fragments in the literature, including the propositions advance by Turner, Killian, and the Langs, into Smelser's bag, denouncing the whole mixed lot as extensions of LeBon's "irrationalism." In effect, a new conception of collective behavior was created by its critics. It was in the image of the pathological approach that modern members of the Chicago school had much earlier identified and rejected.

The theme that collective behavior had been immature, lacking any central theoretical scheme, was soon miraculously replaced by an illusion of tight, dogmatic, and destructive coherence. In 1975, in *The Strategy of Social Protest*, William Gamson referred to "the collective behavior paradigm." He declared, furthermore, "The study of social protest has only recently emerged from the straightjacket of collective behavior" (1975, 30). But who were now said to be the exponents of the paradigm? LeBon and Smelser were, of course, but so were William Kornhauser, Theodore Adorno, Lewis Feuer, and—get this—Eric Hoffer! As the Resource Mobilization perspective or perspectives (or, if we must sound scientific, "paradigms") crystallized, the collective behavior paradigm or "classical collective behavior theory" was shaped into a contrasting conception. Whereas Perrow (1979) neatly distinguished between two variants of Resource Mobilization theory, RM I and RM II, no "CB I" (the Chicago school) and "CB II" (Smelser) were acknowledged by adherents to either RM perspective. Turner and Killian, Lang and Lang, and Smelser were lumped together as the principal proponents of what was now represented as a highly coherent irrationalist paradigm. They were treated as the joint heirs of LeBon's blatant ideological biases and his erroneous irrationalist theories, transmitted, apparently without modification or criticism, through Park and Blumer. In *Theories of Civil Violence* (1988), James B. Rule brackets collective behavior theory with "the irrationalists' doctrine" although noting some differences between them.

Rarely mentioned in characterizations of the Chicago school is the work of Joseph Gusfield, author of *Symbolic Crusade* (1963) and editor of *Protest, Reform and Revolt: A Reader in Social Movements*, published in 1970. Gusfield did distinguish social movements from elementary collective behavior and, at times, from the field of collective behavior, introducing the term *collective action*. Yet he did not scorn his roots in the Chicago school, saying of Blumer, his teacher, "His emphasis on processes of development and the rise of a new social order out of collective behavior has remained a major source of my understanding of collective action" (1970, vii). Gusfield did not seek to separate the study of collective behavior and the analysis of social movements, but rather sought to integrate them, stating, "By using the term 'collective action' we emphasize that our concern with collective behavior is its implications for social movements and social change" (1970, 2).

In 1973 two comprehensive theoretical treatments of social movements appeared, John Wilson's *Introduction to Social Movements* and Anthony Oberschall's *Social Conflict and Social Movements*. Neither sociologist is a member of the Chicago School, and Oberschall is almost always mentioned as one of the founders, if not *the* founder, of Resource Mobilization theory. Neither of these two theorists seem to have difficulty in accepting the contributions of collective behavior theory, particularly CB I to their approaches.

Wilson said, "Social movements have traditionally been included in that subfield of sociology known as collective behavior" (1973, 7). He then referred to the "standard trinity," Turner and Killian, the Langs, and Smelser, but his subsequent statement distinguishing collective behavior from everyday behavior incorporated what is clearly an emergent norm perspective. Smelser's central notion that collective behavior is action mobilized under a generalized belief is nowhere in evidence.

Oberschall (1973) wrote at greater length about "the Chicago school and Collective Behavior." He quite clearly distinguished between this school and the earlier French school and, particularly, mass society theory that is so often cited as evidence of the sins of the classical collective behaviorists. In his first chapter, after a critical analysis of the pathological approach advanced by Hippolyte Taine, LeBon, Trotter, Freud, and to some extent Tarde, he portrayed the Chicago school, beginning with Park, as achieving a desirable integration of "the public opinion and collective behavior approaches" (1973, 14). He wrote, "Although Park accepted the accuracy of LeBon's mistaken description of crowd behavior..., he was also mindful of Gabriel Tarde's conception of the public and the tradition of rational public opinion formation kept alive among British writers.... More important, for Park collective behavior was an integral part of the normal operation of social change." Then Oberschall added, "Fortunately, these basic truths were never forgotten in the best subsequent scholarly treatment of collective behavior" (1973, 14).

Later, calling Ralph Turner "the foremost American theorist of collective behavior" (1973, 15) he wrote of him, "He stresses the continuities between normal, institutionalized process and collective behavior" (1973, 15) and added, "This is the program of my book as well" (1973, 16).

Yet since Oberschall wrote these words numerous sociologists writing about social movements have charged that the "classical theory" of collective behavior has been flawed by failure to recognize the continuities between preexisting institutions and norms and collective behavior; by emphasis on the irrationality of social movement participants; by reliance on strain, discontent, grievances, and psychological discomfort as the causes of social movements; and by postulating spontaneity and emergence as characteristic of social movements along with planning and continuity. Sometimes the epithet used to condemn the theory is "social psychological," as if every theory of group behavior does not rest on some concept of human nature and individual action, explicit or implicit.

The broadest, most damning indictment remains, however, that collective behaviorists subscribe to a theory of irrationality and emotionality, whereas their critics have shown that movements are rational, based on decisions by "actors rationally estimating their chances of success by using social movements to attain desired goals" (Wood and Jackson 1982, 36). How diverse the perceptions of sociologists themselves can be is seen in the fact that Wood and Jackson regard rational-calculation theories as "psychological theories emphasizing rationality," classifying them with psychological theories emphasizing irrationality, not with social-structural theories or with social psychological theories.

It is perplexing, to say the least, that Turner and Killian, as members of the trinity of classical collective behaviorists, should be regarded as "irrationalists" in view of their early and repeated attempt to show that the issues of irrationality and emotionality are pseudo-issues. How much more clearly could they have stated it than they did in 1957: "Another recurring problem in collective behavior is the tendency to single out for study only those collective phenomena of which the observer disapproves and to depict the processes of collective behavior in value-laden terms. . . . On this basis, collective behavior is often erroneously contrasted to 'rational' behavior by being designated 'irrational' or 'emotional'" (1957, 16). They then went on to show that whether the external criterion of efficiency in achieving a goal or the internal criterion of the weighing of all possible alternatives by the individual actor is used, we cannot distinguish institutional behavior from collective behavior. By the first criterion "much institutional behavior is irrational and much collective behavior is rational" (1957, 17). By the second, "most institutional behavior is irrational, since social norms narrow the range of alternatives which the individual can effectively consider" (1957, 17). I do not find any disagreement between this position and that stated by Michael Schwartz, "People who join protest organizations are at least as rational as those who study them," and "Despite this rationality, individuals and groups of individuals can adopt and act upon totally erroneous and counterproductive theories and ideas" (1976, 135). Whatever "rational" may mean, to act rationally does not mean to act "correctly"—there seems to be agreement on this. If there is disagreement, it concerns the sources of "error" in a cognitive process that includes both rational and nonrational components.

Oberschall, of course, saw rationality as central to collective behavior, writing, "The rational component in social conflict and collective behavior is present much in the same way as it is present in other choice and decision-making situations in everyday life" (1973, 25). But he added, "There are emotional and nonrational components in collective behavior, but the same can be said for everyday behavior." He says of both the agents of social control and their opponents, "they respond to concrete problems and choices in complex ways, with a mixture of outrage, anger, puzzlement, and shrewd, informed calculation." Then, he said, "The information and conceptions they possess can be found out by means of empirical study and can then be fed back into the model of man, the decision-maker, whose main features are rationality with respect to means and ends and selfishness with respect to motives" (1973, 25).

This model is, however, derived from economics and the Scottish moralists and is not the model of what Ralf Dahrendorf (1973) called "homo sociologicus." Randall Collins, after noting that we humans pride ourselves that "pretty much everything we do is based on rational thought processes," declared, "Against all this common sense belief in rationality, however, sociology stands out as a dissenter. One of the central discoveries of sociology is that rationality is limited and appears only under certain conditions. More than that: society itself is ultimately based not upon reasoning or rational agreement but upon a nonrational foundation" (1982, 3–4). In *A Sociology of Sociology*, Robert Friedrichs showed that even the logic of science is not purely rational—he spoke of "the irrational roots of the rational" (1970, 151). He said of sociology, "The activities of sociologists as scientists . . . are always and in principle nested in a larger frame that includes the intrasubjective, the unique, and the existential" (1970, 299).

Although some sociologists may freely choose, subjectively, to base their theory and research on a strong, undiluted rationality assumption, this does not qualify them as umpires on the game-field of social movements with the right to thumb collective behaviorists off the field. Data and theory about social movements have been enriched by the work of resource mobilization theorists of whatever variety, organizational theorists (because an essential characteristic of a social movement is organization), the sociologists of religion, political sociologists, anthropologists such as Gerlach and Hine (1970), and even psychologists. To use the term that I would like to discard, I still will ask, "Is it *rational* or fair to define the collective behavior perspective as solely misleading, unproductive, and even destructive?"

I return now to my thesis that the Chicago school, "Collective Behavior I," has never treated collective behavior as purely or primarily irrational; it has insisted instead that it is purposive, resting on a cognitive and not an affective base. Indeed, Quarantelli has observed privately that collective behaviorists' analysis of panic became so highly cognitive in emphasis that we seemed in danger of forgetting that people do get scared. John Lofland's recent effort (1981) to classify forms of elementary collective behavior on the basis of the dominant emotion appears to be a reaction to what may have become the too rational assumptions of the Chicago school.

Although the pre–1975 diagnosis that the field lacked a coherent theory was hardly flattering to the work of earlier collective behaviorists, it is still closer to the truth than the charge that any particular paradigm has ever dominated it. Diversity, not purity, has characterized theory of collective behavior. This diversity has been enhanced in the case of social movement theory because this area has never been exclusively or even primarily the domain of sociologists. Historians, political scientists, anthropologists, psychologists, and journalists have contributed data as grist for the mill of theorists.

This disclaimer notwithstanding, I boldly offer some generalizations about "Collective Behavior I." Collective behaviorists, in contrast to theorists and polemicists with pathological or psychodynamic perspectives, have never defined collective behavior in any of its forms as being fundamentally and essentially irrational. They have defined it, instead, as being behavior in which people jointly create new norms, new structures, or as Blumer put it, a new social order. The field has been of interest chiefly as a major approach to understanding social change, not social disorganization or individual pathology. Another common element of the perspective is the proposition that collective behavior is extra-institutional—not unrelated to previously existing structures and norms but transcending, opposing or modifying them and in so doing generating new forms.

The collective behavior approach, from Park and Blumer through the current work of such scholars as the late Louis Zurcher, David Snow, Norris Johnson, and John Lofland, has led to a broad perspective on social movements. It has included in its purview political and religious movements, movements with strategies of societal manipulation or personal transformation, movements arising in different societies and widely separated in time. Although treating the crowd as an elementary form worthy of analysis in its own right, the relationship of crowds to social movements has been considered even more important, as Gusfield pointed out.

In the application of collective behavior to social movements, there has been a primary emphasis on ideology, not as false consciousness but as beliefs sincerely held no matter how irrational or inaccurate they may appear to others, and motivating most but not necessarily all of the members. Nonrational elements have also been included for analysis because of the central assumptions that the membership is heterogeneous and the motives of individuals mixed. If anything, the cognitive elements have been treated too much as self-evident so that their modification by nonrational influences has loomed larger than was intended.

With regard to ideology, collective behaviorists have never assumed that either the participants in social movements or the social scientists who study them could construct objectively "correct" or "accurate" definitions of reality, whether in regard to the sources of grievances and the sense of injustice, the solutions to social problems, or the success or failure of a movement. Implicitly or explicitly, the approach has been phenomenological.

Collective behaviorists have also assumed structure to be an intrinsic feature of social movements and even of crowds. What has concerned them is not the absence of structure, but the difference between the structure of collectivities and that of established groups. In the study of social movements, instrumental actions and strategic planning have been accorded importance, but spontaneity and expressive behavior have also been accepted as having their place, just as they do in institutionalized behavior.

Cognition, affect, heterogeneity, and interaction have been key concepts in this perspective. Collective behavior has been viewed as consisting of purposive, self-conscious, cognitive activity in constructing a revised version of reality when established norms and structures cease to afford acceptable directions or supply satisfying lines of action. It has been recognized that although participants, being heterogeneous, may include coldly calculating professionals, adventurers, or agents provocateur, the more typical members experience an emotional involvement that is an important part of motivation: anxiety, fear, anger, indignation, grief, or elation. Finally, in the interaction in collectivities, members display more than ordinary responsiveness to cues from other actors as they engage jointly in the construction of a definition of the situation. Furthermore, membership itself is not institutionalized or formally defined but is highly dependent on how well the actor appears to conform to the emergent definition of the situation constructed by the collectivity.

Such an approach is sociological as well as social psychological. It assumes the necessity of analysis on both the macrosociological and microsociological levels, just as it assumes the importance of both cognitive and affective elements in human behavior. If anyone wishes to define this approach as "irrational" that is their own definition of this aspect of reality.

There are approaches so one-sided as to explain social movements as the product of impulses to action activated by what Walter Coutu (1949) used to call "little men" inside individuals—instincts, unconscious motives or psychological states, or conditioned responses. These theories minimize the cognitive element of action; and may be called "irrationalist." To locate the "little men" outside the actors in the social structure, as did Smelser in his reliance on "structural strain," or in the level of resources available for mobilization, as postulated in some Resource Mobilization theories, also suggests a type of irrationality, minimizing the cognitive element of action in another way. This is reminiscent of the cultural determinism represented in William F. Ogburn's theory that inventions emerged because, through cultural accumulation, the cultural base was "ripe" for them. The cognitive activity of the inventor was of little importance; the inventor was compared to the fly on the wheel of the chariot who thought he made the wheel go round. In reaction to this viewpoint I wrote, in 1964, "men and groups have so often been regarded by sociologists as the creatures rather than the creators of social change" (1964, 426). In the same essay, writing as a collective behaviorist—maybe by that time already "classical"—I said of our perspective, "attention is focused on the interaction of conscious, striving human beings as part of an emergent collectively which is the social movement" (1964, 427). Today, the same criticism may be made of theories that emphasize structure so much they become deterministic. A collective behaviorist perspective is still justified, not as the only way to study social movements, but as a respectable and worthy alternative and complement to other approaches.

An Elite-Sustained Movement: Women's Rights in the Post–World War II Decades

VERTA TAYLOR

The standard historical account of the American women's movement holds that between the decline of the suffrage movement in the 1920s and the resurgence of the contemporary women's movement in the 1960s, there was virtually no organized activity on behalf of women's rights. This chapter challenges that view by focusing on women's rights activities from 1945 to the 1960s, a period known for its antifeminism. The end of World War II ushered in strong antifeminist sentiment that preached domesticity, motherhood, and suburbia, or conformity to what Betty Friedan (1963) labeled *the feminine mystique*. Feminist protest, associated with nineteenth-century suffragists, had become "as quaint as linen dusters and highbutton shoes" (Coughlan 1956). There were, nevertheless, organizations that survived and continued to promote women's rights between the suffrage victory and the rise of the contemporary movement; what really happened to change the nature of the feminist challenge was that the mass base of the women's movement dissipated.

Based on this case study, I introduce the concept of an "elite-sustained movement" to characterize a kind of movement that functioned, in the absence of mass indigenous support and substantial external support, by drawing organizational strength primarily from the resources provided by elites from within the ranks of the movement. By examining a social movement in a stage of continuance and survival, this analysis extends existing theoretical formulations that depict either success and institutionalization or failure and decline as the only possible outcomes of social movements. The analysis also demonstrates the ways that elite-based movement infrastructures and moderate tactics promote the continuation and survival of a movement in a hostile environment, but also result in the dissipation of a movement's mass base. Finally, the chapter documents the impact that the women's rights movement of the postwar period had on the goals, composition, and strategies of the contemporary women's movement, thus providing evidence of ways that movements in stages of continuance and survival can affect later rounds of activism by the same challenging group.

Theoretical Context and Propositions

Researchers have generally been more interested in the beginnings of social movements than in movement survival and continuation. Nevertheless, because of the dynamic nature of movements, a long conceptual tradition exists in the social movement literature for analyzing the course and outcome of movements. Early work influenced by classical collective behavior theory assumed that social movements inevitably pass through a standard "life cycle" from their emergence to their decline either as a result of institutionalization and bureaucratization or failure and collapse (Edwards 1927; Hopper 1950; Blumer 1951; Lang and Lang 1961). In recent years, resource mobilization theorists have challenged this monolithic view of social movement development (Zald and Ash 1966; Oberschall 1978; Perrow 1979; Zald and McCarthy 1979; Jenkins 1983; Miller 1983). They argue that the course of any social movement is a product of an interplay between factors internal to the movement—such as its organization, base of support, leadership, and type of goals—and factors in the outside world that support or constrain its aims and actions. Researchers have applied the concepts of resource mobilization theory to analyze the organizational and tactical characteristics of the contemporary American civil rights movement in different stages of development, but little comparable work has been done on other recent American social movements (McAdam 1982, 1983; Morris 1981, 1984; Jenkins and Eckert 1986). This chapter follows in the tradition begun by scholars of the civil rights movement by applying resource mobilization concepts to analyze the women's rights movement from 1945 to the mid-1960s. I argue that these years mark a period of continuance and survival in the history of the American women's movement, not a total decline of women's rights activity, as most accounts of the period hold (O'Neill 1971; Flexner 1975). The analysis is guided by four propositions.

Proposition One

A social movement will take different organizational forms and adopt different strategies depending on whether it is in a stage of formation, mass mobilization, continuance and survival, or decline. This proposition directs our attention to the evolution of social movement organization and action and to the ways these aspects affect and are affected by general processes of social change. A central tenet of resource mobilization theorists is that, as a movement evolves, various organizational forms and tactical choices are possible, depending on the way in which a movement's internal features interact with the support and opposition a movement encounters from the social environment (Zald and Ash 1966). This proposition has received support from a number of empirical studies (Bromley and Shupe 1979, 1983; Lofland 1977, 1979; Morris 1981; McAdam 1983, 1984; Jenkins and Eckert 1986; Whittier 1991).

Proposition Two

In a social climate hostile to its aims, the survival and continuation of a social movement depends on its internal organizational strength and ability to reduce the intensity of the threat posed by the movement. Researchers, for the most part, have not addressed the question of how a social movement challenge is sustained but have assumed that the same factors that give rise to a movement are also linked to movement persistence. McAdam (1982) emphasized two factors set in motion after a movement emerges that are necessary to sustain and continue any insurgent challenge. First, a movement must develop some type of formal organizational structure capable of maintaining an adequate flow of resources. This assumption is central to resource mobilization theory and has received substantial empirical support (Oberschall 1973; Gamson 1975; C. Tilly, L. Tilly, and R. Tilly 1975; Morris 1981; Tierney 1982; Staggenborg 1988, 1989). The second dilemma that a movement must resolve is the response of other groups that oppose or seek to control its actions. According to McAdam (1982), the nature of the goals and tactics embraced by a movement are of crucial importance, because together they largely determine the degree of threat/opportunity posed by the movement.

Social movement scholars have not applied these factors specifically to analyze movements attempting to mobilize in political environments that are not receptive to pressure from the challenging group. This proposition assumes, however, that in unfavorable political climates that offer few sources of external support, a social movement will rely on those internal resources over which it has the greatest measure of control, specifically its organizational structure and its goals and tactics.

Proposition Three

Elite-based infrastructures and institutionalized tactics promote movement survival but also contribute to the loss of a movement's ties to its indigenous base. A central tenet of the McCarthy and Zald formulation of resource mobilization theory is the significance of outside elites and other types of external support for mobilization (1973, 1975, 1977). Although some researchers have confirmed that elite support is essential for social movement growth and success (Lipsky 1971; Carden 1977; Gelb and Palley 1982), others have found that elite patronage just as frequently contributes to the demise of a movement by channeling it into moderate goals and institutionalized actions and transforming it into formal organizations that depend too heavily on ties to external sponsors rather than on indigenous support (Piven and Cloward 1977; Helfgot 1981; McAdam 1982; Haines 1984; Marger 1984; Hoffman 1989). The debate over the consequences of elite patronage seems thus far to have placed undue emphasis, however, on the role of outside elites as compared with elites from within the indigenous group.

Resource mobilization theorists and others discuss the fact that not only leaders but also activists in a social movement are likely to be drawn from certain privileged groups and social strata because of the risks that participation entails and the resources, especially discretionary time and money, required for participation (Lenin 1970; Oberschall 1973; McCarthy and Zald 1973; Pinard and Hamilton 1988). This proposition suggests, then, that elite involvement may prove beneficial for movements at certain points in their history. Elite-based movements can provide an environment in which already-committed activists can continue to pursue their goals when the larger social and political context provides little support for their cause. But this proposition also makes explicit the fact that an elite movement infrastructure that relies primarily on institutionalized tactics can contribute further to the demise of the mass or indignenous base of a movement.

Proposition Four

The organization and tactics of a social movement in a stage of continuance and survival can have consequences for later rounds of mass mobilization of the challenging group. With a few exceptions (Mueller 1987), social movement consequences most often have been analyzed in terms either of success or failure in the achievement of the movement's stated goals, or in terms of patterns of decline that result from the inherent instability and transitory nature of movements (Gamson 1975; Freeman 1983; Miller 1983). Based on the data presented here, I argue that a more fruitful way of analyzing social movements, particularly those based on fundamental and persistent social cleavages, is as entities that persist through time, mobilizing resources with which they pursue strategies that have consequences not only in their own time but also that affect later stages of mobilization and outcomes (Taylor 1989). This proposition addresses the question of the historical continuity of the women's movement by maintaining that the organization and actions of the elite-sustained movement of the 1940s and 50s shaped the nature and direction of the revitalized movement that took up the fight for gender equality in the late 1960s.

Data

The data for this study come from documentary material in public and private archival collections and 57 interviews with individuals involved in women's rights activities from 1945 to the 1960s. Fuller description of the movement and complete documentation are available in Rupp and Taylor (1987).

1. Archival data included the papers of the National Woman's Party and the League of Women Voters, the two major factions of the suffrage movement, and the papers of the President's Commission on the Status of Women (1961–163), a group whose activities facilitated the resurgence of the women's movement in the mid-1960s. Other materials examined were unofficial and official organizational documents, publications, personal letters, and memos available in public and private archival collections, most of which were housed at the Schlesinger Library at Radcliffe College or the Library of Congress. The papers of individual women were an important source of information, not only about feminists' organizational careers, but also about the activities of diverse women's organizations.
2. The second source of data was open-ended, semistructured, tape-recorded interviews with leaders and key members of the most central groups involved in women's rights activities in the period. Twelve women active at the national level were interviewed, and thirty-three women were interviewed who were involved at the local level. A final source of interview data was twelve transcribed interviews conducted by other researchers and available in archival collections.

A Stage of Continuance and Survival: An Elite-Sustained Movement

After the suffrage victory in 1920, the women's movement fragmented into groups concerned with such issues as peace, birth control and family planning, child protection laws, civil rights, union organizing, and other social reforms (Cott 1987). Only one group, the National Woman's Party (NWP), continued to identify itself explicitly as feminist. Although the NWP was unsuccessful in promoting its vision and strategies among other women's organizations and in widening the base of feminist support, there were other groups that joined with it to form an elite-sustained women's movement that pressed for women's rights on a variety of fronts.

I use the term *elite-sustained* to characterize the women's rights movement that existed from 1945 to the emergence of the contemporary wave of feminism because it was comprised of: (1) a small cadre of the discontented group attempting to "speak for" a broader constituency; (2) a membership that was highly educated, especially politically conscious, and had the discretionary time, specialized expertise, and economic independence and other resources necessary for sustained participation; and (3) a very small or nonexistent indigenous or mass base.

By the post-1945 period, the organized women's rights movement consisted of a small group of organizations and individuals that saw themselves as the heirs of the suffrage movement. Its major goals consisted of the Equal Rights Amendment (ERA), the appointment of women to policy-making positions, and recognition of women's history. Three criteria set apart those groups and networks of women who identified as part of the women's rights movement in this period from those that did not: commitment to feminist ideology, working in concert toward common aims, and some degree of interaction, if only on occasion, with the self-identified feminist NWP.

The center of this elite-sustained movement was the NWP (see Table 19). Best known for its use of disruptive tactics such as picketing the White House during the suffrage campaign, the NWP was never popular with mainstream suffragists because of its militant feminism. In the years following the suffrage victory, the NWP further broke with other groups when it launched, under the leadership of Alice Paul, its relentless and unswerving campaign for the ERA. By the end of World War II, the NWP's membership was aging and had dwindled from approximately 60,000 in the last years of the suffrage movement to about 4,000, only 600 of whom paid dues. In addition to losing and failing to attract new and younger members, the NWP suffered serious internal conflicts that hampered its effectiveness and contributed further to its reputation as a group of amusingly eccentric and anachronistic old feminists.

The core movement organizations clustered around the NWP supported the Equal Rights Amendment. Most of these groups existed prior to the 1940s and 1950s and had purposes broader than passage of the ERA. Still, their activities on behalf of the amendment were central to the functioning of the movement in this period, especially because support for the ERA bloomed after World War II as large national women's organizations joined the previously short list of endorsers and the amendment made headway in Congress. Preexisting organizations that formed the core of the movement included groups of professional women such as the National Federation of Business and Professional Women's Clubs (BPW), the National Association of Women Lawyers, and the American Medical Woman's Association. In addition, a loose network of women dedicated to winning recognition for the pioneers of the suffrage movement can be considered part of the core.

Not all of the groups interested in women's rights could be traced directly to the suffrage movement. Several new groups emerged in the 1940s and 1950s and enlarged the core of the movement. The Connecticut Committee for the Equal Rights Amendment, founded in 1943, was a state group active on the ERA and other feminist issues. It spawned other state groups, the most successful of which was the Massachusetts Committee, established in 1955. The Lucy Stone League, a women's rights group established in New York in 1950, was a revived and broadened organization based on the Lucy Stone League of the 1920s, which fought for the right of a married woman to retain her own name. The NWP also organized new groups, including the Women's Joint Legislative Committee, set up by Alice Paul in 1943 as a coalition lobbying group; the Industrial Women's League for Equality, established in 1944 by NWP leaders to counter arguments that women workers did not support the ERA; and the St. Joan Society, a group of pro-ERA Catholics founded in 1943 to influence Catholic opposition to the amendment.

On the periphery of the movement were a variety of other preexisting and emergent groups that supported the goals of the women's rights movement and cooperated periodically with core movement organizations to pursue common aims, but not with the same level of interest and consistency as the core groups. Included among these were the General Federation of Women's Clubs, the National Council of Women, the American Association of University Women (AAUW), the National Association of Colored Women, the Federation of Women Shareholders, a network of women in politics, and service organizations such as Zonta International, Soroptimist International, and Altrusa International. Women in World Affairs and the Multi-Party Committee of Women formed in 1946 to support the appointment of women to policy-making positions.

An analysis of 108 leaders and activists indicated that the majority of participants were white, middle or upper class, well-educated, employed in professional or semiprofessional occupations, and were in their fifties, sixties, or seventies by 1950.[1] In most respects, they differed sharply from the general population of American women. For instance, 81 percent were employed, 88 percent were over the age of fifty-five, and 41 percent were not married. Leaders of national organizations made up an important component of the membership. But despite their shrinking numbers, feminist activists, for the most part, directed little effort toward expanding the base of

Table 19. Women's Movement Organizations, 1945 to the 1960's

	Core	Periphery
Pre-existing	National Woman's Party Business and Professional Women's Clubs (BPW) National Association of Women Lawyers American Medical Women's Association Pioneers of the Women's Movement Network	General Federation of Women's Clubs National Council of Women American Association of University Women (AAUW) National Association of Colored Women Women in Politics Networks Zonta Soroptimist Altrusa Federation of Women Shareholders
Emergent	Connecticut Committee Massachusetts Committee Lucy Stone League Women's Joint Legislative Committee Industrial Women's League for Equality St. Joan Society	Women in World Affairs Multi-Party Committee of Women Assembly of Women's Organizations for National Security

support for women's rights. Rather, they functioned primarily through channeling preexisting group and individual commitment.

Thus, although it is true that women's activists were unable to mobilize broad-based sentiment on behalf of their aims in the post–1945 years, there were groups and networks of women who pressed for women's rights in government and the political parties and saw their activities as a continuation of their participation in the early twentieth-century suffrage movement. Many used the term *women's movement* or its nineteenth-century equivalent, *woman movement*, to refer to their activities. Tilly (1978, 61) pointed out that it "takes confidence, even arrogance, to override a group's own vision of its interests in life." Conceptualizing the activities of women's rights activists as an elite-sustained movement makes it possible to recognize their efforts as a stage that promoted the continuance and survival of the American women's movement.

Survival in a Hostile Environment

Organizational Strength and Movement Infrastructure

Groups and individuals involved in the struggle for women's rights from 1945 to the 1960s recognized that they were working in a social and political climate that was generally hostile to feminism. They expected and found little external support—whether political access, public attention, or favorable sentiment—for their cause. Rather, the movement depended on preexisting organizational and personal networks of women already committed to feminism.

Social movement scholars generally agree that the mobilization potential of a movement is affected by the degree of preexisting organization among dominated groups (Oberschall 1973; Tilly 1978; Freeman 1983; Jenkins 1983; McAdam 1983; Morris 1984; White 1988). If the social base of a movement is already linked together in some sort of primary network so that people have common values and moral commitments, interpersonal ties, and communication on a regular basis, then this prior group solidarity can serve as a basis for mobilization (Aveni 1978).

The infrastructure or organizational base of the post–1945 women's rights movement can be traced to three main sources: a shared bond of participation in the suffrage struggle, business and professional ties, and friendship and other close personal relationships. It should not be surprising that participation in the suffrage struggle would hold together a community of women across the years. Women who had demonstrated in front of the White House during World War I and suffered imprisonment and forced feeding were not likely to forget the sacrifices they had made for "the cause." Lamenting the passage of that period in her life, Florence Kitchelt, the force behind the Connecticut Committee for the ERA, asked a fellow suffragist in 1950 whether or not she ever felt, "as I do, that the 'modern' woman is missing something very thrilling, uplifting as well as unifying in not being able to take part in a suffrage campaign? Those were the days!"[2]

Not only were women who had themselves participated in the suffrage struggle likely to remain committed to women's rights in this period, but also women with illustrious suffrage mothers, grandmothers, and other relatives were often recruited to do feminist work in various organizations. Such women might even use their ancestry as a credential when vying for leadership positions. By connecting the fight for the ERA to the struggle for the vote, the NWP sought to make women who were not involved in the suffrage movement feel as though they were a part of a historical movement for women's rights. One member described the sense of community she felt through participation as "the warm glow that always comes when old Suffrage ties are renewed."[3] Katharine Hepburn, mother of the actress, in a speech to the Connecticut Committee described her experiences in the suffrage struggle. "That whole period in my life I remember with the greatest delight." she said. "We had no doubts. Life was a great thrill from morning until night."[4]

Professional ties also played a part in recruiting women to feminist activity and sustaining their commitment. Because many activists held professional positions, it was not uncommon for them to have become involved in women's rights work through professional associations, such as the BPW, the AAUW, or the National Association of Women Lawyers. Many women in Democratic party politics who held posts in government became fast friends and worked together to improve political opportunities for women. Margaret Hickey, a president of the BPW, an editor of the *Ladies' Home Journal*, and later a member of the President's Commission on the Status of Women, attributed her interest in the women's movement to her contact with a network of powerful women lawyers involved in women's rights work.

Friendship and primary group networks, then, were at the root of most women's involvement in the movement. Whether these grew out of a common suffrage heritage, professional ties, or other strong emotional and social bonds, the community of women that formed the core of the movement was held together by close and loving friendships. Correspondence among women rang with expressions of intimacy and friendship, indicating the intensely personal nature of the political work, and close friendships frequently were what led the few new women recruited to the movement in this period to participation in feminist activity.

Intense devotion to a woman leader and the tight friendship circle that formed around her was another reason given by women for joining the struggle for women's rights, and these leader-follower relationships functioned to integrate members into the feminist community. Alice Paul, often described as a charismatic leader, was the object of intense love and admiration, although she was not the only woman leader to inspire the loyalty and devotion of other women. Indicative of the superhuman characteristics attributed to Paul is the fact that she was addressed in one letter as "My Beloved Deity."[5] Not surprisingly, both adoration and denunciation of Alice Paul tended to focus on the same qualities: her complete absorption with but one thing and her perennial lack of gratitude for the effort others spent on behalf of women's rights.

Another kind of relationship found among core activists was a long-time committed couple relationship between women (Rupp 1980). As far as can be determined from the data, these women would not have identified their relationships as lesbian, although charges of lesbianism, especially in conjunction with "red-baiting," were occasionally whispered. Having another woman as a life partner seemed to facilitate feminist work because these women's personal lives meshed well with their political commitments, thus removing one of the obstacles to participation that women with families or nonsupportive husbands sometimes experienced (Cook 1977). The woman-centered environment at Belmont House, the NWP headquarters, is suggested by the fact that one member wondered if being married would prevent one's acceptance there.[6] Women involved in the movement in this period created a homosocial world through which they carried out their political work and maintained the kind of female public sphere that Freedman (1979) argued is so essential to the survival of feminism.

In addition to providing solidary incentives for participation, these networks of women, however marginal their cause, had access to important resources that made it possible to maintain the struggle for women's rights. NWP's Belmont House, located on Capitol Hill in Washington, D.C., was used by members and nonmembers as a feminist hotel, meeting place, and center of organizational, political, and social activities. Leaders of women's organizations had access to politicians and elites in government as well as to existing networks of women who could be mobilized for lobbying or letter-writing campaigns on a particular issue.

Because the women who were active on behalf of feminism were mainly middle and upper class with the financial means and time to devote to the cause, many were able to make the movement their life work. Individuals such as Rose Arnold Powell rebelled against the "utter man-mindedness" she saw around her by devoting her life to winning recognition for the pioneers of the suffrage movement.[7] From the 1920s until her death in 1961, Powell mounted a campaign to honor the memory of suffrage pioneer Susan B. Anthony by working to have her face carved on Mount Rushmore, have schools named after her, make her birthday a national holiday, and have calendar companies mark Anthony's birthday. Politician India Edwards sent out her 1959 Christmas cards bearing the printed message: "May there be more women in politics and more 'little women' in Girls Clubs in 1960!"[8]

Even as the political and cultural wave that had once carried feminists forward receded, there were women who continued literally to devote their lives to feminist activity. One woman described the intense commitment and close personal bonds among members as "ingrown to the point of fanaticism," whereas another woman apologized for her lack of activity, explaining, "I am interested but not with that dedicated drive I so admired."[9]

The movement consisted of a dedicated core of activists held together by a willingness to devote themselves almost single-mindedly to the cause, close interpersonal ties, a shared moral commitment, and strong leadership capable of coordinating and devising multiple strategies and actions. The strong sense of community and the conviction of being "the true feminists" helped to ensure the survival of feminist activity in the face of ridicule and opposition from the larger society.

Constraints and Tactical Choices

Despite an increase in female labor force participation and female enrollment in institutions of higher education, support for women's rights and opportunities declined sharply following World War II. Although women's organizations succeeded in having 236 bills pertaining to women's rights introduced into Congress in the 1950s, only 14 passed (Klein 1984). Even the one government agency established explicitly to improve the conditions of working women, the Women's Bureau of the Department of Labor, was generally unfriendly to groups most likely to identify as part of the women's rights movement (Harrison 1988). For instance, the assistant director of the Bureau, in planning a conference on womanpower to be held in 1954, sought a title that would "avoid the old battle of the sexes idea and pitch a conference around the idea of men and women working together and maintaining the Nation's economy."[10] Many large women's organizations disassociated themselves from the women's movement. The League of Women Voters, successor to the mainstream suffrage organization, shifted its focus after passage of the suffrage amendment to educating voters about political issues and candidates. The distance that the league maintained from women's rights advocates is evident in the comments of one member who contrasted "those primarily interested in women (the old line 'feminists') and those primarily interested in seeing women play their part in world affairs."[11]

To avoid increasing the intensity of opposition directed against them, women's rights activists relied on institutionalized tactics intended to achieve greater access to the political arena. Their principal strategies were persuasive rather than conflictive and involved lobbying and letter writing aimed at affecting national issues and actions with the support of elites in government and politics. Unlike in the suffrage years, women's rights groups avoided actions such as public demonstrations and protests that required the involvement of large numbers of participants and were likely to provoke unfavorable public sentiment and alienate political leaders.

Given the privileged social and occupational status of most activists and the limited support and other resources feminists could realistically expect to mobilize from the larger society, the movement emphasized strategies that drew on the specialized skills and expertise of its participants. A large number of the core activists were lawyers. For instance, lawyers who were members of the NWP instigated a prohibition on sex discrimination in Title VII of the 1964 Civil Rights Act through their contacts with conservative Representative Howard Smith (Brauer 1983). Women's organizations worked for the ERA by lobbying Congress, seeking endorsements for political office and from other organizations, establishing coalitions to support the amendment, and educating the public through newspaper and magazine articles, letters to the editor, and radio and television appearances.

The ERA came up for a vote in the Senate for the first time in 1946 and failed. Year after year, women's rights activists worked to have the ERA introduced in Congress, questioned candidates on their position on the amendment, built a list of Congressional sponsors, fought to keep the ERA in all party platforms (the Republican platform first included it in 1940, the Democratic platform in 1944), lobbied Senators and Representatives, sought the endorsements of other organizations, pressured the appropriate subcommittees, pressured the full committees, and lobbied again after each failure of the amendment to pass, and they did this without losing hope. Commenting on the persistence of feminists' lobbying efforts, one Representative from Connecticut wondered in 1945 "whether or not the Congress meets for any other purpose but to consider the Equal Rights Amendment."[12]

Women organized letter-writing campaigns on every issue the movement supported: not only the ERA, but also the election of pioneer feminists like Susan B. Anthony to the Hall of Fame and the appointment or election of women candidates. Sometimes these campaigns were undertaken by a single individual, but attempts almost always were made to make the efforts of individuals appear to represent a wider constituency by, for example, printing up organizational letterhead even if only one individual wrote all the letters and constituted the entire membership. One woman did almost all of the work of the Connecticut Committee for the ERA, as did another for the Spokane branch of the NWP.

Consistent with what other scholars have found, then, the strategies and tactics of the women's rights movement in this period were influenced not only by its limited opportunities for collective action but by its base of mobilization and its position in the power structure (Useem and Zald 1982). The movement operated from a small social base of women of privileged social status with strong ties to established institutions, most of whom had developed their commitment to feminism during the suffrage era. This, together with the unlikelihood of generating the kinds of wide support necessary for more disruptive tactics, made the general strategies of legal change and institutional reform almost inevitable.

The tactics that a social movement adopts influence the way that other groups in the political environment react to it (Turner 1970; Turner and Killian 1972; McAdam 1982; Barkan 1984; Cable 1984). If a movement relies primarily on institutionalized tactics or "proper channels" to express its grievances, established groups are less likely to view it as threatening than if it uses noninstitutionalized means.

Absence of Ties to Mass Base

Ironically, the same factors that made it possible for a small minority of women's rights activists to continue their struggle in the postwar decades contributed also to their inability to generate wide indigenous support. First, the movement was rooted in networks of elite women, held together by strong interpersonal relationships, who relied primarily on preexisting network ties rather than mass mobilization strategies to enhance the movement's claims to legitimacy. Second, women's rights activists used strategies to win support for the ERA from external groups that isolated them from a broader constitutency.

Elite Base of Movement

Social movements strive for two kinds of legitimacy (Useem and Zald 1982). Legitimacy of numbers is achieved by mobilizing a large number of people committed to challenging the existing distribution of power (Tilly 1979; Oliver and Marwell 1988). The rationale behind this strategy is that if a challenging group is able to amass a sizeable number of supporters, the polity will be forced to accede to its demands. Legitimacy of means is achieved by convincing the public, through its aims and strategies, that the movement is the most appropriate vehicle for achieving the goals of its constituency (Turner 1969; Gamson 1975). That is, a movement must make certain that its grievances are seen as credible and serious enough to warrant the challenge it poses the prevailing power structure (Turner 1969; McCarthy and Zald 1977).

After 1945, the number of active supporters of women's rights had dwindled so that the movement could not claim a large and active base of supporters. From the perspective of the mass media, the public, and others that the movement hoped to influence, the small numbers, age, and the social and economic status of those who comprised the movement made it easy to question whether or not those active on behalf of women's rights legitimately represented the interests of American women. Furthermore, because the kinds of women involved in the struggle for women's rights could be seen as having access to other more institutionalized means of influence, the movement had difficulty convincing others that it truly represented a disadvantaged group for whom a social movement was the appropriate vehicle for expressing its grievances. An example of the way in which the motives of women's rights activists were challenged is found in a letter to the *Washington Star*, complaining about a feature on a major woman leader of the period. "Whenever I see these smug pictures of women who have abdicated their normal functions and entered the field of politics and the like I instinctively say failure and slacker. Such women have flunked at their own jobs and yet pretend to tell men what they should do in their normal field."[13]

Recognizing that it was important that they appear to represent a larger sector of women than they actually were able to mobilize, feminists used various strategies to give the appearance of having a mass constituency. Among the small core of intensely committed movement activists were many well-known and admired women who headed national women's organization, several of whom had been leaders in the suffrage struggle. These women typically were able to command a following when support for a particular goal or strategy was needed, and most of the larger organizations maintained newsletters as a way of keeping members informed of their programs and actions. Thus, when a national organization supported an issue such as the ERA, movement leaders claimed the entire membership of that organization in listing numbers of women who supported the movement's position. The Business and Professional Women's Clubs' (BPWs') "Buttonhole Campaign" in 1953–54, in which state and local members were urged to "buttonhole" their senators and representatives to ask for passage of the ERA, is an example of the way in which the national leadership of women's oranizations mobilized their membership.

Organizations and individuals sought occasionally to bring young women into equal rights work but never had much success. They used strategies that ranged from publishing material in college newspapers designed to appeal to young women to holding luncheons and square dances to try and attract younger members. A "junior" organization of the NWP sprang up in California in 1945 but was described by one member, in a manner revealing her own class bias, as "made up of young business women ... active in defense work ... not to the manor born but definitely well informed."[14] On the whole, women's rights activists made few efforts to mobilize younger women but considered them "hopeless," or, in the words of one NWP member, "not interested in anything but cosmetics, T.V. and modern amusements."[15]

Other tactics used to inflate the movement's numbers were seeking paper membership by listing as members those who paid one-time nominal dues and seeking the establishment of state branches by a single member, a favorite strategy of the NWP. Multiple organizational memberships were common among activists. Many NWP members belonged to the BPW and the AAUW, and some even to the anti-ERA League of Women's Voters, and they used their multiple memberships to try to win or maintain endorsements for the ERA. For example, a number of NWP members who belonged to the BPW attended a BPW convention and, concealing their Woman's Party membership, mingled with the delegates and gathered at intervals in a central room to map out strategy for winning wider support in the organization for the ERA.

To counter the image that the women's rights movement was a small, exclusive, and eccentric groups whose claims of injustice might not be warranted, movement leaders established coalition groups among previously committed networks of women that would give the impression of a large and broad-based membership, such as the Women's Joint Legislative Committee for the ERA. But the most important strategies devised to establish legitimacy of means were those that attempted to give the appearance that the movement had wide grassroots appeal among less privileged women, such as working-class, union, and black women.

The support of women from labor was extremely important because a primary source of opposition to the ERA in this period was organized labor, which favored the protective legislation that the amendment threatened. To persuade the public and elite groups that women in labor supported the ERA, the NWP used the few working-class and union women who were active in the movement as spokespersons whenever possible. As one member cynically described the dealings of the NWP with a New York subway worker, "they paid her to leave her job and paid her more so that she could go around and work, to show that they weren't all patricians."[16] The NWP also founded and financed the Women's Industrial League for Equality to give the appearance of a coalition between labor and the women's rights movement.

The same kind of tokenism was evident in the way black women were dealt with by the movement. Organizations such as the National Council of Negro Women perceived most of their interests as antagonistic to those of the self-identified women's rights movement, largely because of the long tradition in the movement of defining "women's rights" as property rights, the right to vote, and other legal and political rights relevant mostly to white middle-class women. Although women's organizations made no large-scale efforts to open their membership to black women in the forties, fifties, and sixties, they did encourage the participation of highly visible women like Mary Church Terrell, founder of the National Association of Colored Women, to obtain her group's endorsement of the ERA. Throughout this period, feminists tended to perceive the burgeoning civil rights movement as a competitor likely to win rights for black women and men that were still denied to white women. In 1944, for example, a women's rights leader who was also a member of the National Association for the Advancement of Colored People wrote a letter to the *New York Times* publicizing a case in which black teachers in South Carolina won equal pay with white teachers under the 14th Amendment to point out that the Constitution protected black women and men but not white women. When the National Association of Colored Women objected to the letter by pointing out that black women received the lowest wages of any group in society, a NWP member explained that "when it comes to the question of racial equality, the negroes have no sense of humor whatever."[17]

The kinds of actions taken by women's rights movement organizations with regard to union women and black women suggest that feminists were not necessarily committed to creating truly meaningful political alliances with other constituencies. Rather, they tended to use their limited ties to union women and black women for their own ends, to give the impression that the women's movement represented a truly aggrieved group that was broad-based and, thus, legitimate in making its demands for change.

Isolationist Strategies

The loss of indigenous support can be understood also as a by-product of the tactical choices made by the movement, specifically relying on institutionalized tactics and pursuing reform goals that required the support of established elites in government and politics (McAdam 1982). Put simply, the more a movement relies on external groups, the more likely it is to have its efforts channeled in directions that suit the interests of these groups (McAdam 1982; Jenkins and Eckert 1986). Thus, despite the fact that the goals of feminism went against the broad social trends of the times, the movement as a whole sought respectability in its relationship to almost all other social and political issues besides women's rights as narrowly defined by the movement. Because the ERA was in this period opposed by labor, liberals, left-wing organizations, and most women's groups, feminists did not generally coalign their efforts with other progressive interests. When they did, it was out of self-interest, as was the case with the struggle by members of the NWP to include sex discrimination in civil rights legislation in 1957 and 1964, actions that were viewed with suspicion by many civil rights groups.

In effect, some feminists were willing to use almost any strategy, no matter how contradictory to the basic principles of social equality, to further their goals as long as it did not involve mass protest and other disruptive tactics. Frequently, this meant using tactics that took advantage of divisions among political elites and other powerful groups, which is a common strategy of reform movements (McAdam 1982). For example, the NWP attempted to win support for the ERA by using the fact that the Communist Party opposed the amendment to threaten vulnerable groups and individuals; the NWP pointed out that organized labor, by being against the ERA, was on the side of the Communists. This tactic unintentionally alienated socialist women from the organized women's rights movement.

In summary, the factors that contributed to the survival of the women's rights movement in such an unfavorable social climate almost ensured that the movement would remain small, encapsulated, and sectarian, and undoubtedly reduced its overall impact. Consistent with what other scholars have found, the strategies and tactics of the movement were influenced not only by its limited opportunities for collective action but also by its base of mobilization and its position in the power structure (Useem and Zald 1982;

Tarrow 1983; Garner and Zald 1987; Jenkins 1987; Klandermans 1991). Most of the organizations and networks of women that comprised the movement pursued limited goals consistent with their own interests and the kinds of resources they were able to obtain from the high status of movement participants, such as financial support, informal connections to decision makers in politics and government, legal expertise, and communication networks with mainly middle-class professional and semiprofessional women. Furthermore, the core movement organization, the NWP, evaluated its ties to any external individual or group solely on the basis of support or opposition to the ERA and, thus, was as willing to obtain the support of McCarthyites as Communists, anti-Semitic groups as Jewish groups, and racist groups as black women's organizations. Paradoxically, this tactical choice prevented the women's rights movement from developing meaningful ties with other organized constituencies of women in labor and civil rights, further isolating feminist groups and preventing the movement from launching a broad-based challenge.

Consequences for the Resurgent Movement

The women's movement that survived from 1945 to the mid-1960s did not accomplish its agenda, although it did make some limited progress toward its three major goals: passage of the ERA, placement of women in policy-making positions, and recognition of women's history. But a more important question for understanding the historical continuity of the women's movement is what consequences, if any, the actions of the elite-sustained women's movement of this period had for the revitalized movement that took up the fight for gender equality in the late 1960s. The founding of the National Organization for Women (NOW) in 1966 serves as a useful signpost in mapping the transition from the movement of the 1940s and 1950s to the resurgence of the contemporary women's movement. NOW brought together labor union activists, government employees, and longtime feminist activists and quickly took leadership of what is generally called the liberal branch of the women's movement (Freeman 1975; Ferree and Hess 1985). At about the same time, women involved in the Civil Rights and New Left movements formed the more locally organized radical branch of the women's movement (Evans 1979). Studies have not generally recognized the connections between the existing women's rights movement and the resurgent one.

To understand the impact that the elite-sustained movement had on the feminist challenge that followed, it is useful to think of this phase of the movement as providing crucial resources that molded the strategies and possibilities for action of the subsequent cycle of the movement. Following Freeman (1979), we can view the resources that a movement depends on as both tangible and intangible. Tangible resources include the money, space, and publicity a movement needs to exist and to make its program known, whereas intangible resources refer to such things as the time and commitment of supporters, specialized expertise, access to networks and decision makers, status in the larger society and within the movement.

One of the most important tangible resources the movement of the 40s and 50s provided to the resurgent movement was space, which resulted from the fact that women active in the movement in this period were affluent, capable of paying membership dues and leaving bequests that supported the activities of their respective organizations and networks. For example, Belmont House, the NWP headquarters, was used as a centralized location for ERA lobbying for groups other than the NWP, for celebrations of the women's movement's history, and as a temporary residence for scholars and students engaged in feminist research in Washington. Belmont House, in fact, remained symbolically important to the women's movement throughout the 1970s, so much so that in 1981 a group of younger feminists named themselves the Congressional Union and attempted, under the leadership of Sonia Johnson, to take control of the NWP and its headquarters. Other organizations, too, such as the BPW and AAUW, had headquarters in Washington that were focal points of activity, information, and support as the movement coalesced around the ERA and other activities in the 1970s.

Second, the elite-sustained women's rights movement also had access to different sources of publicity than were available to the emerging movement. The radical branch of the contemporary women's movement relied on the network of underground newspapers of the New Left to publicize its ideas and actions (Freeman 1975). In contrast, the emerging liberal branch lacked any established means of communication. However, groups that had been involved in women's rights work in the forties and fifties did have channels that they could activate on behalf of the ERA and other issues. The newsletters of the BPW, NWP, professional women's organizations, and women's service organizations continued into the 1980s to reach a sector of women that otherwise would never have been reached by the publications of NOW, much less by the newsletters and writings of the more radical groups of younger women. In addition, the elite-sustained movement had access to established media that helped to publicize feminist activity when new groups such as NOW had yet to form such ties. For example, the *Christian Science Monitor* and the *New York Herald Tribune* both supported the ERA from the 1940s on, and journalists looked to groups such as the NWP and the BPW for information about the ERA before NOW gained visibility and wider support.

Far more important than the tangible resources, however, were the intangible ones that preexisting feminist groups provided new organizations that emerged in the late 1960s. First, women gave time and commitment to the cause. Many of the same women who had taken an active role in women's rights work in the 1940s and 1950s continued to be activists into the 1980s. Some focused on trying to convince emerging feminist organizations, such as NOW and even coalitions of radical young women who attended the National Conference on New Politics in Chicago in 1967, to support the ERA. For instance, a NWP member went to hear Kate Millett speak at Purdue University in the early 1970s and took ERA literature to hand out to the crowd. Contrasting vividly the old suffragists of the women's rights movement with the new feminists of the 1970s, she noted that she was the only one there in a hat and that everyone else, including Millet, had long hair.[18]

Second, women's rights supporters, active for many years on behalf of the ERA, brought specialized expertise in the form of regular procedures for lobbying Senators and Representatives and a system of record keeping on supporters and opponents of the amendment. Feminists also had some access to key decision makers in government and politics who proved useful in the campaign for the ERA and the struggle to include sex discrimination in Title VII of the Civil Rights Act of 1964. Title VII is the only section of the civil rights bill to include sex as well as race discrimination, and it remains the major piece of legislation safeguarding women's employment rights. The issue of its nonenforcement by the government was an early rallying cry for the liberal branch of the women's movement in the late 1960s.

Most interpretations of the process by which Title VII came to include gender-based as well as racial discrimination focus on the fact that opponents of the civil rights bill introduced and supported the amendment prohibiting sex discrimination as an attempt to ridicule and defeat the legislation by dividing its supporters. In fact, the story behind the inclusion of sex discrimination under Title VII is more complex (Zelman 1982; Brauer 1983; Rupp and Taylor 1987; Harrison 1988; Freeman 1991). It illustrates how the NWP used its connections and expertise on behalf of a legal change that eventually served as a cornerstone of the contemporary women's movement. NWP members wrote to Representative Howard Smith suggesting that he introduce an amendment forbidding sex discrimination, which he did. Then the NWP lobbied to support passage of the amendment. Although in actuality a number of women in government assisted the NWP in the fight for the amendment, Alice Paul insisted, characteristically, that the NWP "bore the entire burden of this battle with absolutely no help of any kind from other women's organizations."[19] Another member of the NWP involved in the struggle said: "I seriously doubt that women could have gotten Title VII without just a very small number of us."

Women's rights activists had a third resource that was useful to contemporary feminists. They had access to networks within a variety of women's organizations that might otherwise have remained isolated from feminism as it increasingly became identified in the 1970s as an opposition movement. By 1970, most national women's organizations supported the ERA, and many embraced some of the newer goals of the movement.

A fourth resource that the elite-sustained movement brought to the women's movement in subsequent years was status within the feminist world. Alice Paul, in particular, became a legendary figure who was elevated by later feminists to the role of heroine. With the founding of NOW in 1966, a new group of women who had not identified with the movement in the 1940s and 1950s took up the feminist cause. Not surprisingly, some members of the NWP were among the founders of NOW and, most important, NOW's endorsement of the ERA in 1967 came largely through NWP's prompting. In Alice Paul's mind, "we captured the NOW people."[19] Although the two organizations did cooperate for a time, the Woman's Party's attitude toward NOW hardened as NOW grew into a vigorous organization with a multi-issue platform and took leadership of the ERA struggle. One longtime activist disapproved of NOW's attention to "problems of sexual or psychological supremacy," which she felt debased the struggle for equal rights to the "battle of the sexes."[20]

Fifth, although the women's rights movement after 1945 did not always reach the wider public with its program, it did have some influence, even if limited, on the larger society. Probably the best evidence is that John F. Kennedy in 1961 appointed the President's Commission on the Status of Women. Part of the reason for establishing the commission was to devise an alternative to the ERA to placate the amendment's opponents, including Democratic Party women and organized labor. That the position of feminists on the ERA entered into the deliberations of the commission is clear from the attitude expressed by one member, who hoped that "we could steer a complete path between not opposing the equal rights amendment and not bringing down the wr[a]th of the very influential Woman's Party ... upon our fresh green heads."[21] It was to no one's surprise that the commission's final report did not favor the ERA and that its deliberations did forestall, temporarily, further action on the amendment. However, as scholars have pointed out, the real significance of the President's Commission on the Status of Women was that, through recommending the establishment of state commissions, it ultimately gave birth to the organizational infrastructure that served as a stimulus for the growth of a new and more broad-based cycle of the women's movement (Harrison 1988).

Sixth, and perhaps most important, was the impact of Title VII of the Civil Rights Act and the ERA on the revitalized movement. Title VII and the issue of its enforcement by the Equal Employment Opportunity Commission helped to spark the founding of NOW, shaped its earliest actions, and rendered the political system vulnerable to the feminist challenge by facilitating collective action around gender discrimination in the workplace (Freeman

1975). Whatever else can be said about the impact of the elite-sustained stage of the movement on the women's movement that followed, one thing is clear: The Equal Rights Amendment, authored in 1923 by the NWP, was handed to a new generation of women and became, for almost a decade, one of the major issues around which the modern women's movement mobilized. To be sure, in the early years of NOW, the emphasis on the ERA and the strategy of attempting to mobilize what Paula Giddings called "the unreconstructed elements of the Feminine Mystique" alienated black women and union women (Giddings 1984). This suggests another way in which the elite-sustained movement influenced the cycle of the women's movement that followed it: The contemporary women's movement, by adopting the goals of earlier feminist activists, was unable to shake the class and race limitations of its predecessors.

In summary, the activities of feminists from 1945 to the mid-1960s had consequences in their own times as well as for the resurgent women's movement. Although new organizations emerged in the late 1960s and 1970s and took up a broader range of feminist issues, mobilized a larger and more diverse constituency, and developed new and varied strategies, the women's rights movement of the 1940s and 1950s, by serving as a source of mobilizable resources, affected the goals, composition, and strategies of the round of feminist activism that followed it.

Conclusions

Drawing on the resource mobilization perspective, I introduced four propositions to analyze the nature and outcome of the women's rights movement that survived in a period noted by most scholars for its antifeminism and general lack of support for women's rights.

Proposition One maintains that the structures and strategies of social movements will be influenced by whether the movement is in a stage of formation, success, continuance and survival, or decline. The women's rights movement that survived from 1945 to the mid-1960s was largely, although not exclusively, a remnant of the early twentieth-century suffrage movement and provided an environment in which a small and sectarian group of committed feminists could carry on the struggle for women's rights until the equation of forces changed, making it possible for a later generation of women to resume more overt mass-based protest. A particular form of social movement organization, an "elite-sustained movement," allowed feminists to keep the faith and continue the struggle in the absence of mass indigenous support or substantial support from powerful groups and individuals outside the movement. Previous conceptualizations of social movement outcomes depict only two consequences, success and institutionalization or failure and collapse. This analysis extends social movement theory by introducing continuance and survival as a possible outcome of social movements. Elsewhere I offer the concept of abeyance to characterize the holding process by which movements sustain themselves in nonreceptive political environments (Taylor 1989; Taylor and Whittier 1992).

Consistent with Proposition Two, the data suggest that the continuation of the women's movement can be attributed to its organizational strength and the nature of its goals. The movement consisted of personal networks held together by strong personal ties. As a result of the movement's distinctive membership base and the lack of opportunities for mass actions, women's rights activists pursued moderate goals that reflected the interests and resources obtainable from high-status participants.

This chapter introduces the concept of an elite-sustained movement to describe the type of social movement organization that characterized the women's rights movement from 1945 until the emergence of the modern women's movement. Existing theoretical formulations depict two types of social movement organization and action, the classical social movement and the professional social movement. The former is an indigenous effort that draws its existence from leadership, a mass-base of support, ideas, and tactics generated from within the dominated group (Morris 1981, 1984). The professional social movement, in contrast, is a top-down reform effort that depends on a core of trained entrepreneurs for its leadership, has a small or "paper" membership, obtains most of its resources from elites outside of the dominated group, and attempts to "speak for" rather than mobilize the direct beneficiaries of its actions (McCarthy and Zald 1973). However, Jenkins and Eckert (1986) pointed out that researchers have not adequately distinguished between two kinds of professional movement organizations, those that originate in and are supported by indigenous movements and those created by elite patrons and movement entrepreneurs that have no roots in indigenous movements. This chapter clarifies this distinction by introducing the concept of an elite-sustained movement to depict a form of social movement organization that owes its support to elites from within the challenging group rather than to outside elites, as is the case with the professional social movement.

Consistent with Proposition Three, I presented evidence that suggests that, in addition to promoting movement survival in a hostile social environment, an outcome of elite-based organization is the further erosion of ties to a movement's indigenous base of support. The fact that women's rights advocates lacked sufficient ties to other constituencies, such as labor unions, socialists, and black women, prevented the movement from launching a unified and broad-based challenge. To the extent that American women were differentiated by class, political, ethnic, and cultural identities and interests, the strategy of pursuing the ERA frequently meant compromising the interests of a sizeable portion of one sector of the movement's potential constituency for the support of another. Hence the movement's exclusivity, together with its cooptability, resulted in the further isolation and encapsulation of an already small and sectarian group of women's rights activists.

Despite its small size, limited success in reaching its goals, and isolationism, the women's rights movement of the 1940s, 1950s, and early 1960s did have an impact on the women's movement that followed. It provided a range of tangible and intangible resources that affected the goals, strategies, and composition of the mass women's movement that emerged in the latter half of the 1960s. I have presented evidence, then, for Proposition Four, that the organization and tactics of a social movement in a stage of continuance and survival can have consequences for later rounds of mass mobilization.

This case study suggests a different model for understanding recurrent collective action around a common set of interests than does the dominant one proposed by scholars of social movements. A preference for studying the emergence of movements rather than continuity has led to analyses of cycles of growth and decline as though they were independent of prior stages of collective action around the same grievances. Making a similar point, Gusfield (1981, 325) stated that "It is a matter crying for study that movement remain dormant and reemerge with their rhetoric shining and seemingly unused." Social movements based on longstanding grievances do not have simple beginnings and endings, but are better understood in terms of an ebb and flow.

Notes

1. This profile of the leadership and core membership is based on a careful reading of archival material, particularly correspondence, as well as research in biographical sources. For 108 women identified as leaders and core members, information was recorded about race, class, education, occupation, age, place of residence, political affiliation, political views, marital status, presence and number of children, living situation, and time of first involvement in the women's movement. In addition, any comments made by participants about the social characteristics of the membership were noted.

2. Florence Kitchelt to Katharine Ludington, 14 August 1950, Kitchelt papers, box 6 (175), Schlesinger Library (SL), Radcliffe College.

3. Caroline Babcock to Grace Cook Kurz, 8 July 1946, NWP papers, reel 89.

4. Katharine Hepburn, speech to the Connecticut Committee, n.d. [1946], Kitchelt papers, box 6 (153), SL.

5. Lavinia Dock to Alice Paul, 9 May 1945, NWP papers, reel 86.

6. Elizabeth Osgood to Caroline Babcock, 4 September 1945, NWP papers, reel 87.

7. Rose Arnold Powell to Mary Beard, 23 June 1948, Powell papers, box 2 (27), SL.

8. India Edward's Christmas card, 1959, Emma Guffey Miller papers, box 4 (62), SL.

9. Florence Kitchelt to Margaret Bruton, 11 July 1947, Kitchelt papers, box 4 (93), SL; Julia Pemberton to Alice Paul, [November 1955], NWP papers, reel 101.

10. Winifred G. Helmes to Helen Loy, 15 October 1954, Alice Leopold papers, box 2 (8), SL.

11. Dorothy Kenyon to Anna Lord Strauss, 5 October 1946, League of Women Voters papers, box 696, Library of Congress, Washington, D.C.

12. Joseph E. Talbot to Florence Kitchelt, 12 February 1945, Kitchelt papers, box 8 (234), SL.

13. Letter to the editor of the *Washington Star*, 3 November 1946.

14. Neenah Hastings Lessemann to Alice Paul, 16 May 1945, NWP papers, reel 86; [Nora Stanton Barney] to Alice Paul, 7 May 1945, NWP papers, reel 86.

15. Lillian H. Kerr to Alice Paul, 12 April 1956, NWP papers, reel 102.

16. Interview; Anita Pollitzer to Agnes Wells and Mildred Palmer, 17 May 1951, NWP papers, reel 98.

17. Gaeta Wold Boyer to Florence Kitcheit, 11 May 1944, Kitchelt papers, box 7 (201), SL.

18. Mary Kennedy to Alice Paul, 11 February 1971, NWP papers, reel 112.

19. Memo from Alice Paul, 10 July 1964. NWP papers, reel 109.

20. Alice Paul, "Conversations with Alice Paul: Woman Suffrage and the Equal Rights Amendment," Interview conducted in 1972 and 1973 by Amelia R. Fry, Regional Oral History Office, University of California, 1976.

21. Miriam Holden to [Mary Glenn] Newell, 11 February 1967, NWP papers, reel 110.

22. Transcript of Proceedings, Committee on Political and Civil Status of Women, 24 August 1962, President's Commission on the Status of Women papers, box 14, John F. Kennedy Library, Boston, Massachusetts (formerly in Waltham, Massachusetts).

Fragmentation and Cohesion in American Society

GARY T. MARX

> Only connect. That was the whole of her message.
> —E. M. Forster, *Howard's End*

Much of Henry Quarantelli's professional life has been devoted to the study of community, although not as the term is conventionally defined in sociology. Following Shakespeare's advice to "by indirection find directions out" he has looked to the rupture (or anticipated rupture) of community and established social patterns during situations of crisis. In the rich tradition of deviant or contrasting case analysis, he has given us insights into processes of ordinary behavior by looking at extraordinary situations. Through his extensive worldwide efforts he has also helped create new institutions and communities. In this chapter I continue this interest in community by considering some issues around fragmentation and cohesion in American society. In doing this I also touch on other issues that have occupied him such as mass communications, popular culture, and collective behavior.

In the 1950s and early 1960s, American society was criticized for its conformity and homogeneity. Observers such as David Riesman in *The Lonely Crowd* (1950), William Whyte in *The Organization Man* (1956), and C. Wright Mills in *The Power Elite* (1956) decried what they saw as the leveling and stifling effect of bureaucracy and mass society. The individual was seen to be overly responsive to the group and unduly timid. In the 1960s, many idealistic and optimistic young persons heeded John F. Kennedy's charge to "ask not what your country can do for you, but what can you do for your country?" But in recent years these concerns have rarely been heard. They have been replaced by what is in some ways an opposite concern over the pronounced differences and indifferences in contemporary society. Worry over the tyranny of community has been replaced by worry over whether or not there is any community to begin with, or what kind of community it is.

I am grateful to Susan Eckstein, Claude Fischer, Russ Neuman, Susan Silbey, David H. Smith, and Jim Wood for their suggestions.

Consider the following:

—the rise of strident single issue political groups
—the popular tax revolt
—a negative politics that earmarks candidates to be defeated rather than elected
—a retreat from public institutions and the public provision of health and welfare services (e.g., increased attendance at private schools and emphasis on the role of private organizations in meeting social needs)
—the dispersion of political power from the national to the state and local level and among a plethora of seemingly gridlocked interests and bureaucracies
—an apparent resurgence of ethnicity and increased ethnic diversity as new groups from Asia, Latin America, and the Caribbean grow in size
—an increase in adversarial and litigious behavior and a tendency to treat an ever larger number of relationships in legal terms
—an apparent increase in tax cheating, littering, running stop signs, vandalism, the failure to pay fines and other indicators that may suggest civic decline
—high rates of predatory crime and the appearance of new forms such as drive-by shootings
—inventions such as the personal radio and video games that can isolate the person from direct interaction with others
—the enhancement of specialized channels of communication such as cable television that permit "narrowcasting," appealing to highly differentiated groups, rather than the more diffuse "broadcasting"
—an increase in the geographical separation of groups based on life-style or special needs (e.g., whole towns, neighborhoods, or buildings for the retired, young adults without children, the physically handicapped, and homosexuals)
—the appearance of medievallike walled or "gated" towns (e.g., in California and Florida) with access to the town itself gained only by passing elaborate security and surveillance measures

It is possible to look at such examples and conclude, as many popular and scholarly accounts have, that social bonds are weakening and that American society is becoming less cohesive.

Is differentiation based on life-style, race and ethnicity, language, single-issue political commitments, occupation, leisure interests, or geographical location increasing? Are the selfish concerns of the individual lessening concern for the needs of others? In this more private and privatized society are social ties fewer and weaker? Is trust in others and in basic institutions decreasing? Is the society becoming more intolerant, contentious, factionalized, and parochial? Is there a decline in the willingness to compromise and related conditions that are essential for a civil society, or is the nature of the social bond simply changing? Are some of the old forms weakening and being replaced by new forms? Does increased involvement in a local group or increased cultural diversity necessarily imply fragmentation relative to the larger society? These questions are not easily answered. Yet they are vital.

The examples just considered are quite diverse. If we ask how they can be conceptualized in more general and abstract terms, each in its own way can be seen to involve a form of fragmentation.

Fragmentation is an inclusive multidimensional concept that touches familiar social concepts such as: atomization, isolation, privatization, alienation, anomie, disintegration, segmentation, and tribalism. Conceptual overlap, trends, interrelationships, processes, and basic causes in each of these should be analyzed. But my purpose here is merely to identify issues, summarize some data and suggest some possible interpretations on a broad level.

As used here, the concept of fragmentation is not fully captured by the dictionary definition, "a part broken off." Some examples deal with parts that have recently been added (e.g., new Asian immigrant groups). Other examples involve parts that, rather than being detached, are simply loosely attached (e.g., the socially isolated). Other examples involve the level at which attachment, linkage, or integration occurs, for example, local or national. In other cases what is of interest is a feeling of disconnectedness rather than external behavior or structures. Fragmentation is not a perfect concept, but its breadth permits capturing some related phenomena.

There is no simple way to assess fragmentation. A variety of types of individual and group data are required to measure it: (a) behavioral indicators of social attachment and isolation; (b) attitudes and feelings regarding trust and tolerance; (c) indicators of group heterogeneity and homogeneity; and (d) indicators of group structure. Here because of space limitations, I emphasize the first and the last with brief mention of the other two.

Social Attachment-Isolation

One obvious indicator is the extent and form of connections with others. I consider: (a) measures of direct involvement; (b) measures of indirect involvement such as exposure to mass media; (c) measures of pro- and antisocial behavior such as charity and crime.

Direct Involvement

One factor suggesting an increase in isolation is the greater prevalence of single-person households. The percentage of persons living alone increased from 16 percent in 1970 to 22 percent in 1987 (in 1950 this figure was only 11 percent).[1] Married couples accounted for less than 63 percent of all households. Yet the marriage rate has not changed—it was 10.6 per thousand population in 1970 and the same in 1980. Divorce, however, increased from 3.5 to 5.2 per one thousand. However, in 1981, for the first time in twenty years, the number of divorces actually declined by 3 percent relative to the preceding year. Rates of remarriage remain high; about three out of four divorced persons remarry, most within a few years after their divorce.

Work is a major way that individuals are connected to society. Labor force participation has been increasing for women. The 1980 census found that for the first time a bit more than half (52 percent) of all women were working outside the home. In 1987 this figure had increased to 56 percent. At the same time, the labor force participation of adult males is declining. The latter went from a high of 87 percent in 1951 to 76 percent in 1987. Early retirement has become a reality for many workers. In 1987, 30 percent of males, aged fifty-five to sixty-four, were not in the work force.

Although the image of an America turning increasingly inward and persons withdrawing from others and broader social participation may have dramatic appeal, there is no trend data to support it. The National Opinion Research Center (NORC) has regularly asked representative national samples about the frequency of spending "a social evening with relatives," with "someone who lives in your neighborhood," and with "friends who live outside the neighborhood." There is no consistent pattern of change between 1974 and 1983. For example, in 1974 about 56 percent of the sample visited with relatives several times a week or more (11 percent reported visits of once a year or less), 41 percent visited with friends living outside their neighborhood (19 percent reported visits of once a year or less). In 1983, these percentages were 52 percent spending a social evening with relatives (12 percent reported visits of once a year or less, 43 percent reported visits with friends who live outside the neighborhood, 17 percent reported visits once a year or less). NORC polls in 1973 and 1978 found approximately 70 percent of Americans reporting "a great deal" or "a very great deal" of satisfaction with their friendships.

One of the characteristics that the astute nineteenth-century French observer of American society, Alexis de Tocqueville, found most interesting was the proliferation of voluntary organizations. The United States may not differ from Europe today in this regard, but it is clear that a majority of the American public belongs to organizations and that this has not changed much in recent years. According to NORC data in 1974, 25 percent of Americans reported that they belonged to no formal groups and organizations (e.g., fraternal, service, labor, hobby, professional, and church affiliated groups). However, 50 percent of the population belonged to two or more such groups. In 1983, these figures were about the same.

These overall patterns can obscure differences across types of organization. For example, participation in labor and religious organizations has been declining, although this is far truer of the former than the latter. The unionized sector of the U.S. work force has declined from a high of 37 percent in 1940 to 27 percent in 1970 to 24 percent in 1980.

There has been a slight decline in religious participation judged in terms of the percentage of the population belonging to a religious body (this general trend masks increased membership in more fundamentalist and newer religious groups). In 1960, 64 percent of the population reported belonging to a religious organization, as against 63 percent in 1970 and 59 percent in 1980. In 1972 half the population reported attendance at religious services a few times a month or more, whereas in 1978 this figure dropped to 44 percent.

Beyond asking individuals about their membership in groups, we can ask about the availability of groups for people to join. Opportunities for participation appear to have increased. For example, the number of organizations listed in the Gale Encyclopedia of Associations increased from 10,734 in 1970 to 17,644 in 1984. The vast majority of these are national nonprofit membership organizations. This listing represents only a fraction of all voluntary organizations because those with minimal formal organization and those at the local level are generally excluded.

Involvement in leisure and cultural activities has increased significantly over the last two decades, and new forms of participation have appeared. With respect to sports, both participation and attendance have increased. For example the number of persons playing softball went from 16 million in 1970 to 37 million in 1987. The number of recreation boats owned went from 9 to 15 million. In 1970, 36 million people reported they had gone fishing and/or hunting, whereas in 1980 this figure was 58 million.

The more passive attendance at major sports events also increased. Thus, attendance at major league baseball games went from 29 million in 1971 to 53 million in 1987. Significant increases also occurred for football, basketball, hockey, and even greyhound racing.

Attendance at cultural events also increased. For example, the number of opera companies went from 648 in 1970 to 1,019 in 1981, with attendance increasing from 4.6 million to 11.1 million. The number of symphony orchestras increased from 1,441 to 1,572. The number of concerts went from 6,599 to 19,327, and attendance went from 13 to 23 million.

Attendance at rock concerts also appears to have increased. Forms of entertainment on a massive scale such as Woodstock and the 1982 "US" festival appeared. At the US concert, for example, more than two hundred thousand persons gathered in a field near San Bernardino, California for three days of almost nonstop performance by major rock stars. Although hardly the equivalent of a close-knit community where one is personally known, these nonetheless are a type of social participation. Concert organizers sought to create "a new kind of unity" through music. Themes of community and peace were self-consciously put forth. There was little conflict or disorder. Many participants reported a strong feeling of solidarity. For some persons such concerts represent a secular form of religious participation.

Shopping malls represent another new arena for participation. Whether initially intended or not, these have become social centers. Indeed, shopping malls are the third most visited places in America. Although clearly not offering the intimacy of a club or visit to a private home, neither are they as impersonal as city streets. For some young people, visiting the mall has become what driving up and down the main drag was to an earlier generation of youths. Malls have a social as well as an economic function. For many Americans going to the mall is an event in and of itself, apart from shopping. Malls offer a protected environment with a variety of diversions and a place to see and be seen.

Rates of geographic mobility in 1980 closely resembled those for 1970. In 1970, 47 percent of the population had moved in the last five years, whereas in 1980, the figure was about the same. Between 1975 and 1980 25 percent of the population was living in a different house but in the same county; 10 percent were in a different county; 10 percent were in a different state; and 2 percent were living abroad. In 1980, 18 percent of all households moved. Although relative to a broader trend, it appears that such mobility has been dropping since the mid-nineteenth century. Still the average American will move fourteen times in his or her lifetime.

Another special measure that traditionally suggested isolation, but probably is less important now, is the density of the area in which a person lives. The percentage of the population living in Standard Metropolitan Statistical Areas (the Census Bureau's measure of metropolitan areas) was the same in 1980 as in 1970 (77 percent). Yet there has been a slight deconcentration of population and growth outside of incorporated places. In a shift without precedent, population has disproportionately increased in rural areas and smaller urban clusters. During the 1970s metropolitan areas grew by 10 percent, whereas nonmetropolitan areas grew by 17 percent.

Mass Media and Electronic Communication

Face-to-face interaction and group membership are not the only forms of social attachment. Exposure to the mass media and interaction that occurs via telephone, mail, telegraph, fax, audio and videocassette recorders, videotext, videodisc, teletext, television, computer, and other emerging technologies also merit consideration. Various portable paging devices, along with cordless telephones, answering machines, devices for the remote reception of messages and shortwave radios are like an invisible spider's web reaching over vast areas and making possible instantaneous interaction.

People are receiving and sending more messages than ever before, from a greater variety of sources. This increase generally goes well beyond that which can be attributed to population increases. The chance to receive messages beyond a face-to-face context has significantly expanded, and to a lesser but still marked extent, so too has the chance to send messages.

To take a familiar example, mail received per capita increased from 411 to 632 pieces between 1970 and 1987, and the number of first class letters increased from 245 to 326. Word processing, which makes possible personalized mailings at a fraction of the traditional cost, is likely to significantly increase mail communication.

Telephone use has increased even more dramatically. The percentage of households with telephones steadily expanded from sixty-two in 1950 to ninety-three in 1988. The number of telephones per thousand population increased from 582 in 1970 to 789 in 1981. The number of daily phone conversations has come close to doubling each decade in recent times. Domestic long-distance calls increased from about 3 to 9 percent of all calls between 1970 and 1981. The number of overseas calls annually went from 3.3 million in 1970 to 199.6 million in 1980.

Beyond an increase in the amount of phone communication, types of interaction over the phone have become more varied. Terms such as *hot line* and *warm line* have entered the vocabulary. The variety of phone counseling, information, and recreational services available seems to be continually expanding, from those dealing with drug abuse, suicide, consumer, and parenting issues (a group that includes "phonefriend," an after-school telephone help line for children alone while parents are at work) to dial-a-prayer, dial-a-joke, and dial-a-sexual conversation lines.

The merging of television, radio, telephone, and computer networks in the home, perhaps supplemented by a television camera and a printer, can mean vastly expanded communication possibilities. When phones are connected with computers and with videotext systems, the possibilities for information retrieval and specialized contact are greatly expanded. In 1983, there were an estimated 4 million home computers and in 1987, 16 million. Beyond strictly instrumental uses, such as for banking, ordering goods and services, and obtaining information, interactive computer networks permit playing video games, sending messages, and getting involved with electronic bulletin boards.

Such computer communication can be the equivalent of the ham radio and even the singles bar. Depending on the format, participants may leave messages or participate in instantaneous "conversations" with others through a central computer. Some commercial dating services help subscribers make initial contact through home computers. The exchange of electronic messages, often using a code or alias, may eventually lead to direct social encounters. This is a more active form of participation than computer-aided dating services that merely show a videotape and/or generate a list of persons with particular characteristics.

Various forms of "teleconferencing" such as video telephone calls may serve as substitutes for travel and be socially richer than mere print or verbal communication. With the spread of fiber optics, video phone conversations in the home may become widespread soon.

One experiment in interactive cable television linked elderly persons living alone to neighborhood community centers, relatives, and friends. An evaluation found that the experiment had reduced their sense of isolation. Such new forms of communication and information retrieval can be particularly helpful for those who have physical disabilities.

The computer may serve to reduce isolation in other important, if less personal ways. Thus, the interactive cable television Qube project in Columbus, Ohio, offered a variety of uses. Programs on ten community channels sought viewers' opinions. The system also offered a motion-sensing burglar alarm and a smoke detector. For example, sensing smoke, an electronic message would automatically be sent to a central computer and the address, number of persons in the household, and nearest fire hydrant and police station would be displayed.

Radio and television ownership have also become more prevalent. Radio possession has gone from 2.1 per household in 1950 to 5.1 in 1970, to 5.6 in 1989. Ninety-seven percent of cars have radios. The percentage of households with television sets increased from 9 in 1950 to 95 in 1970, to 98 in 1988. The average number of sets per household has also steadily increased. Half of American households have multiple sets. The average viewing time per day has increased from 5.1 hours per household in 1960, to 6.1 in 1970, to 7.1 in 1988. Cable television subscribers went from 4.5 million in 1970 to 21 million in 1982 to 43 million in 1988.

The only form of communication not showing a significant increase is newspaper circulation. Thus, although the population increased almost 20 percent between 1970 and 1987, daily newspaper circulation remained at about 62 million. Only one city in four now has a daily newspaper.

Surprisingly, these increases have not come at a cost of more traditional forms of communication such as reading. Americans are reading as much or more than ever, although reading has become more specialized. About 50,000 books were published in 1983, compared to 40,000 in 1970 and only 11,000 in 1950. Book sales per capita and the number of retail book stores have been increasing. Library circulation has increased at twice the rate of population growth in the last four decades.

Pro- and Antisocial Behavior

Some inferences about cohesion and fragmentation can also be drawn from trends in pro- and antisocial behavior. One sign of caring about others is charitable contributions, although this is also effected by the state of the economy and the tax system. Over the last two decades, United Way contributions have just about kept up with inflation, going from $457 million in 1960, to $788 million in 1970, to $1,530 million in 1980. Using dollars adjusted for gross national product and population size, private philanthropy by individuals, foundations, corporations, and charitable bequests increased from $9 billion in 1960 to $20 billion in 1970, to $54 billion in 1981 to $94 billion in 1987.

An interesting form of prosocial participation lies in the expanding self-help movement. In 1983, an estimated 15 million persons belonged to approximately 500,000 self-help groups. Twenty-four regional clearinghouses were available to refer callers to appropriate organizations. Traditional self-help groups such as Alcoholics Anonymous have been joined by Gamblers Anonymous, Weight Watchers, neighborhood crime and arson watches, food and other cooperatives, community development organizations, a rich array of support groups concerned with women's, men's, and family-related issues such as pregnancy and birth, parenting, divorce, jealousy, juvenile delinquency, child abuse, retarded children, aging parents, retirement, widowhood, and groups concerned with a variety of other mental and physical health issues.

Crime rates are one indicator of antisocial behavior. Although crime rates increased in the late 1960s and early 1970s, they have remained relatively constant over the last decade. In 1982, the number of people victimized by serious crime declined by 4.1 percent. This was the largest one-year decline since the National Crime Survey began ten years before. To be sure, both white-collar and street crime are serious problems in American society, but data on the latter do not suggest an escalation in rates consistent with sensationalized media reporting.

Although not antisocial in the way that predatory crime is, suicide represents the extreme form of social withdrawal. Suicide rates for those over forty-five have not changed much in recent decades, but this is not the case for younger persons. In the last two decades the suicide rate among young persons has increased sharply. For persons fifteen to twenty-four this figure was 5.2 per 100,000 in 1960, 8.5 in 1980 and 10.2 in 1986. Homicide rates have also increased among the young, particularly among minority group members.

It would be nice to have measures of change in white-collar crime such as political corruption, embezzlement, price-fixing, and violations of consumer protection laws, but adequate trend data on these are not available, partly because of their low visibility. Prosecution rates have increased, but that need not reflect absolute increases in the violations themselves.

Attitudes

The afore mentioned data hardly suggest an unraveling of social bonds. With respect to most traditional measures of social involvement, the United States has not changed much in recent decades, and new forms of participation have appeared.

However, it might be argued that the heart of fragmentation is not so much any overt behavior, but the way people feel about themselves and society. Looked at in this way, fragmentation implies malaise and demoralization. In a famous speech in 1979, President Jimmy Carter talked of "a crisis in confidence [that] strikes at the very heart and soul of our national will."

The subjective aspect is of obvious importance, but it may be only weakly or indirectly linked (particularly in the short run) to more objective factors. Objective social conditions need not be automatically accompanied by feelings of attachment or perceptions of a cohesive society.

A society lacking in cohesiveness is likely to be one where persons do not care about, or worse, are intolerant of others and believe others to feel the same way. This may lead to feelings of mistrust, suspiciousness, cynicism, and pessimism. These attitudes may as well extend to basic social institutions.

The case for fragmentation is perhaps strongest with respect to the attitudinal data. As Lipset and Schneider (1983) reported, the survey research data does not suggest a very trusting society. Yet this is hardly new. The United States has always been mistrustful relative to much of Europe. There has not been much change in the last decade, although changes relative to the 1960s are more pronounced. There is some evidence that trusting attitudes are increasing relative to the decline that followed Watergate. Tolerance for diversity has been increasing over the last decades. The data does not suggest a badly demoralized, pessimistic population. Space limitations permit consideration of only one area: tolerance.

With respect to tolerance for and acceptance of diversity, the trend data suggest that Americans are becoming less prejudiced (at least at the level of what they will admit to a poll taker) and more supportive of Constitutional protections for all persons. One indicator of this is whether or not a person would consider voting for a minority group member for president. For example, according to Gallup polls in 1958, the percent of the population who would consider voting for a black, a woman, a Jew, or a Catholic for president was 38, 52, 62, and 68 respectively, whereas in 1983 these figures had increased to 77, 80, 88, and 92 percent. Other measures of support for nondiscrimination in housing, jobs, accommodations, restaurants, public service, and social relations suggest that the American public is becoming increasingly tolerant, at least as measured by surveys. The belief in racial inferiority has declined. Reviewing evidence from the last four decades of public opinion, Smith (1982) concluded that there has been a general shift toward liberalism. This is particularly the case for attitudes toward abortion, civil rights, race relations, and religion.

Heterogeneity and Homogeneity

Let us turn from these measures characterizing individuals to some aggregate measures that characterize groups or cultures. Among the most important traditional axes of group diversity have been race and ethnicity, religion, sex, age, and region. These in turn are often associated with social class, life chances, life-styles, and a variety of attitudes and behavior.

Perhaps the simplest approach is to ask for each of these whether the different elements are proportionally increasing or decreasing and whether new elements are appearing and/or old ones disappearing. Our attention is restricted to three areas: age, race and ethnicity, and life-style differences.

The youth culture that gained prominence in the 1960s may be balanced by a "mature" or "gray" culture that will become increasingly important in the next decades. This reflects the fact that the U.S. population is aging. The median age is expected to rise from 30 in 1981 to 36 by 2000 to 42 by 2050. Census projections put the proportion of the population over 65 (which was 11 percent in 1982) at 13 percent by the year 2000 and at 22 percent by 2050. The "old, old," over 85, increased by 56 percent in the last decade. The percentage of the population over 85 is expected to go from 1 percent in 1981 to 1.9 percent in 2000 and to 5 percent in 2050. This group has different needs than the "younger old" in areas such as health-care, housing, and transportation.

As the population ages, the work force may shrink in size, and ethnic diversity may be encouraged as a result of the increased need for immigrant workers. Under more liberal immigration laws, special provisions for refugees, and changed economic conditions, the last several decades have seen significant legal and illegal immigration to the United States. According to some estimates, almost nine million persons came here between 1970–80, a figure close to the peak immigration period of 1900–10. The 1980 census found 11 percent of the population speaking a language other then English at home. Immigration, of course, bears on the general issue of ethnic and racial diversity. Between 1970–80, the white population increased by 6 percent. Rates of increase for nonwhite groups were more pronounced. Thus, the rate of population increase for blacks was 17 percent, for persons of Hispanic origin 61 percent, for Indians and Eskimos 72 percent, and for persons of Asian origin 127 percent. If present trends with respect to birth rates and immigration continue, by the year 2000, Hispanics are likely to be the largest ethnic group in the United States with 12 percent of the population (35 million persons). This would put them slightly ahead of the blacks and the two largest groups today, whites of Anglo-Saxon and German background. The Asian population may rise to 10 million.

It is possible to imagine an American society that is increasingly balkanized along Anglo, Afro, Hispanic, and Asian lines. The spread of bilingual education (which for some persons may mean little or no effective education in English), a requirement that under certain conditions ballots be printed in a language other than English, the economic and political advantages (and technical ease) of foreign language mass media communication and marketing, and the continual influence of the home country, through immigration or patterns of going back and forth, suggest this.

However, the current data point away from such a conclusion, although bilingualism and strong ethnic communities at points of entry such as Miami, El Paso, and San Francisco are likely to remain strong. Relative to European immigrants who came here in the early twentieth century, the more recent immigrants from Asia and the Western Hemisphere are closer to the American population in social characteristics. They are more geographically dispersed between regions and within cities. The ghettoization that characterizes blacks seems unlikely to occur on a large scale. Groups such as Asians and Cubans have experienced significant social mobility in a short time. Rates of intermarriage for the second generation are very high (e.g., for Asians, over 50 percent).

The United States as an industrial state with its historic values, economic interdependence and opportunities, higher education, geographic and social mobility, and rapid communication exerts stronger pressure toward assimilation, the mixing of cultures, and pluralism, than toward separatism. Many of the correlates of ethnicity, particularly those associated with racial definitions, are of course strong, yet in many areas they have diminished. We are far from being a unisex or nonracial society where knowing a person's sex or race permits you to predict no more about them than would knowing the color of their eyes. Yet we are also far from the traditional patrician and racist society where gender and race were highly correlated with almost everything else. The declining significance of ethnicity is part of a broader move toward a society where ever more choices are open to the individual.

One of the major consequences of the 1960s was increased tolerance for, and legitimacy of, life-style diversity. Persons today have more choices about how they will live than ever before. Those whose nontraditional preferences were previously hidden or not expressed find it easier to locate each other and form supportive groups. These in turn further the group's concerns and also help it to draw new adherents.

Diversity has been encouraged by greater resources and new forms of communication. The United States has always been highly diverse, but previously this was not valued and legally protected to the extent that it is today. Ethnic immigrants, for example, were expected to shed their old ways and fully adopt the dominant culture. Homosexuality and prostitution were hidden and stigmatized. The physically disadvantaged were often segregated and invisible. The merits of continuing to speak a non-English language were little appreciated. In recent times, pluralism has become more legitimate.

The standards and life-style associated with being white, male, Anglo-Saxon, Protestant, youthful, eastern, urban, and middle class, although still very strong, are not as dominant as they once were. As traditional ascriptive sources of division have weakened, the range of groups that a person can affiliate with and life-style choices have broadened. More than seven out of ten Americans feel they have "more freedom of choice" with respect to how they will live their lives than their parents had. Whether in living arrangements, work, architecture, food, clothes, music, or recreation, cultural diversity appears to have increased within the United States, at least within some broad confines.

Family and work patterns have also become more varied. In the early 1950s, 70 percent of households were composed of a father at work and a mother at home with a child or children. Today this is the case for only 15 percent of households. A variety of other living arrangements and patterns have become more prominent—from hetero-and homosexual cohabitation to group homes shared by unrelated persons. In 1988, 29 percent of total households were made up of "nonfamily" persons whereas in 1970, this figure was 20 percent. By 1988, 8 percent of total households were made up of unmarried persons of the opposite sex sharing living quarters, a group that has increased significantly since 1970.

The detached single-family dwelling has been supplemented by cooperatives, condominiums, and multiple-dwelling units. In some cases, as with housing designed for the retired, the physically handicapped, and the young unmarried, these represent efforts to build community through architecture and spatial separation.

Work force involvement has become more diversified. Greater variety in work schedules and career patterns can be noted. The recent decade has seen an increase in flexible work hours and days (including compressed work weeks), shared part-time jobs, time off from work, whether for maternity (or in a few cases, paternity) leaves, and sabbaticals. In 1982, more than one worker in five was on a flextime work schedule, a compressed work week (e.g., forty hours in four days), or part-time employment. It is possible that more adults will desire part-time work and that career shifts, whether at mid-life or earlier, will become more prominent.

Whether on balance work is becoming more differentiated is unclear. A new social division may be emerging between the highly skilled and rewarded specialists who have undergone extensive training and other unskilled, poorly paid, interchangeable workers.

One indication of expanded life-style and leisure choices is the increased diversity and expansion of communications media. Persons have an astounding number of choices when it comes to deciding what to watch, read, and listen to.

Russell Neuman and Ithiel de Sola Pool (1984) estimated that the supply of mass communication (TV, radio, newspapers, magazines, books, and motion pictures) increased at a rate of 6.7 percent between 1960 and 1980, although media use grew at only 2.1 percent per year. Looked at another way, in 1960, the mass media offered three million words per capita per day, whereas in 1980, this had grown to 11 million words, an increase of 367 percent! Users have increased control over what they can watch not only because more is available to watch, but also because inventions such as videotaping permit viewing at one's leisure apart from standardized transmission.

The standardized days of the *Saturday Evening Post, Life, Look*, and *Colliers*, three national television networks, and a few national radio networks are gone. They have been supplemented by media catering to much more specialized concerns.

In various areas of popular culture, markets have become more differentiated. Let us consider music and food. In the past in most areas of the country those seeking to hear classical, jazz, blues, folk, or country and western music were lucky to find these on one radio station, and then for only a limited time. Such programming is much more accessible now, as stations that specialize in one type of music have appeared. Even within a broad type of music, there is much more choice. In the case of what was once simply called popular music, there are now soft rock, hard rock, rhythm and blues, punk, funk, and disco stations.

In the case of food, beyond the plethora of international, gourmet, diet, and health foods increasingly available in frozen or fresh form, there has been a proliferation of diverse restaurants. In larger urban areas those eating out can choose between Italian, Greek, Chinese, Japanese, Korean, Mexican, Indian, Vietnamese, Thai, Indonesian, and Middle Eastern restaurants. In some urban areas, even within the cuisine of one country such as China, a diner may have a choice of Cantonese, Mandarin, or Szechuan.

Marketing specialists who once treated consumers as if they had homogeneous buying tastes now compete for more clearly defined segments of the market. Differentiated products are increasingly aimed at selected groups. In commenting on this "segmentation" one specialist observes "every market is breaking into smaller and smaller units, with unique products aimed at defined segment." This targeted marketing both responds to, and helps further the splintering of, consumer tastes and needs. It pays to create products for diverse markets, and diversity itself has become a more important value. This runs counter to one aspect of the mass society argument regarding the leveling and homogenizing effect of industrial society.

Group Structure

This broad factor refers not to the behavior of individuals but to organizational structure and relations, particularly aspects of the national society and relations among groups. A major aspect of modernization is the incorporation of diverse localities into a national entity. Whereas this need not involve a highly centralized political system, it is associated with transportation and communication.

We have seen the steady incorporation of rural into metropolitan areas, of villages into towns, towns into cities, and cities into sprawling contiguous urban areas (e.g., the Boston-Washington and San Diego-San Francisco coastal corridors). The spanning of the United States by a network of superhighways, the continued building and expansion of airports, and improved telecommunications help create a national society. Federal Express and related delivery services, able to move goods anywhere in the country overnight, are a good example, as is the success of the daily national newspaper, *U.S.A. Today*. The strengthening of the physical elements of a national infrastructure has encouraged some forms of economic centralization, as suggested by the rise of large chain stores, conglomerates, and multinational corporations, in place of small locally owned businesses.

But equally apparent are examples of economic decentralization—from the AT&T breakup to new forms of enterprise based on the production and distribution of information. Although industrial production needs economies of scale and, hence, concentrates production, information production can be dispersed, because communication with outlying areas can be instantaneous.

The number of small businesses has grown significantly in recent decades—from 93,000 in 1960 to 600,000 in 1980. Rather than working in huge, impersonal environments, over three-quarters of Americans work in settings of twenty or fewer persons. Smaller businesses create a majority of the new jobs that are added to the economy. At the same time, the Fortune 500 companies employed 1.2 million fewer people in 1983 than they did in 1970.

Although bureaucratic organization is not about to disappear, its limitations have been more clearly appreciated. The turn toward decentralization, the simplification of paperwork and procedures, and participatory management are efforts to overcome some of these limitations. This is also the case with more explicit organizational efforts to nurture and create networks that transcend formal organizational lines. In many settings, the line between formal and informal organization has become more blurred. Informal relations have, of course, always been present in organizations but generally they were viewed as something to be stamped out or ignored.

As recent Japanese economic successes have helped make the problems of traditional bureaucratic structure more apparent, efforts to incorporate elements of the informal structure and to back off from rigid bureaucratic forms have appeared. The new emphasis on networking is a good example. Because social networks cross organizational boundaries and hierarchies, communication tends to flow more rapidly and honestly. Networks may serve both the individual and the group. The individual may feel more a part of the group, and diverse groups are linked horizontally together via individual network participants.

Apart from their increased presence (or at least visibility) in formal organizations, networks appear to be increasing in importance as elements of informal and voluntary associations, whether for leisure or self-help. The increased ease of locating and communicating with persons with highly specialized interests through electronic means has aided this. Along with greater leisure and resources, there seems to be a new boldness and new norms regarding interaction with strangers who share a special interest. Informal horizontal linkages based on shared leisure, health, political, and consumption interests have increased in importance relative to traditional sources of group participation based on the extended kin group, religion, and neighborhood.

Although it might be surprising to some of the conservative social theorists who, since the French Revolution, predicted and feared an inexorable march toward an ever more powerful, liberty-destroying, centralized nation-state, the last decade has seen some move toward decentralization of political power. Evidence of this can be seen in increased political participation at the local level, including more referenda and initiative, weakened federal regulatory agencies and greater state powers.

There has been a decline in voting for nationally elected officials. For example, in 1964, 62 percent of the population voted in the presidential election, whereas in 1980 this figure was 53 percent. In off-year Congressional elections the decline was from 53 percent to 35 percent. Voting at the local level, however, has increased. Spirited local issues sometimes draw more than three-quarters of the voters. This greater local action reflects intensely felt political concerns, as well as resources provided by interest groups and the appearance of professional entrepreneurs skilled at political mobilization.

The increase in local political activity has not always been at the expense of broader issues. As awareness of national and international interdependence has spread, many national issues have become local. As one activist observed, "the exciting thing, for me, is that the average American, on the grassroots level, is connecting his or her local problems with global issues . . . they're starting to ask questions about where their tax money is going that they never asked before. They are getting active in the village square on global issues." Local referenda on issues of foreign policy and nuclear arms are an example.

Conclusion, Complications, Implications

What are we to conclude from this review of developments relevant to fragmentation and cohesion in American society? In a society as large and diverse as the United States, for almost any trend a countertrend (or at least a contrary anecdote) can be found. And as a Yiddish proverb holds "for instance is not proof."

As prestidigitators well know, what you see depends on where you look. What is seen as increased fragmentation at the national level may mean increased cohesion at the local level. This is likely to be the case with political decentralization. Or if we look at increased conflict *between* groups engaged in collective behavior, we may miss an increase in solidarity *within* them. Looked at broadly, the construction of a superhighway linking cities with each other and with suburban and rural areas can be seen as integrative. Yet taking a micro focus, it may be disintegrative, because it can break up what were once socially and physically close knit communities (e.g., by bringing in outsiders, eliminating housing in the highway's path as a result of construction, or creating physical barriers between sections of a neighborhood).

The temporal frame of reference is also important. Inferences about general societal trends based on only a limited time may prove deceptive. For example, some recent indicators of fragmentation, such as decreased concern with welfare needs and the move toward privatization, may be a function of the economic slowdown that will be reversed if times improve.

The presence or absence of an external crisis is also relevant. For example, in times of a broadly supported war, internal cohesion and altruism are likely to be greatest. The external threats may serve to counterbalance forces tending toward fragmentation. However, when these threats are absent, centrifugal forces are given greater play.

Awareness of broader time frameworks can call attention to longer range unifying consequences. The various civil rights struggles of recent decades have clearly meant increased conflict between groups. But looked at in the longer run, they may create national unity, as the society becomes more open and inclusive of previously excluded groups.

Looked at in the short run, some of what appears to be directional change will be shown in the long run to be a temporary blip or part of a random pattern. Or it may represent periodic oscillation and the dynamic nature of social systems. Like natural systems, social units experience a cycle of growth, decay, and reformulation. This seems to be the case with centralization-decentralization. When the limits of either are perceived to be reached, social forces come into play that generate their opposite.

An important insight from Henry Quarantelli's disaster research is that humans are rarely frozen into collective inactivity in the face of crisis. Instead they innovate and act back on the environment. Identifying and publicizing a trend may trigger countervailing efforts that then reverse it. It may be that beyond a certain limit fragmentation generates counter-efforts, as persons search for ways to mediate its harsher consequences. Collective behavior in the form of political challenges, the self-help movement, and the search for more fulfilling human relationships may be part of this reaction.

Similarly, although there is an apparent increase in social conflict and in the tendency to see relationships in legal terms, there may be an equivalent increase in the willingness to compromise. This is suggested by the rise of alternative processes of dispute resolution such as arbitration and mediation carried out in neighborhood justice centers or informal courts.

Even if we are clear about the level of analysis, the time frame, cycles, and the validity of an empirical pattern, persons may disagree on the meaning of a given indicator or on the value implications of an indicator whose meaning is agreed upon.

Some measures are not easily interpreted:

—Should a low rate of political participation be seen as a sign of fragmentation, indifference, resignation, and detachment from the social order, or as a measure of contentment with it? Can't high rates of political participation also indicate fragmentation when they are accompanied by rancorous conflict?

—Does the fact that many Americans are on the move geographically (whether within or between communities) suggest rootlessness, isolation, and the lack of community? Or should this mobility be seen as conducive to economic development and integration, as persons have a greater breadth of experience, come to understand and appreciate diversity and are able to some degree to choose where and how they live?

—Does the move away from central cities toward small urban clusters and more rural areas mean increased integration into smaller close-knit communities in the face of the impersonal city, or does it mean increased isolation as people do not live in as close proximity to each other?

—Should the decline of the local daily newspaper and the increase in "umbrella" newspapers serving a number of localities be seen as fragmentation, or does it suggest greater inclusion, as a larger number of persons are sharing the same communication?

—Should the move toward decentralization, networks, and increased local associational involvements and diversity be welcomed as a liberating response to a society that has grown too centralized and standardized, or should it be feared as the bearer of anarchy, disorganization, inequality, and rancorous conflict among people with little that binds them together?

—Is a diminution of the national-state's role in the provision of services necessarily undesirable, if local groups step in to fill the need? Will greater local involvement and power mean more efficiency, but less fairness with respect to minority interests?

—How should we view people scurrying by on a busy sidewalk all wearing headsets and seemingly oblivious to each other? Does this take the haunting "lonely crowd" to a new low? Or do we need to know more about what they are listening to, to reach any conclusions? How would our assessment vary depending on whether they are able to listen to a broad range of specialized material that uniquely suits them, or only a narrow range of standardized material that has been produced and sanitized by big government or business? In the case of the former, granted diversity and choice are positive values, but is there a point beyond which these are undesirable? What if people are not listening to the same language, or the messages they receive share nothing with those received by others, and even denigrate them? In a related example, does the lonely person watching television indicate the height of fragmentation, or is television an important integrative device for a person who otherwise would be isolated? How does the assessment change if the person is watching general and/or national networks versus highly specialized and/or local networks?

Future Trends

We began by noting examples suggesting that American society is becoming increasingly fragmented. Where is our society headed? The data do not point clearly in one direction. A detailed prediction of what the next two decades will be like with respect to fragmentation and cohesion requires distinct predictions for multiple indicators. There is no single overarching conclusion. The current data present a mixed and complicated picture, yet a consistent case for increased fragmentation cannot be made.

If we were to create an elaborate typology combining dimensions of social involvement, attitudes, diversity, and structure (each with a high and low value relative to fragmentation-cohesion), at the extremes we find two models of particular interest. One involves a heterogeneous, decentralized society, high in social isolation and feelings of distrust, composed of unrelated groups with few, if any, unifying interests or beliefs. At the other extreme we would find a homogeneous, centralized society, low in social isolation and feelings of distrust, composed of tightly linked groups with a great deal in common. Contemporary American reality does not easily fit into either of these extremes.

Yet I find three trends of particular interest: (l) the increased tolerance for diversity and the emergence of a new pluralism increasingly based on choice rather than ascription; (2) the increased importance of electronically based interaction and the weakening of physical place as a factor in cohesion; and (3) the emergence of new social forms that will combine and make less meaningful some of our traditional distinctions, for example, between centralization and decentralization.

Apart from some objective measures of whether or not cultural diversity is actually increasing, diversity is being redefined away from something to be merely tolerated to something more positive, as indicated by laws protecting minority rights and by subtle changes in values. This greater legitimacy, when mixed with increased affluence, entrepreneurs seeking to serve and create markets, and the ease of communication, encourages diversity.

The element of choice seems to be more important with respect to cultural diversity. Traditionally, many behavioral differences were simply expected as a result of the race, ethnic background, or gender one inherited. All behavior correlated with these will not disappear, but to a greater extent than ever before, where there are differences, it will be as result of conscious choice, not the unreflective acceptance of social definitions. For example some distinctive ethnic behavior will remain, but more as something proudly sought, than as something unwillingly thrust upon an individual. Elective groups will play a more important role in societal differentiation and as sources of cohesion and group identification.

Although there are important new elements favoring diversity, there are also strong elements operating to incorporate diversity into the broader society. This greater legitimacy for diversity ironically also serves to shape the broad national culture that persons share. The same technology and values that encourage this diversity also may contribute to increased national integration. As diverse groups touch, they may blend, as well as grate. Contact may mean conflict over scarce resources and heightened group identification. But it can also mean the appearance of altogether new forms such as Chinese restaurants serving kosher food, or the appearance of a new language—"Spanglish" among many persons living near the Mexican–United States border. And rather than remaining distinctive, local cultural elements may be incorporated into an ever-changing national culture. This countertendency toward the incorporation of diversity can be seen in examples such as the "nationalization" of country and western music, southern barbecue food, Californian hot tubs, and the slang from drug and computer subculture groups that has spread far beyond its origins. Similar dynamics at the international level, such as multinational firms and improved communications, were to diffuse what had once been insulated cultural forms.

We are moving away from a society whose major resources of solidarity/cohesion were importantly determined by geography—whether the face-to-face presence of others such as kin and friends, neighborhoods or regions, to a society where "community" is ever less restricted to physical proximity. Related to this is a change in temporal dimensions. In what amounts to "freeze dried" interaction, electronic media store relevant data and are always ready to offer it up at the leisure of the user. The further weakening of spatial and temporal barriers to interaction (begun by the telegraph and telephone) will likely have profound implications.

The meaning of loneliness and isolation may be redefined. Out of sight need not mean out of mind. Rather than being defined by how much persons interact on a face-to-face basis, or are part of a geographically defined community, integration will increasingly be defined by how much persons are involved with each other electronically and are part of some community of interest, rather than of place. Face-to-face interaction can decline, even as societal participation increases. More specialized social ties will become increasingly important and easier to make.

What is most interesting about the present paradoxical period is that new forms are coexisting rather than displacing old forms. We are seeing a reassertion of some processes that were in decline during the industrializing era. We may be in a transitional period where the balance will eventually tip one way or the other. But it is likely that opposing trends will continue in the coming decades. The great processes of centralization, bureaucratization, standardization, and specialization may be altered or weakened in some spheres, but in others they will grow stronger. What is new is a greater duality, as trends involving their opposites—decentralization, networks and lateral forms of organization, heterogeneity and generalization—gain in strength.

Value Implications

Beyond social science issues there are questions involving value judgments. The value implications of fragmentation and increased integration are also complex and contradictory.

The mere fact that some traditional forms of social participation and sources of social solidarity are being replaced does not insure that the new forms are better, or even equivalent. Are vicarious and symbolic forms of participation as gratifying as direct participation? Is interaction via machines and over long distances as satisfying as the direct human contact that occurs in face-to-face settings? Will the possibility of greater efficiency, convenience, predictability, and anonymity create a preference for interaction with machines rather than persons? How close can machines come to using the discretion and judgment of the wise human being? Do computerized forms, letters, and machine produced phone messages debase communication by their standardization and impersonality? Will social skills be lessened and emotional needs unfulfilled as more interaction occurs directly with, or through, machines? Can we have the advantages offered by a thoroughly "wired" society without the ominous feeling of being watched and violations of privacy and civil liberties?

Interaction on the basis of highly specialized, shared, elective interests has obvious advantages for the cohesion of the group in question. But do other needs then go unfulfilled, because interaction is based on only a tiny facet of the person rather than on the total personality? Furthermore, what happens to those who are left behind and have few choices to make—the elderly, the physically and mentally handicapped, the poor and isolated and marginal persons? In a society whose cohesiveness is based on ascription, such persons are more likely to be taken care of, because their position is not dependent on what they do, or choose, and others cannot as readily choose to abandon them.

Even with adequate safety nets and provisions for those who cannot really choose, the vast array of choices may carry a psychic and political burden for some persons. This can vary from immobilization as a result of sensory overload, to increased receptiveness to totalitarian direction that promises that one will not have to choose, as Erich Fromm argued in *Escape from Freedom*.

The seemingly rich array of choices offered by the market can make for societal diversity and feelings of individuality, as persons combine their myriad options for consumption, work, and leisure in distinctive ways. Yet questions can be raised about how meaningful some of these choices are. Some are in reality "pseudo-choices" masking the absence of more fundamental choices. For example, the absolute number of choices open to a car purchaser is enormous with respect to manufacturers, models, colors, and accessories, but the larger choice of an effective mass transit system, as against the personal auto, is not offered.

Will the new forms of communication make for a more democratic, less hierarchical, more participatory and generalist society, as persons are able to quickly gain knowledge that previously was available only to experts? Is this a logical extension of a democratizing tendency that began with the invention of the printing press and the spread of mass literacy? Or will we see instead an even less egalitarian, more divided society, as access to these means of communication is restricted to the computer literate and affluent? Will the gap between the information rich (and its likely correlate, the economically rich) and the poor increase and further enhance the life chances and opportunities for interaction and social participation of the privileged?

Questions can also be posed about the balance between two-way communication and the traditional mass media in which the person is simply a passive recipient. In this sense the printing press, the mimeograph machine, and the typewriter may be far more democratic, because the fixed costs associated with obtaining them are far less than for broadcast networks or highly sophisticated computer resources. On the other hand, by permitting instant mass feedback, computers may increase involvement and democratic participation as it becomes easier for persons to offer direct responses.

Those concerned over fragmentation in American society often make assumptions about the integrative consequences of homogeneity and the disintegrative consequences of diversity. At the heart of this may lie romantic notions about the supposedly homogenous rural area thought to be united by shared kinship, religious, ethnic, class, occupational, political, leisure, and geographic factors.

Yet even for most persons who do not fall into the error of idealizing a peaceful, pastoral past that never was, fragmentation probably carries a negative tone. It may be associated with disunity, lack of coherence, conflict, redundancy, inefficiency, indifference, selfishness, confusion, and powerlessness. We associate health with wholeness and pathology with separation.

However, there is another side to this. Cohesion can come at a high price, being associated with totalitarianism, oligarchy, monopoly, standardization, and the stifling of dissent and creativity. Second, apart from any perverse aspects of cohesion, depending on the social context, fragmentation may have positive elements. It may be associated with privacy, liberty, individualism, autonomy, freedom, and an enriching democratic diversity. It may mean greater flexibility and adaptability. A degree of skepticism and mistrust is not a bad thing for a healthy democratic society with an active citizenry. Skepticism and vigilance can be virtues in the face of political inequality and the expansive tendencies of power. A certain distance toward or distrust of strangers is prudent in an impersonal urban society.

We have argued that social forces are present that make for diversity and division as well as increased similarity. Yet even if the trends were much stronger in the former direction, this would not necessarily mean a less cohesive society, or at the extreme, a fragmented society tottering on self-destruction. The consequences depend on how diversity is patterned, whether or not there are structures to control and mediate conflict, and whether or not unifying values are present that transcend group differences.

Groups may be distinctive yet interwoven. As Durkheim argued, modern society is partly held together by the interdependence found among its highly differentiated parts. A related factor is how measures of diversity for the society as a whole overlap with respect to group formation. Even where the society shows many sources of heterogeneity, if these are distributed between, rather than within groups, cohesiveness may be enhanced. Diversity cutting across groups may increase integration. As S. M. Lipset argued, multiple criss-crossing group memberships and identities can increase understanding and appreciation of the need for tolerance.

Yet even where within-group homogeneity and commitment to particularistic groups is high, this need not mean a divided society lacking common concerns. To be sure, it may mean greater conflict and competition with other groups. However, as Georg Simmel and Lewis Coser observed, these, too, represent forms of social interaction. When conflict is carried out according to limiting rules, it can be integrative.

This is central to the idea of pluralism. In spite of the historic diversity of American society, at a broader level there has generally been consensus about the legitimacy of the society. Some consensus on values and beliefs has brought a degree of unity to highly diverse groups (aided, of course, by police and the military). At a broader level, there is consensus on the desirability of diversity. People can civilly agree to disagree. Increased commitment to one's own group (whatever the base—class, age, race, and ethnicity, religion, sex, recreational, or occupational interests) need not be divisive. When it is accompanied by tolerance for the rights of others, it may contribute to social cohesion.

There is not much evidence that this tolerance is declining. Indeed, as education has increased, survey research suggests that Americans are becoming more tolerant of diversity and more accepting of the rules of the democratic game, however uneven the path. Difference can add richness and variety and out of the clash of divergent ideas can come new forms. Of course, there are limits. Diversity from poverty and powerlessness is highly undesirable, and some of the examples with which this chapter opened are cause for concern. However, given the positive and negative aspects of both fragmentation and its opposites, conclusions must be drawn guardedly. That the traditional sources of these tendencies are changing is clear, even if the facts and value implications are less clear. There is a need for social research on the form and consequences of these changes and for philosophical and policy analysis of the values and choices at stake.

Note

1. Unless noted, all data are drawn from The Statistical Abstract of the United States for 1982 and 1989.

Bibliography

Adams, D. S. 1983. Selection, socialization, and four types of members in the American Red Cross chapter: A typology of voluntary association members. *Journal of Voluntary Action Research* 12: 31–45.

Aguirre, B. E., and E. L. Quarantelli. 1983. Methodological, ideological, and conceptual-theoretical criticisms of the field of collective behavior: A critical evaluation and implication for future studies. *Sociological Focus* 16: 195–216.

Aguirre, B. E., E. L. Quarantelli, and J. L. Mendoza. 1988. The collective behavior of fads: The characteristics, effects, and career of streaking. *American Sociological Review* 53 (August): 569–84.

Aldrich, H. 1979. *Organizations and environments.* Englewood Cliffs, NJ: Prentice-Hall.

Aldrich, J. E. N.d. X-ray pioneers in Nova Scotia, 1896–1917. MG 100 Vol. 58 No. 77. Public Archives of Nova Scotia.

Alexander, D. 1986. Disaster preparedness and the 1984 earthquakes in central Italy. Working Paper no. 55. Boulder: Institute of Behavioral Science, University of Colorado.

Alexander, D. 1990. Behavior during earthquakes: A Southern Italian example. *International Journal of Mass Emergencies and Disasters* 8 (March): 5–29.

Alexander, J. C. 1982. *Theoretical logic in sociology, volume 1: Positivism, presuppositions, and current controversies.* Berkeley and Los Angeles: University of California Press.

Allport, G. W., and L. Postman. 1974. *The psychology of rumor.* New York: Henry Holt.

Altheide, D. L. 1985. *Media power.* Beverly Hills, CA: Sage.

Amato, P. R., R. Ho, and S. Partridge. 1984. Responsibility attribution and helping behavior in the Ash Wednesday bushfires. *Australian Journal of Psychology* 36, no. 2: 191–203.

Ambrosini, P. J. 1983. Clinical assessment of group and defensive aspects of rumor. *International Journal of Group Psychotherapy* 33: 69–83.

American Psychiatric Association. 1980. *Diagnostic and Statistical Manual of Mental Disorders.* 3d ed. Washington, DC: American Psychiatric Association.

Anderson, W. A. 1969. Disaster warning and communication processes in two communities. *Journal of Communication* 19, no. 2: 92–104.

Anderson, W. A., R. R. Dynes, and E. L. Quarantelli. 1974. Urban counterrioters. *Society* 11: 50–5.

Andrews, G., C. Pollack, and G. Stewart. 1987. The determination of defensive style and questionnaire. *Archives of General Psychiatry* 25: 455–60.

Antonovsky, A. 1987. *Unravelling the mystery of health: How people manage stress and stay well.* San Francisco: Jossey-Bass.

Arminger, G., and G. Bohrnstedt. 1987. Making it count even more. In *Sociological Methodology*, ed. C. Clogg. New York: Jossey-Bass.

Aveni, A. 1977. The not-so-lonely crowd: Friendship groups in collective behavior. *Sociometry* 49: 96–99.

Aveni, A. 1978. Organization linkages and resource mobilization: The significance of linkage strength and breadth. *Sociological Quarterly* 19: 185–202.

Aysam, Y., and P. Oliver. 1987. *Housing and culture after earthquakes*. Oxford: Oxford Polytechnic.

Baker, E. J. 1979. Predicting response to hurricane warnings: A reanalysis of data from four studies. *Mass Emergencies* 4: 9–24.

Baker, J. E. N.d. Letter MG27 No. 3. Public Archives of Nova Scotia.

Barber, B. 1983. *The logic and limits of trust*. New Brunswick, NJ: Rutgers University Press.

Barkan, S. E. 1984. Legal control of the Southern Civil Rights Movement. *American Sociological Review* 49: 552–69.

Barton, A. 1969. *Communities in disaster: A sociological analysis of collective stress situations*. New York: Doubleday & Company.

Bartone, P., K. Ursana, K. Wright, and L. Ingraham. 1989. The impact of a military air disaster on health of assistance workers. *The Journal of Nervous and Mental Disease* 177: 317–28.

Bates, F. L., ed. 1982. *Recovery, change and development: A longitudinal study of the Guatemalan earthquakes*. Athens: Department of Sociology, University of Georgia.

Bates, F. L. 1991. The social network. Manuscript. Athens: INTERLAB, University of Georgia.

Bates, F. L., and C. C. Harvey. 1988. *The structure of social systems*. 3d ed. Maliber, FL: Krieger Press.

Bates, F. L., and W. G. Peacock. 1987. Disasters and social change. In *Sociology of disasters*, ed. R. R. Dynes, C. Pelanda, and B. A. DeMarchi, 291–330. Milan, Italy: Franco Angeli.

Bates, F. L., and W. G. Peacock. 1989. Conceptualizing social structure: The misuse of classification in structural modeling. *American Sociological Review* 54 (August): 565–77.

Bauer, R. A., and D. B. Gleicher. 1953. Word of mouth communication in the Soviet Union. *Public Opinion Quarterly* 17: 297–310.

Bauman, D. 1983. Determination of the cost effectiveness of flood hazard information. Papers and proceeding of the Applied Geography Conference 6: 292.

Becker, H. S. 1963. *Outsiders: Studies in the sociology of deviance*. Glencoe, IL: Free Press.

Becker, H. S. 1967. Whose side are we on? *Social Problems* 14: 239–47.

Bell, F. M. 1918a. *The romance of the Halifax disaster*. Halifax: H. A. Marshall, Limited.

Bell, F. M. 1918b. Interim medical report re Halifax disaster. RG 24 6359 HQ71-26-99 Vol. 1, 225–53.

Bellamy, L. J. 1987. Evacuation data. Paper presented at the European Conference on Emergency Planning for Industrial Hazard, November, at Villa Ponti, Varese, Italy.

Benson, J. K. 1977. Innovation and crisis in organizational analysis. *Sociological Quarterly* 18 (Winter): 3–16.

Berah, E. F., H. J. Jones, and P. Valent. 1985. The role and reactions of a mental health team involved in the Victorian bushfire aftermath. Report of proceedings of a research workshop on Human Behavior in Disaster in Australia, Mt. Macedon, Australia.

Berk, R. 1988. Causal inference for sociological data. In *Handbook of sociology*, ed. N. Smelser. Beverly Hills, CA: Sage.

Bird, M. J. 1962. *The town that died.* Toronto: McGraw-Hill.

Blalock, H. 1961. *Causal inferences in nonexperimental research.* Chapel Hill: University of North Carolina Press.

Blalock, H. 1979. *Social statistics.* New York: McGraw-Hill.

Blalock, H. 1982. *Conceptualization and measurement in the social sciences.* Beverly Hills, CA: Sage.

Blalock, H. 1984. *Basic dilemmas in the social sciences.* Beverly Hills, CA: Sage.

Blanshan, S., and E. L. Quarantelli. 1981. From dead body to person: The handling of fatal mass casualities in disasters. *Victimology* 6: 275–97.

Blau, P. 1970. A formal theory of differentiation in organizations. *American Sociological Review* 35 (April): 201–18.

Blumer, H. 1948. Public opinion and public opinion polling. *American Sociological Review* 13 (October): 542–54.

Blumer, H. 1951a. Collective Behavior. In *New outline of the principles of sociology*, ed. A. M. Lee, 166–222. New York: Barnes & Noble.

Blumer, H. 1951b. Social movements. In *Principles of sociology*, ed. A. M. Lee, 199–220. New York: Barnes & Noble.

Blumer, H. 1957. Collective behavior. In *Review of sociology: Analysis of a decade*, ed. J. B. Gittler, 127–58. New York: Wiley.

Blumer, H. 1971. Social problems as collective behavior. *Social Problems* 18 (Winter): 301–13.

Bolin, R. 1982. *Long-term family recovery from disaster.* Institute of Behavioral Science Monograph Series. Boulder: University of Colorado.

Bolin, R. 1989. *Temporary sheltering after the Whittier Narrows earthquake.* Final report to the National Science Foundation. Department of Sociology and Anthropology, New Mexico State University.

Bolin, R. 1990. *The Loma Prieta earthquake: Studies of short-term impacts.* Monograph Series. Boulder: Institute of Behavioral Science, University of Colorado.

Bolin, R., and P. Bolton. 1986. *Race, religion, and ethnicity in disaster recovery.* Monograph Series, no. 42. Boulder: Institute of Behavioral Science, University of Colorado.

Bolin, R., and L. Stanford. 1990. Shelter and housing issues in Santa Cruz county: Impacts of the Loma Prieta earthquake. In *The Loma Prieta earthquake: Studies of short-term impacts*, ed. R. Bolin. Boulder: Institute of Behavioral Science, University of Colorado.

Bolton, P. 1979. Long-term recovery from disaster: The case of Managua. Ph.D. diss. University of Colorado.

Bolton, P. 1988. Preliminary project report to the National Science Foundation. Seattle: Battelle HARC.

Boman, B. 1985. The Vietnam veteran as a disaster victim. Report of proceedings of a research workshop on Human Behavior in Disaster in Australia, Mt. Macedon, Australia.

Bosworth, S. L., and G. A. Kreps. 1986. Structure as process: Organization and role. *American Sociological Review* 51: 699–716.

Bramson, L. 1961. *The political context of sociology*. Princeton, NJ: Princeton University Press.

Brauer, C. M. 1983. Women activists, southern conservatives, and the prohibition of sex discrimination in Title VII of the 1964 Civil Rights Act. *Journal of Southern History* 49: 37–56.

Braverman, H. 1974. *Labor and monopoly capitalism*. New York: Monthly Review Press.

Breslau, N., and G. C. Davis. 1987. Postraumatic stress disorder: The stressor criterion. *Journal of Nervous and Mental Disease* 166, no. 5: 307–16.

Britton, N. R. 1985. The state/territorial emergency service as a constrained organization within the Australian Disaster Management System. Ph.D. diss. James Cook University, Townsville, Australia.

Britton, N. R. 1989a. Reflections on Australian disaster management: A critique of the administration of social crisis. Working paper, no. 3. Disaster Management Studies Center. Lidcombe, NSW: Cumberland College of Health Sciences, University of Sydney.

Britton, N. R. 1989b. Sociological studies in disaster. *The Macedon Digest: The Australian newsletter of disaster management* 4, no. 3: 4–6.

Britton, N. R. 1990. Permanent disaster volunteers: An outlier-and hence forgotton—group? In *Towards the 21st century: Challenges for the voluntary sector*, ed. D. Billis, 90–116. Center for Voluntary Organisation, London: London School of Economics.

Britton, N. R. 1991a. Permanent disaster volunteers: Where do they fit? *Nonprofit and Voluntary Sector Quarterly* 20, no. 4.

Britton, N. R. 1991b. Organisational and community response to a technological emergency: Case study of a major incident within a metropolitan Australian city. Occasional paper, No. 6. Armidale, NSW: Centre for Disaster Management, University of New England-Armidale.

Britton, N. R. 1991c. Hazardous industries and risky technologies: A new role for insurance. In *Natural and technological hazards: Implications for the insurance industry*, ed. N. R. Britton and J. Oliver, 201–18. Armidale, NSW: Centre for Disaster Management, University of New England.

Britton, N. R. 1991d. Constraint of effectiveness in disaster management. *Canberra Bulletin of Public Administration* 64 (May): 54–66.

Britton, N. R., C. C. Moran, and B. J. Correy. 1990. The permanent volunteer emergency worker: Exploring relevant factors in stress coping. Occasional paper series, no. 3. Disaster Management Studies Centre, Cumberland College of Health Sciences, Lidcombe, Australia: University of Sydney.

Britton, N. R., and R. Wettenhall. 1991. The first fifteen years: Australia's natural disasters organization. Occasional paper, no. 4. Armidale, NSW: Centre for Disaster Management, University of New England.

Brodsky, C. M. 1988. The psychiatric epidemic in the American work place. *Occupational Medicine: State of the Art Review* 3: 653–62.

Bromley, D. G., and A. D. Shupe. 1979. *Moonies in America: Cult, church, and crusade*. Beverly Hills, CA: Sage.

Bromley, D. G., and A. D. Shupe. 1983. Repression and the decline of social movements: The case of the new religions. In *Social movements of the sixties and seventies*, ed. J. Freeman, 336–47. New York: Longman.

Bronson, H. 1918. Some notes on the Halifax explosion. Transcript of the royal society of Canada. Section 3, 1–4.

Brough, G. et al 1989. Hillsborough: The truth. *The Sunday Times*, 23 April 1989, A13.

Brown, M., and A. Goldin. 1973. *Collective behavior: A review and reinterpretation of the literature*. Pacific Palisades, CA: Goodyear.

Bucher, R. 1957. Blame and hostility in disaster. *American Journal of Sociology* 62: 467–75.

Bucher, R., C. E. Fritz, and E. L. Quarantelli. 1956a. Tape-recorded interviews in social research. *American Sociological Review* 21: 359–64.

Bucher, R., C. E. Fritz, and E. L. Quarantelli. 1956b. Tape-recorded research: Some field and date processing problems. *Public Opinion Quarterly* 20: 427–39.

Buckner, H. T. 1965. A theory of rumor transmission. *Public Opinion Quarterly* 65: 54–70.

Butler, G., and A. Mathews. 1983. Cognitive processes in anxiety. *Advances in Behavior Research and Therapy* 5: 51–62.

Cable, S. 1984. Professionalization in social movement organizations: A case study of Pennsylvanians for biblical morality. *Sociological Focus* 17: 287–304.

Calhoun, C. 1982. *The question of class struggle: Social foundations of popular radicalism during the industrial revolution*. Chicago: University of Chicago Press.

Campbell, D. T. 1975. Degrees of freedom and the case study. *Comparative Political Studies* 8 (July): 178–93.

Campbell, D. T., and J. S. Stanley. 1966. *Experimental and quasi-experimental designs for research*. Chicago: Rand McNally.

Cantril, H. 1941. *The psychology of social movements*. New York: Wiley.

Caplan, N. S., and J. M. Paige. 1971. A study of ghetto rioters. *Scientific American* 219: 115–21.

Caplow, T. 1947. Rumors in war. *Social Forces* 25: 298–302.

Caplow, T., H. M. Bahr, and B. A. Chadwick. 1984. *Readiness of local communities for integrated emergency management planning*. Final report. Charlottesville, VA: United Research Services.

Caporale, R. 1989. Toward a cultural-systemic theory of natural disasters: The case of the Italian earthquakes of 1968 in Belice, 1976 in Friuli and 1980 in Campania-Basilicata. In *Preparations for, response tor and recovery from major community disasters*, ed. E. L. Quarantelli and C. Pelanda, 253–71. Newark: Disaster Research Center, University of Delaware.

Carden, M. 1977. *Feminism in the mid-1970s*. New York: Ford Foundation.

Carr, A. T. 1974. Compulsive neurosis: A review of the literature. *Psychological Bulletin* 81: 311–8.

Carroll, G. R. 1987. *Ecological models of organizations*. Cambridge, MA: Ballinger Publishing.

Carter, T. M. 1980. Community warning systems: The relationships among the broadcast media, emergency service agencies, and the National Weather Service. In *Disasters and the mass media*, 214–28. Washington, DC: Committee on Disasters and the Mass Media, National Academy of Science.

Charles, M. T., and J. C. K. Kim, eds. 1988. *Crisis management: A casebook*. Springfield, IL: Charles C. Thomas.

Chen, Z. 1987. Social responses to the earthquake rumor going round Zhanghou city, the Southern Fujian province during February 1987 and the countermeasures against it. Trans. Q. Yang and W. Wang. *Recent Developments in World Seismology* 104: 19–22.

Cisin, I. H., and W. B. Clark. 1962. The methodological challenge of disaster research. In *Man and society in disaster*, ed. G. W. Baker and D. Chapman, 25–53. New York: Basic.

Clifford, R. A. 1956. The Rio Grande flood: A comparative study of border communities. *National Research Council Disaster Study 7*. Washington, DC: National Academy of Sciences.

Cohen, J. 1985. Strategy or identity: New theoretical paradigms and contemporay social movements. *Social Research* 52, no. 4.

Collins, R. 1981. On the micro foundations of macro sociology. *American Journal of Sociology* 86: 984–1015.

Collins, R. 1982. *Sociological insight*. New York: Oxford University Press.

Collins, R. 1985. *Three sociological traditions*. New York: Oxford University Press.

Collins, R. 1987. Interaction ritual chains, power, and property: The micro-macro connection as an empirically based theoretical problem. In *The micro-macro link*, ed. J. C. Alexander, R. M. Giesen, R. Munch, and N. J. Smelser, 193–206. Berkeley, Los Angeles, and London: University of California Press.

Condie, B. 1989. Time to harness powerful emotions for the general good. *Daily Telegraph*, 22 April 1989, Sport 4.

Connelly, W. J. 1987. The Halifax (Nova Scotia) explosion of 1917: An epilogue. *Journal of the Royal Society of Medicine* (December): 774–75.

Conrad, P., and J. W. Schneider. 1980. *Deviance and medicalization: From badness to sickness*. St. Louis: C. V. Mosby.

Cook, B. W. 1977. Female support networks and political activism: Lillian Wald, Crystal Eastman, Emma Goldman. *Chrysalis* 3: 43–61.

Correy, B. J. 1989. The NSW Bush Fire Bridgades Organization: A comparative study of three areas in the Sydney region. Honors thesis. Department of Geography, Sydney, Australia: University of Sydney.

Coser, L. 1974. *The functions of social conflict*. New York: Free Press.

Costner, H. L. 1969. Theory, deduction and rules of correspondence. *American Journal of Sociology* 75: 245–63.

Cott, N. F. 1987. *The grounding of modern feminism*. New Haven: Yale University Press.

Couch, C. 1968. Collective behavior: An examination of some stereotypes. *Social Problems* 15: 310–22.

Couglan, R. 1956. Changing roles in modern marriages. *Life*, 24 December 1956.

Coutu, W. 1949. *Emergent human nature*. New York: Knopf.

Currie, E., and J. H. Skolnick. 1970. A critical note on conceptions of collective behavior. *Annals of the American Academy of Political and Social Science* 391: 34–55.

Cutter, S. L., and K. Barnes. 1982. Evacuation behavior and Three Mile Island. *Disasters* 6, no. 2: 116–24.

Dahrendorf, R. 1973. *Homo sociologicus*. London: Routledge & Kegan Paul.

Danzger, M. H. 1975. Validating conflict data. *American Sociological Review* 40: 570–84.

Danzig, E. R., P. W. Thayer, and L. R. Galater. 1958. *The effects of a threatening rumor on a disaster-stricken community*. Disease study, no. 10. Washington, DC: Disaster Research Group, National Academy of Sciences.

Davis, I. 1975. Disaster housing: A case study of Managua. *Architectural Design* (January): 42–47.

Davis, I. 1977. Emergency shelter. *Disasters* 1, no. 1: 23–40.

Davis, W., and E. J. Phares. 1967. Internal-external control as a determinant of information-seeking in a social influence situation. *Journal of Personality* 35: 547–61.

Dawson, D. 1988. An investigation into the personality characteristics associated with job performance in ambulance officers. Paper presented at the conference, Dealing with Stress and Trauma in Emergency Services, Melbourne, Australia.

de Tocqueville, A. 1954. *Democracy in America*. New York: Random House.

DeFleur, M. L. 1962. Mass communication and the study of rumor. *Sociological Inquiry* 32: 51–70.

DeMarchi, B. A. 1986. Italian sociology and the study of social movements. In *Comparative methods and research on collective behavior and social movements*, ed. G. Marx. *International Journal of Mass Emergencies and Disasters* 4, no. 2: 177–244. Special issue.

Demerath, N. J. 1957. Some general propositions: An interpretive summary. *Human Organization* 16: 28–9.

Diggory, J. C. 1956. Some consequences of proximity to a disease threat. *Sociometry* 19 (March): 47–53.

Dollard, J., L. W. Doob, N. E. Miller, O. H. Mowrer, and R. R. Sears. 1939. *Frustration and aggression*. New Haven: Yale University Press.

Dombrowsky, W. R., and J. K. Schorr. 1986. Angst and the masses: Collective behavior research in Germany. In *Comparative methods and research on collective behavior and social movements*, ed. G. Marx. *International Journal of Mass Emergencies and Disasters* 4, no. 2: 61–90. Special issue.

Drabek, T. E. 1968. *Disaster in aisle 13: A case study of the coliseum explosion at the Indiana State Fairgrounds, October 31, 1963*. Columbus: College of Administrative Science, Ohio State University.

Drabek, T. E. 1969. Social processes in disaster: Family evacuation. *Social Problems* 16 (Winter): 336–49.

Drabek, T. E. 1970. Methodology of studying disasters: Past patterns and future possibilities. *American Behavioral Scientist* 13 (January): 331–43.

Drabek, T. E. 1979. Communication: Key to disaster management. *Insight* 3 (July): 3–4.

Drabek, T. E. 1985. Managing the emergency response. *Public Administration Review* 45: 85–92.

Drabek, T. E. 1986. *Human system responses to disaster: An inventory of sociological findings*. New York: Springer-Verlag.

Drabek, T. E. 1987a. Emergent structures. In *Sociology of disasters*, ed. R. R. Dynes, B. A. DeMarchi, and C. Pelanda, 259–90. Milan, Italy: Franco Angeli.

Drabek, T. E. 1987b. *The professional emergency manager: Structures and strategies for success*. Boulder: Institute of Behavioral Science, University of Colorado.

Drabek, T. E. 1989a. Disasters as nonroutine social problems. *International Journal of Mass Emergencies and Disasters* 7 (November): 253–64.

Drabek, T. E. 1989b. Taxonomy and disaster: Theoretical and applied issues. In *Social structure and disaster*, ed. G. A. Kreps, 317–45. Newark: University of Delaware Press.

Drabek, T. E. 1990. *Emergency management: Strategies for maintaining organizational integrity*. New York: Springer-Verlag.

Drabek, T. E. 1991. *Implementation of microcomputer technology in emergency management agencies*. Boulder: Institute of Behavioral Science, University of Colorado.

Drabek, T. E., and K. Boggs. 1968. Families in disaster: Reactions and relatives. *Journal of Marriage and the Family* 30 (August): 443–51.

Drabek, T. E., and G. J. Hoetmer, eds. 1991. *Emergency management: Principles and practice for local government*. Washington, DC: International City Management Association.

Drabek, T. E., and W. Key. 1984. *Conquering disasters: Family recovery and long-term consequences*. New York: Irvington.

Drabek, T. E., and E. L. Quarantelli. 1967. Scapegoats, villains, and disasters. *Trans-Action* 4 (March): 12–7.

Drabek, T. E., and J. S. Stephenson. 1971. When disaster strikes. *Journal of Applied Social Psychology* 1, no. 2: 187–203.

Drabek, T. E., H. L. Tamminga, T. S. Kilijanek, and C. R. Adams. 1981. *Managing multiorganizational emergency responses: Emergent search and rescue networks in natural disasters and remote area settings*. Boulder: Institute of Behavioral Science, University of Colorado.

Drabek, T. E., R. Braito, C. C. Cook, J. R. Powell, and D. Rogers. 1982. Selecting samples of organization: Central issues and emergent trends. *Pacific Sociological Review* 25 (July): 377–400.

Draft Advertisement. N.d. Rg 24 4548 86-1-11. National Archives of Canada.

Dubin, R. 1978. *Theory building*. New York: Free Press.

Duncan, O. D. 1975. *Introduction to structural equation models*. New York: Academic Press.

Duncan, O. D. 1984. *Notes on social measurement: Historical and critical*. New York: Russell Sage.

Dunning, C. M. 1985. Prevention of stress. *Role stressors and supports for emergency workers*. Rockville, MD: National Institute of Mental Health.

Durham, T. W., S. L. McCammon, and E. J. Allison. 1985. The psychological impact of disaster on rescue personnel. *Annals of Emergency Medicine* 14, no. 7: 664–68.

Durkheim, E. 1938. *The rules of sociological method*. Trans. S. A. Soloway and J. H. Mueller. Chicago: University of Chicago Press.

Durkheim, E. 1951. *Suicide: A study in sociology*. New York: Free Press.

Durkheim, E. [1893] 1966. *On the divison of labor in society*. Trans. G. Simpson. New York: Free Press.

Dynes, R. R. 1970. *Organized behavior in disaster*. Lexington, MA: Heath Lexington Books.

Dynes, R. R. 1983. Problems in emergency planning. *Energy* 8, no. 8–9: 653–60.

Dynes, R. R. 1990. Community emergency planning: False assumptions and inappropriate analogies. Preliminary Paper, no. 145. Newark: Disaster Research Center, University of Delaware.

Dynes, R. R., and B. E. Aguirre. 1979. Organizational adaptation to crises: Mechanisms of coordination and structural change. *Disasters* 3, no. 2: 71–74.

Dynes, R. R., and E. L. Quarantelli. 1968a. Redefinition of property norms in community emergencies. *International Journal of Legal Research* 3: 100–12.

Dynes, R. R., and E. L. Quarantelli. 1968b. Group behavior under stress: A required convergence of organizational and collective behavior perspectives. *Sociology and Social Research* 52 (July): 416–29.

Dynes, R. R., and E. L. Quarantelli. 1971. The absence of community conflict in the early phases of natural disasters. In *Conflict resolution: Contributions of the behavioral sciences*, ed. C. G. Smith. South Bend, IN: University of Notre Dame Press.

Dynes, R. R., and E. L. Quarantelli. 1973. Editor's introduction: Issue on urban civil disturbances, organizational change and emergence. *American Behavioral Scientist* 16: 305–11.

Dynes, R. R., and E. L. Quarantelli. 1980. Helping behavior in large-scale disasters. In *Participation in social and political activities*, ed. D. H. Smith, J. Macaulay, and Associates. San Francisco: Jossey-Bass.

Dynes, R. R., and E. L. Quarantelli. 1981. *Disruption on the campuses of Ohio colleges and universities*. Newark: Disaster Research Center, University of Delaware.

Dynes, R. R., and E. L. Quarantelli. 1986. Role simplification in disaster. In *Role stressors and support for emergency workers*, 23–37. Washington, DC: Center for Mental Health Studies of Emergencies.

Dynes, R. R., B. DeMarchi, and C. Pelanda. 1987. *Sociology of disasters*. Milan, Italy: Franco Angeli.

Dynes, R. R., E. L. Quarantelli, and G. A. Kreps. 1972. *A perspective on disaster planning*. Newark: Disaster Research Center, University of Delaware.

Dynes, R. R., E. L. Quarantelli, and D. E. Wenger. 1990. *Individual and organizational response to the 1985 earthquake in Mexico City, Mexico*. Newark, DE: Disaster Research Center, University of Delaware.

Dynes, R. R., A. H. Purcell, D. E. Wenger, P. E. Stern, R. A. Stallings, and Q. T. Johnson. 1979. Report of the Emergency Preparedness and Response Task Force Washington, DC: President's Commission on the Accident at Three Mile Island, Executive Office of the President.

Editorial Office of RDWS. 1989. Carry out stipulations for issuing earthquake predictions seriously to avoid the spread of rumors and misdissemination of earthquake news. Trans. Q. Yang and W. Wang. *Recent Development in World Seismology* 130: 14–19.

Edwards, L. P. 1927. *The natural history of revolution*. Chicago: University of Chicago Press.

Eisenstat, R. A., and R. D. Felner. 1983. Organizational mediators of quality of care: Job stressors and motivators in human service settings. In *Stress and Burnout in the Human Service Profession*, ed. B. A. Farber. New York: Pergamon Press.

Ekiert, G. 1989. Collective identities and collective protest. Rebellions in the Soviet concentration camp system. Unpublished manuscript.

Emergency Management Planning Division, Department of Public Safety, Consolidated City of Indianapolis. 1988. The Park Fletcher air crash, Indianapolis, Indiana, 20 October 1987: A case study. Indianapolis, IN: Emergency Management Planning Division, Department of Public Safety, Consolidated City of Indianapolis.

Emergency Management Planning Division, Department of Public Safety, Consolidated City of Indianapolis. 1990. Marion County Management Plan: Medical service section. Indianapolis, IN: Emergency Management Planning Division, Department of Public Safety, Consolidated City of Indianapolis.

Emergency Management Planning Division, Department of Public Safety, Consolidated City of Indianapolis; Indiana State Fair Administration, State of Indiana; Indiana State Police, State of Indiana; State Emergency Management Agency, State of Indiana. 1989. Indiana State Fair Major Incident Coordination Plan: Reference and action sections. Indianapolis, IN: Emergency Management Planning Division, Department of Public Safety, Consolidated City of Indianapolis.

Enoch, J. K. 1988. Origins, social networks, and restructuring: An analysis of three methodologies. Master's thesis. Williamsburg: College of William and Mary.

Erikson, K. T. 1976. *Everything in its path.* New York: Simon and Schuster.

Estepp, M. H. 1990. Combination fire departments roots in the past: Visions in the future. Technical paper presented at the South Pacific Fire Conference, Auckland, New Zealand.

Ettedgui, E., and M. Bridges. 1985. Posttraumatic stress disorder. *Psychiatric Clinics of North America* 8, no. 1: 89–103.

Evans, R. R., ed. 1969. *Readings in collective behavior.* Chicago: Rand McNally.

Evans, S. 1979. *Personal politics: The roots of women's liberation in the Civil Rights Movement and the new left.* New York: Knopf.

Farmer, E. Z. 1989. Structure, organization, and social movements. In *Social structure and disaster*, ed. G. A. Kreps, 135–66. Newark: University of Delaware Press.

Faust, H. S., and L. B. Brilliant. 1981. Is the diagnosis of mass hysteria an excuse for incomplete investigation of low-level environmental contamination? *Journal of Occupational Medicine* 23: 22.

Ferree, M. M., and B. B. Hess. 1985. *Controversy and coalition: The new feminist movement.* Boston: Twayne.

Ferris, J. M. 1988. The use of volunteers in public service production. *Social Science Quarterly* 69, no. 1: 3–23.

Festinger, L., D. Cartwright, K. Barber, J. Fleischl, J. Gottsdanker, A. Keysen, and G. Leavitt. 1948. A study of rumor: Its origin and spread. *Human Relations* 1: 464–86.

Fireman, B., and W. A. Gamson. 1977. Utilitarian logic in the resource mobilization perspective. In *The dynamics of social movements: Resource mobilization, social control, and tactics*, ed. M. N. Zald and J. D. McCarthy. Cambridge, MA: Winthrop.

Fireman, B., W. A. Gamson, S. Rytina, and B. Taylor. 1979. Encounters with unjust authority. *Research in Social Movements, Conflicts, and Change* 2: 1–34.

Firth, R. 1956. Rumor in a primitive society. *Journal of Abnormal and Social Psychology* 53: 122–32.

Fischer, C. 1975. The study of urban community and personality. *Annual Review of Sociology* 1.

Flexner, E. 1975. *Century of struggle*. Cambridge: Harvard University Press.

Flynn, C. B. 1979. *Three Mile Island telephone survey: Preliminary report on procedures and findings*. Washington, DC: Nuclear Regulatory Commission.

Flynn, C. B., and J. A. Chalmers. 1980. *The social and economic effects of the accident at Three Mile Island*. Washington, DC: Nuclear Regulatory Commission.

Form, W. H., and S. Nosow. 1958. *Community in disaster*. New York: Harper & Bros.

Forman, R. E. 1963. Resignation as a collective behavior response. *American Journal of Sociology* 69: 285–90.

Forrest, T. R. 1972. Structural differentiation in emergent groups. Ph.D. diss. Columbus: Department of Sociology, Ohio State University.

Foster, T. A. 1917–18. The Halifax disaster. *Journal of the Maine Medical Association*: 199–203.

Francis, P., and G. A. Kreps. 1984. *Disaster and the social order: Organization and social network. Final report*. National Science Foundation, Grant no. CEE-9121135.

Freedman, E. 1979. Separatism as strategy: Female institution building and American feminism, 1870–1930. *Feminist Studies* 5: 512–29.

Freeman, J. 1975. *The politics of women's liberation*. New York: David McKay.

Freeman, J. 1979. Resource mobilization and strategy: A model for analyzing social movement organization actions. In *The Dynamics of social movements*, ed. M. N. Zald and J. D. McCarthy, 167–89. Cambridge, MA: Winthrop.

Freeman, J. 1983. *Social movements of the sixties and seventies*. New York: Longman.

Freeman, J. 1991. How "sex" got into Title VII: Persistent opportunism as a maker of public policy. *Law and Inequality* 9: 163–84.

French, S., C. Ewing, and M. Isaacson. 1984. *Restoration and recovery following the Coalinga earthquake of May 1983*. Working paper, no. 50. Boulder: Institute of Behavioral Science, University of Colorado.

Friedan, B. 1963. *The feminine mystique*. New York: Dell.

Friedrichs, R. 1970. *A sociology of sociology*. New York: Free Press.

Friedsam, J. J. 1962. Older persons in disaster. In *Man and society in disaster*, eds. G. W. Baker and D. W. Chapman, 151–84. New York: Basic.

Friedson, E. 1970. *Professional dominance: The social structure of medical care*. Chicago: Aldine-Atherton.

Friedson, E. 1986. *Professional powers: A study of the institutionalization of formal knowledge*. Chicago: University of Chicago Press.

Fritz, C. E. 1961. Disasters. In *Contemporary social problems*, ed. R. K. Merton and R. A. Nisbet. New York: Harcourt.

Fritz, C. E., and E. S. Marks. 1954. The NORC studies of human behavior in disaster. *The Journal of Social Issues* 10, no. 3: 26–41.

Fritz, C. E., and J. H. Mathewson. 1957. *Convergence behavior in disasters*. National Research Council Disaster Study no. 9. Washington, DC: National Academy of Sciences.

Fritz, N. J., and D. L. Altheide. 1987. The mass media and the social construction of the missing children problem. *The Sociological Quarterly* 28: 473–92.

Fromm, E. 1963. *Escape from freedom.* New York: Holt, Reinhart & Winston.

Gale Research Associates. 1984. *Encyclopedia of associations.* Detriot.

Gallino, L. N.d. The interplay of culture and social structure in the mind. Preliminary paper. University of Turin, Department of Sociology.

Gamson, W. A. 1975. *The strategy of social protest.* Homewood, IL: Dorsey Press.

Gamson, W. A., and A. Modigliani. 1989. Media discourse and the public opinion on nuclear power: A constructionist approach. *American Journal of Sociology* 95: 1–37.

Gans, H. 1962. Urbanism and suburbanism as ways of life. In *Human behavior and social process*, ed. A. Rose. Boston: Houghton Mifflin.

Garfinkel, H. 1956. Conditions of successful degradation ceremonies. *American Journal of Sociology* 61: 420–4.

Garner, R. A., and M. N. Zald. 1987. The political economy of social movement sectors. In *Social movements in an organizational society*, ed. M. N. Zald and J. D. McCarthy. New Brunswick, NJ: Transaction Books.

Gehlen, F. L. 1977. Toward a revised theory of hysterical contagion. *Journal of Health and Social Behavior* 18: 27–35.

Gelb, J., and M. L. Palley. 1982. *Women and public policy.* Princeton: Princeton University Press.

Gergen, K., M. Gergen, and K. Meter. 1972. Individual orientations to prosocial behavior. *Journal of Social Issues* 28, no. 3: 105–30.

Gibbs, M. S. 1989. Factors in the victim that mediate between disaster and psychopathology: A review. *Journal of Traumatic Stress* 2, no. 4: 489–514.

Giddings, P. 1984. *When and where I enter: The impact of black women on race and sex in America.* New York: Bantam.

Gillespie, D. F., M. W. Sherraden, C. L. Streeter, and M. J. Zakour. 1986. Mapping networks of organized volunteers for natural hazard preparedness. Final report. St. Louis, MO: Washington University, George Warren Brown School of Social Work.

Gleser, G. C., B. L. Green, and C. Winget. 1981. *Prolonged psychosocial effects of disaster: A study of Buffalo Creek.* New York: Academic Press.

Goffman, E. [1974] 1986. *Frame analysis: An essay on the organization of experience.* Boston: Northeastern University Press.

Golec, J. 1983. A contextual approach to the social psychological study of disaster recovery. *Journal of Mass Emergencies and Disaster* 1 (August): 255–76.

Gormley, Mrs. F.A. 1918. "The Halifax Disaster." *Canadian Medical Association Journal*, 59–62.

Grey, J., and E. L. Quarantelli. 1984. The behavior of first responders and their initial definition of acute chemical emergencies. *Disaster Management* 4: 6–12.

Gruntfest, E. C. 1977. What people did during the Big Thompson flood. Working paper, no. 32. Boulder: Institute of Behavioral Science, University of Colorado.

The Guardian. 1989. Police name 94 who died. *The Guardian*, 18 April 1989, 2.

Gurney, J. N., and K. J. Tierney. 1982. Relative deprivation and social movements: A critical look at twenty years of theory and research. *The Sociological Quarterly* 23 (Winter): 33–47.

Gusfield, J. R. 1955. Social structure and moral reform: A study of Woman's Christian Temperance Union. *American Journal of Sociology* 61 (November).

Gusfield, J. R. 1963. *Symbolic crusade*. Urbana: University of Illinois Press.

Gusfield, J. R. 1970. *Protest, reform, and revolt: A reader in social movements*. New York: Wiley.

Gusfield, J. R. 1976. The literary rhetoric of science: Comedy and pathos in drinking driver research. *American Sociological Review* 41: 16–34.

Gusfield, J. R. 1981a. *The culture of public problems: Drinking-driving and the symobolic order*. Chicago: University of Chicago Press.

Gusfield, J. R. 1981b. Social movements and social change: Perspectives of linearity and fluidity. In *Research in social movements, conflict and change*, ed. L. Kriesberg. Greenwich, CT: JAI Press.

Haas, J. E., and P. Trainer. 1974. Effectiveness of the Tsunami Warning System in selected coastal towns in Alaska. Proceeding of the Fifth World Conference on Earthquake Engineering, Rome, Italy.

Haas, J. E., R. Kates, and M. Bowden. 1977. *Reconstruction following disaster*. Cambridge: MIT.

Hage, J., and M. Aiken. 1967. Relationship of centralization to other structural properties. *Administrative Science Quarterly* 12 (June): 72–91.

Haicheng Earthquake Study Delegation. 1977. Prediction of the Haicheng earthquake. *Eos* 58: 236–72.

Haines, H. H. 1984. Black radicalization and the funding of civil rights. *Social Problems* 32: 31–43.

Hall, R. H. 1987. *Organizations: Structures, processes and outcomes*. Englewood Cliffs, NJ: Prentice-Hall.

Hannigan, J. A. 1985. Alain Touraine, Manuel Castells, and social movement theory: A critical appraisal. *The Sociological Quarterly* 26, no. 4: 435–54.

Hannon, M. T., and J. Freeman. 1989. *Organizational ecology*. Cambridge: Harvard University Press.

Harris, D. F. N.d. A history of the medical aspects of the disaster in Halifax and Dartmouth, Nova Scotia. M61 Vol. 2124. Public Archives of Nova Scotia.

Harris, D. F., ed. N.d. False impressions as to the part played by local doctors. In *A history of the medical aspects of the disaster and Halifax and Dartmouth, Nova Scotia*. Public Archives of Nova Scotia.

Harris, D. F., ed. N.d. The activities at New Glasgow. In *A history of the medical aspects of the disaster and Halifax and Dartmouth, Nova Scotia*. Public Archives of Nova Scotia.

Harris, D. F., ed. N.d. The formation by Lieut-Colonel McKelvey Bell of the Medical Relief Committee. In *A history of the medical aspects of the disaster and Halifax and Dartmouth, Nova Scotia*. Public Archives of Nova Scotia.

Harris, D. F., ed. N.d. The septicity of the wounds. In *A history of the medical aspects of the disaster and Halifax and Dartmouth, Nova Scotia*. Public Archives of Nova Scotia.

Harris, D. F., ed. N.d. The work of Dr. C. C. Hubly. In *A history of the medical aspects of the disaster and Halifax and Dartmouth, Nova Scotia*, ed. D. F. Harris. Public Archives of Nova Scotia.

Harrison, C. E. 1988. *On account of sex*. Berkeley: University of California Press.

Hartman, E. 1988. *Conceptual foundations of organization theory*. Cambridge, MA: Ballinger.

Hayes, M. R. 1917. Medical Officer on board the U.S.S. Old Colony. Memo to commanding officer: 9 December 1917. MG 27 Vol. 2. Public Archives of Nova Scotia.

Hefez, A. 1985. The role of the press and the medical community in the epidemic of "Mysterious gas poisoning" in the Jordan West Bank. *American Journal of Psychiatry* 142: 833–37.

Heise, D. 1975. *Causal analysis*. New York: Wiley.

Helfgot, J. 1981. *Professional reforming*. Lexington, MA: Lexington.

Henig, J. R. 1982. *Neighborhood mobilization: Redevelopment and response*. New Brunswick, NJ: Rutgers University Press.

Hiroi, O., S. Mikami, and K. Miyata. 1985. A study of mass media reporting in emergencies. *International Journal of Mass Emergencies and Disasters* 3, no. 1: 21–50.

Hirsch, E. L. 1990. Sacrifice for the cause. *American Sociological Review* 55, no. 2: 243–54.

Hirschman, C. 1983. America's melting pot reconsidered. *Annual Review of Sociology* 9: 397–423.

Hodge, D., V. Sharp, and M. Marts. 1981. Contemporary responses to volcanism: Case studies from the Cascades and Hawaii. In *Volcanic activity and human ecology*, ed. P. D. Sheets and D. K. Grayson. New York: Academic Press.

Hoffman, L. M. 1989. *The politics of knowledge*. Albany: State University of New York Press.

Hogg, S. 1980. Reconstruction following seismic disasters in Venzone, Fruili. *Disasters* 2: 173–85.

Holland, C. J. 1989. Effective utilization of victim volunteers in the emergency response. *Proceedings of the International Workshop on Earthquake Injury Epidemiology for Mitigation and Response*, 321–35. Baltimore: Johns Hopkins University.

Hood, C. 1989. Public administration and public policy: Intellectual challenges for the 1990's. *Australian Journal of Public Administration* 48, no. 4: 346–58.

Hood, C., and M. Jackson. 1991. The new public management: A recipe for disaster. *Canberra Bulletin of Public Administration* 64 (May): 16–24.

Hopper, R. D. 1950. The revolutionary process. *Social Forces* 28: 270–79.

Howell, C. D. 1988. *A century of care: A history of the Victoria General Hospital in Halifax 1887–1987*. Halifax: Victoria General Hospital.

Hull, R. B., and D. E. Wenger. 1992. *Relationships among volunteering, evacuation and place attachments: Charleston after Hurricane Hugo*. College Station, TX: Hazard Reduction & Recovery Center, Texas A&M University.

Inglis, S. 1989. Terrace pens turned safe ground to death trap. *Observer Sunday*, 16 April 1989, 3.

Jackman, N. R. 1957. Collective protest in relocation centers. *American Journal of Sociology* 63 (November): 264–72.

Jacobs, N. 1965. The phantom slasher of Taipei: Mass hysteria in a non-western society. *Social Problems* 112: 318–28.

Janowitz, M. 1978. *The last half century*. Chicago: University of Chicago Press.

Jefferson, H. B.,1958. Day of disaster. *Atlantic Advocate* 48, no. 5 (January): 15, 17–18, 21, 23, 24.

Jenkins, J. C. 1983. Resource mobilization theory and the study of social movements. *Annual Review of Sociology* 9: 527–53.

Jenkins, J. C. 1987. Interpreting the stormy sixties: Three theories in search of a political age. In *Research in political sociology, Vol. 3.* ed. R. G. Braungart, 269–305. Greenwich, CT: JAI Press.

Jenkins, J. C., and C. M. Eckert. 1986. Channelling black insurgency: Elite patronage and professional SMOs. Paper presented at the annual meetings of the American Sociological Association, New York, 2 September.

Johnson, D. M. 1945. The phantom anesthetist of Mattoon: A field study of mass hysteria. *Journal of Abnormal and Social Psychology* 40: 175–86.

Johnson, N. R. 1987a. Panic at The Who concert stampede: An empirical assessment. *Social Problems* 34, no. 4: 362–73.

Johnson, N. R. 1987b. Panic and the breakdown of social order: Popular myth, social theory, empirical evidence. *Sociological Focus* 20: 171–83.

Johnson, N. R. 1988. Fire in a crowded theater: A descriptive analysis of the emergence of panic. *International Journal of Mass Emergencies and Disasters* 6: 7–26.

Johnston, D., and N. Johnson. 1989. Role expansion in disaster: An investigation of employee behavior in a nightclub fire. *Sociological Focus* 22: 39–51.

Johnstone, D. 1919. The tragedy of Halifax: The greatest American disaster in the war. Manuscript. Public Archives of Nova Scotia, Circa.

Joyce, D. 1989. Why do police officers laugh at death? The *Psychologist* (September): 379–81.

Kapferer, J. 1990. *Rumors: Uses, interpretations, and images.* Trans. B. Fink. New Brunswick, NJ: Transaction.

Katz, E. 1957. The two-step flow of communication: An up-to-date report on an hypothesis. *Public Opinion Quarterly* 21: 61–78.

Keating, J. P. 1982. The myth of panic. *Fire Journal* (May): 57–62.

Kelley, H. H., J. Condry, A. Dahlke, and A. Hill. 1965. Collective behavior in a simulated panic situation. *Journal of Experimental and Social Psychology* 1: 20–54.

Kemper, T. D. 1980. Altruism and voluntary action. In *Participation in social and political activities*, eds. D. H. Smith, J. Macaulay, and Associates, 306–38. San Francisco: Jossey-Bass.

Kerckhoff, A. C. 1970. A theory of hysterical contagion. In *Human nature and collective behavior: Papers in honor of Herbert Blumer*, ed. T. Shibutani, 81–93. Englewood Cliffs, NJ: Prentice-Hall.

Kerckhoff, A. C., and K. W. Back. 1968. *The June bug: A study of hysterical contagion.* New York: Appleton-Century-Crofts.

Kerckhoff, A. C., K. W. Back, and N. Miller. 1965. Sociometric patterns in hysterical contagion. *Sociometry* 28: 2–15.

Killian, L. M. 1952. The significance of multiple group membership in disasters. *American Journal of Sociology* 57: 309–14.

Killian, L. M. 1956. An introduction to methodological problems of field studies in disasters. Washington, DC: National Academy of Sciences-National Research Council.

Killian, L. M. 1964. Social movements. In *Handbook of modern sociology*, ed. R. E. L. Faris, 426–55. Chicago: Rand McNally.

Kingswood, J. E. 1989. In defence of black humour (Letter 2). *The Psychologist* (November): 488.

Kitz, J. F. 1989. *Shattered city: The Halifax explosion and the road to recovery.* Halifax: Nimbus Publishing Limited.

Klandermans, B. 1984. Mobilization and participation: Social-psychological expansions of resource mobilization theory. *American Sociological Review* 49 (October): 583–600.

Klandermans, B. 1989. Grievance interpretation and success expectations: The social construction of protest. *Social Behavior* 4: 113–25.

Klandermans, B. 1991. Peace movements in Western Europe and the United States. In *International Social Movement Research.* ed. B. Klandermans, 1–39. Vol. 3. Greenwich, CT: JAI Press.

Klapp, O. E. 1956. American villain types. *American Sociological Review* 21, no. 337–40.

Klapp, O. E. 1962. *Heroes, villains, and fools: The changing American character.* Englewood Cliffs, NJ: Prentice-Hall.

Klapp, O. E. 1964. Mexican social types. *American Journal of Sociology* 69: 404–14.

Klapp, O. E. 1969. *Collective search for identity.* New York: Holt, Rinehart & Winston.

Klapp, O. E. 1972. *Currents of unrest, an introduction to collective behavior.* New York: Holt, Rinehart & Winston.

Klapp, O. E. 1978. *Opening and closing: Strategies of information adaption in society.* Cambridge: Cambridge University Press.

Klein, E. 1984. *Gender politics: From consciousness to mass politics.* Cambridge: Harvard University Press.

Knopf, T. A. 1975. *Rumors, race, and riots.* New Brunswick, NJ: Transaction.

Kobasa, S. C., S. R. Maddi, and S. Khan. 1982. Hardiness and health: A prospective study. *Journal of Personality and Social Psychology* 42: 168–77.

Koenig, F. 1985. *Rumor in the marketplace: The social psychology of commercial hearsay.* Dover, MA: Auburn House.

Kreps, G. A. 1978. The organization of disaster response: Some fundamental theoretical issues. In *Disasters: Theory and research*, ed. E. L. Quarantelli, 65–87. London: Sage.

Kreps, G. A. 1981. The worth of the NAS-NRC (1952–63) and DRC (1963–present) studies of individual and social responses to disasters. In *Social science and natural hazards*, ed. J. D. Wright and P. H. Rossi, 91–122. Cambridge, MA: Abt.

Kreps, G. A. 1983. The organization of disaster response: Core concepts and processes. *International Journal of Mass Emergencies and Disasters* 1: 439–67.

Kreps, G. A. 1984. Sociological inquiry and disaster research. *Annual Review of Sociology* 10: 309–30.

Kreps, G. A. 1985. Disaster and the social order. *Sociological Theory* 3: 49–65.

Kreps, G. A. 1989a. *Social structure and disaster.* Newark: University of Delaware Press.

Kreps, G. A. 1989b. Future directions in disaster research: The role of taxonomy. *International Journal of Mass Emergencies and Disasters* 7 (November): 215–41.

Kreps, G. A. 1991a. Organizing for emergency management. In *The principles and practice of emergency management*, eds. T. E. Drabek and G. J. Hoetmer. Washington, DC: International City Management Association.

Kreps, G. A. 1991b. Answering organizational questions: A brief for structural codes. In *Studies in organizational sociology*, ed. G. Miller. Greenwich, CT: JAI Press.

Kreps, G. A., and S. L. Bosworth. 1987. Social networks and organizing: Interdependence as process. Final report, National Science Foundation, Grant no. CEE-8400486.

Kreps, G. A., and S. L. Bosworth. 1989. Response to commentary of Ralph Turner. In *Social structure and disaster*, ed. G. A. Kreps, 213–19. Newark: University of Delaware Press.

Kreps, G. A., J. G. Crooks, and J. R. Linn. 1987. Organizational change: Static and dynamic aspects. Final report, National Science Foundation, Grant no. CEE-8400486.

Kunreuther, H. 1978. *Disaster insurance protection: Public policy lessons.* New York: Wiley.

Kunreuther, H., and E. Ley, eds. 1982. *The risk analysis controversy: An institutional perspective.* Berlin, Heidelberg, New York: Springer-Verlag.

Land, K. C. 1983. Social indicators. *Annual Review of Sociology* 9: 1–26.

Lang, K., and G. E. Lang. 1961. *Collective dynamics.* New York: Thomas Y. Crowell.

LaPiere, R. T. 1938. *Collective behavior.* New York: McGraw-Hill.

Lardry, T., and G. Rogers. 1982. *Warning confirmation and dissemination.* Pittsburgh: Center for Social and Urban Research, University of Pittsburgh.

Lazarsfeld, P., B. Berelson, and H. Gaudet. 1948. *The people's choice.* New York: Columbia University Press.

Lazarus, R., and S. Folkman. 1984. *Stress, appraisal and coping.* New York: Springer.

Leahy, P., and A. Mazur. 1978. A comparison of movements opposed to nuclear power, fluoridation, and abortion. *Research on Social Movements, Conflicts, and Change* 1: 143–54.

Lechat, M. F. 1976. The epidemiology of disasters. *Proceedings of the Royal Society of Medicine*, 69: 421–426.

Lefcourt, H. M. 1976. *Locus of control: Current trends in theory and research.* New York: Wiley.

Lehto, M. R., and J. M. Miller. 1986. *Warnings, vol. I: Fundamentals, designs, and evaluation methodologies.* Ann Arbor, MI: Fuller Technical.

Leik, R. K. et al. 1981. *Community response to natural hazard warnings: Final report.* Minneapolis: University of Minnesota.

Lemert, E. M. 1951. *Social pathology: A systematic approach to the theory of sociopathic behavior.* New York: McGraw-Hill.

Lemert, E. M. 1972. *Human deviance, social problems, and social control.* 2d ed. Englewood Cliffs, NJ: Prentice-Hall.

Lenin, V. I. 1970. What is to be done? In *Protest, reform, and revolt*, ed. J. R. Gusfield, 458–72. New York: Wiley.

Leonard, V. A. 1973. *Police predisaster preparation.* Springfield, IL: Charles C. Thomas.

Lewis, J. M. 1982. Crowd control at English football matches. *Sociological Focus* 15: 417–23.

Lewis, J. M. 1989. A value-added analysis of the Heysel Stadium soccer riot. *Current Psychology* 8: 15–29.

Li, Q. 1988. Socio-psychological analysis of earthquake rumors and earthquake misdisseminations. Trans. Q. Yang and W. Wang. *Recent Developments in World Seismology* 120: 8–14.

Lieberson, S. 1985. *Making it count.* Berkeley: University of California Press.

Lieberson, S., and A. R. Silverman. 1965. The precipitants and underlying conditions of race riots. *American Sociological Review* 30: 887–98.

Lifton, R. J., and E. Olson. 1976. The human meaning of total disaster: The Buffalo Creek experience. *Psychiatry* 39 (February): 1–18.

Lind, A. 1946. *Hawaii's Japanese: An experiment in democracy.* Princeton, NJ: Princeton University Press.

Linder, S. B. 1970. *The harried leisure class.* New York: Columbia University Press.

Lindsey, R. K. 1989. Heartless. *Sunday Mirror*, 23 April 1989.

Linn, J. R., and G. A. Kreps. 1989. Disaster and the restructing of organization. In *Social structure and disaster*, ed. G. A. Kreps, 108–35. Newark: University of Delaware Press.

Lipset, S. M., and W. Schneider. 1983. *The confidence gap.* New York: Free Press.

Lipset, S. M., M. Trow, and J. Coleman. 1962. *Union democracy.* New York: Anchor.

Lipsky, M. 1971. *Protest in the city.* Englewood Cliffs, NJ: Prentice-Hall.

Livermore, D., and J. P. Wilson. 1981. The Mississauga Train Derailment and Evacuation, 10–16 November 1979. *Canadian Geographer* 25, no. 4: 365–75.

Lofland, J. 1977. *Doomsday cult.* New York: Irvington.

Lofland, J. 1979. White-hot mobilization: Strategies of a millinerian movement. In *Dynamics of social movements*, eds. M. N. Zald and J. D. McCarth, 157–66. Cambridge, MA: Winthrop.

Lofland, J. 1981. Collective behavior: Elementary forms. In *Social psychology: Sociological perspectives*, eds. M. Rosenburg and R. H. Turner, 411–46. New York: Basic.

Lofland, J. 1985. *Protest: Studies of collective behavior and social movements.* New Brunswick, NJ: Transaction.

Lofland, J., and V. Johnson. 1990. Citizen surges: A domain in movement studies and a perspective on peace activism in the eighties. Unpublished manuscript.

Lofland, J., and L. H. Lofland. 1984. *Analyzing social settings: A guide to qualitative observation and analysis.* Belmont, CA: Wadsworth.

Loizos, P. 1977. A struggle for meaning: Reactions to disaster amongst Cypriot refugees. *Disasters* 1, no. 3: 231–39.

Lyman, G. H. 1919. *The story of the Massachusetts Committee of Public Safety.* Boston: Wright & Patten.

Mack, R. W., and G. W. Baker. 1961. *The occasion instant: The structure of social responses to repeated air raid warnings.* Disaster Study, no. 15. Washington, DC: National Research Council, National Academy of Sciences.

MacLennan, H. 1958. *Barometer rising.* Toronto: McClelland & Stewart Limited.

Mann, L., T. Nagel, and P. Dowling. 1976. A study of an economic panic: The run on the Hindmarsh Building Society. *Sociometry* 39: 223–35.

Manning, R. 1973. Fifteen years of collective behavior. *The Sociological Quarterly* 14: 279–86.

Marger, M. 1984. Social movement organization and response to environmental change. *Social Problems* 32: 16–30.

Marwell, G., and P. Oliver. 1984. Collective action theory and social movements research. *Research in Social Movements, Conflicts, and Change* 7: 1–27.

Marx, G. T. 1987. Untitled and unpublished book-length manuscript.

Marx, G. T., and J. L. Wood. 1975. Strands of theory and research in collective behavior. *Annual Review of Sociology* 1: 363–428.

Massey, D. 1981. Dimensions of the new immigration to the United States and the prospects for assimilation. *Annual Review of Sociology* 7.

Mauss, A. L. 1975. *Social problems as social movements.* Philadelphia: J. B. Lippincott.

May, P. J. 1985. *Recovering from catastrophes: Federal disaster relief policy and politics.* Westport, CT: Greenwood.

Mayer, M. 1990. Theoretical assumptions in American social movement research and their implications. Paper presented in the Roundtable Social Movement Theory, Session on Social Movements and Collective Behavior of the ISA World Congress, Madrid, Spain.

Mayhew, B. 1980. Structuralism versus individualism: Part 1, shadowboxing in the dark. *Social Forces* 59: 335–76.

McAdam, D. 1982. *Political process and the development of black insurgency, 1930–1970.* Chicago: University of Chicago Press.

McAdam, D. 1983. The decline of the Civil Rights Movement. In *Social movements of the sixties and seventies*, ed. J. Freeman, 279–97. New York: Longman.

McAdam, D. 1984. Tactical innovation and the pace of insurgency. *American Sociological Review* 48: 735–53.

McCammon, S., T. W. Durham, E. J. Allison, and J. E. Williamson. 1988. Emergency workers' cognitive appraisal and coping with traumatic events. *Journal of Traumatic Stress* 1, no. 3: 353–72.

McCarthy, J. D. 1991. Forward to *The myth of the madding crowd*, by C. McPhail. New York: Aldine de Gruyter.

McCarthy, J. D., and M. N. Zald. 1973. *The trend of social movements in America: Professionalization and resource mobilization.* Morristown, NJ: General Learning Press.

McCarthy, J. D., and M. N. Zald. 1975. Organizational intellectuals and the criticism of society. *Social Service Review* 49: 344–62.

McCarthy, J. D., and M. N. Zald. 1977. Resource mobilization and social movements: A partial theory. *American Journal of Sociology* 82: 1212–41.

McFarlane, A. C. 1988. The aetiology of post-tramatic stress disorders following a natural disaster. *British Journal of Psychiatry* 152: 116–21.

McIntosh, C. N.d. MG 1 Vol. 2124, no. 29. Public Archives of Nova Scotia.

McKeen, J. C. 1967. The Halifax explosion: An eyewitness account. 6 December 1967, 1–13.

McLuckie, B. F. 1970. A study of functional response to stress in three societies. Ph.D. diss. Columbus: Ohio State University.

McMurray, Mrs., A. 1974. Former Haliqonian views disaster. *The Chronicle-Herald*, 6 December 1974, 11.

McPhail, C. 1971. Civil disorder participation: A critical examination of recent research. *American Sociological Review* 36: 1058–73.

McPhail, C. 1988. Micro-macro levels and units of analysis in the study of collective behavior and collective action. Unpublished manuscript.

McPhail, C. 1991. *The myth of the madding crowd.* New York: Aldine de Gruyter.

McPhail, C., and D. Miller. 1973. The assembling process: A theoretical and empirical examination. *American Sociological Review* 38: 721–35.

Medalia, N. Z. 1959–60. Who cries wolf? The reporters of damage to police in a pseudo-disaster. *Social Problems* 7: 233–40.

Medelia, N. Z., and O. N. Larsen. 1958. Diffusion and belief in a collective delusion: The Seattle windshield pitting epidemic. *American Sociological Review* 23: 221–32.

Medical Relief Committee. 1917. Daily consolidated report. RG 24 HQ 71-26-89-3 Vol. 1, 108. 19 December.

Medical Relief Committee. 1918a. Weekly consolidated report. RG 24 HQ 71-26-89-3 Vol. 1, 178. 12 January.

Medical Relief Committee. 1918b. Memo to Deputy Minister. RG 24 6159 HQ 71-26-89-3 Vol. 2. 12 December.

Melucci, A. 1985. The symbolic challenge of contemporary movements. *Social Research* 52, no. 4: 789–816.

Merton, R. K. 1957. *Social theory and social structure*. Rev. ed. New York: Free Press.

Metson, G. 1978. *The tragedy of Halifax*. Toronto: McGraw-Hill. (This book contains the official history written by MacMechan.)

Meyers, K. A. 1990. A dialectical analysis of role enactment in established and emergent organizations. Master's thesis. Williamsburg: College of William and Mary.

Mikami, S., and K. Ikeda. 1985. Human response to disasters. *International Journal of Mass Emergencies and Disasters* 3, no. 1: 107–32.

Mileti, D. S. 1974. *A normative causal model analysis of disaster warning response*. Ann Arbor, MI: University Microfilms.

Mileti, D. S. 1975. *Natural hazard warning systems in the United States: A research assessment*. Boulder: Institute of Behavioral Science, University of Colorado.

Mileti, D. S. 1987. Sociological methods and disaster research. In *Sociology of disasters*, eds. R. R. Dynes, B. DeMarchi, and C. Pelanda. Milan, Italy: Franco Angeli.

Mileti, D. S., and E. M. Beck. 1975. Communication in crisis: Explaining evacuation symbolically. *Communication Research* 2 (January): 24–49.

Mileti, D. S., and P. O'Brien. 1991. Public response to the Loma Prieta earthquake emergency and aftershock warnings: Findings and lessons. Fort Collins: Hazards Assessment Laboratory, Colorado State University.

Mileti, D. S., and J. H. Sorensen. 1987. Why people take precautions against natural disasters. In *Taking care: Why people take precautions*, ed. N. Weinstein, 189–207. New York: Cambridge University Press.

Mileti, D. S., and J. H. Sorensen. 1990. *Communication of emergency public warnings: A social science perspective and state-of-the-art assessment*. Washington, DC: Federal Emergency Management Agency.

Mileti, D. S., T. E. Drabek, and J. E. Haas. 1975. *Human systems in extreme environments: A sociological perspective*. Boulder: Institute of Behavioral Science, University of Colorado.

Mileti, D. S., J. Hutton, and J. H. Sorensen. 1981. *Earthquake prediction response and options for public policy*. Boulder: University of Colorado.

Miller, A. 1983. Is confidence rebounding? *Public Opinion* (June): 16–20.

Miller, D. L. 1985. *Introduction to collective behavior*. Belmont, CA: Wadsworth.

Miller, F. 1983. The end of SDS and the emergence of weatherman: Demise through success. In *Social movements of the sixties and seventies*, ed. J. Freeman, 279–97. New York: Longman.

Miller, G. A., D. L. Anderson, and J. Conaty. 1985. Assessing the samples of prior organizational research. *Journal of Management Studies* 22 (July): 369–83.

Mills, C. W. 1956. *The power elite*. New York: Oxford University Press.

Minnon, M. A. 1977. *Miracles and mysteries: The Halifax explosion 6 December 1917*. Hantsport, NS: Lancelot.

Mintz, A. 1951. Nonadaptive group behavior. *The Journal of Abnormal Psychology* 46: 150–59.

Mintzberg, H. 1979. *The structuring of organizations: A synthesis of the research*. Englewood Cliffs, NJ: Prentice-Hall.

Minutes of meeting of the Halifax Relief Committee, 11 A.M., 6 December 1917. Unpublished Public Archives of Nova Scotia.

Mitchell, J. T. 1983. When disaster strikes: The critical incident stress debriefing process. *Journal of Emergency Medical Services* (January) : 36–39.

Mitchell, J. T. 1988. The impact of stress on emergency service personnel: Policy issues in emergency response. In *Managing disaster: Strategies and policy perspectives*, ed. L. K. Comfort, 199–214. Durham: Duke University Press.

Mogil, M., and H. Groper. 1977. NWS's Severe Local Storm Warning and Disaster Preparedness Programs. *Bulletin of the American Meterological Society* 58, no. 1: 318–29.

Mooney, J. A. 1989. Organization and role: Conception and measurement. Master's thesis. Williamsburg: College of William and Mary.

Moore, H. E., F. L. Bates, M. V. Layman, and V. J. Parenton. 1963. *Before the wind: A study of response to Hurricane Carla*. National Research Council Disaster Study, no. 19. Washington, DC: National Academy of Sciences.

Moore, W. B. N.d. Personal narrative—Halifax explosion 6 December 1917. In *A history of the medical aspects of the disaster and Halifax and Dartmouth, Nova Scotia*, ed. D. F. Harris. Unpublished Public Archives of Nova Scotia.

Moran, C. C. 1989. Coping styles in volunteer rescue workers—a preliminary report. Working Paper, no. 2. Disaster Management Studies Centre, Cumberland College of Health Sciences, Lidcombe, Australia: University of Sydney.

Moran, C. C. 1990. Does the use of humor as a coping strategy affect stresses associated with emergency work? *International Journal of Mass Emergencies and Disasters* 8, no. 3 (November).

Moran, C. C., N. R. Britton, and B. J. Correy. 1992. Characterizing emergency responders. *International Journal of Mass Emergencies and Disasters* 10, no. 1 (March): 207–16.

Morris, A. 1981. The black southern sit-in movement. *American Sociological Review* 46: 744–67.

Morris, A. 1984. *The origins of the Civil Rights Movement: Black communities organizing for change*. New York: Free Press.

Morris, A., and C. Herring. 1987. Theory and research in social movements: A critical review. In *Annual Review of Political Science*. ed. S. Long, 137–98. Norwood NJ: Ablex.

Morris, D. 1981. *The soccer tribe*. London: Jonathan Cape.

Morrissey, J. P., R. H. Hall, and M. L. Lindsey. 1982. *Interorganizational relations: A sourcebook of measures for mental health programs*. Rockville, MD: National Institute of Mental Health.

Mueller, C. M. 1987. Collective consciousness, identity transformation, and the rise of women in public office in the United States. In *The Women's movements of the United States and Western Europe*, eds. M. F. Katzenstein and C. M. Mueller, 89–108. Philadephia: Temple University Press.

Murphy, G. N.d. Dr. George Murphy's account of the surgery at the Victoria General Hospital. In *A history of the medical aspects of the disaster and Halifax and Dartmouth, Nova Scotia*, ed. D. F. Harris. MGl Vol. 2124. Public Archives of Nova Scotia.

Myambo, K. 1989. Stress and coping in a voluntary residential relocation program in rural Africa. In *Clinical and Abnormal Psychology*, eds. P. F. Lovibond and P. H. Wilson. Amsterdam: Elsevier.

Myers, K. A. 1990. A dialectical analysis of role enactment in established and emergency organizations. Master's thesis. Williamsburg: College of William and Mary.

Naisbitt, J. 1982. *Megatrends*. New York: Warner Books.

National Opinion Research Center. 1983. *General Social Surveys, 1972–82: Cummulative codebook*. Chicago: University of Chicago Press.

Neal, D. M. 1984. Blame assignment in a diffuse disaster situation: A case example of the role of an emergent citizen group. *International Journal of Mass Emergencies and Disasters* 2 (August): 251–66.

Nehnevajsa, J. 1989. *Volunteering for emergency preparedness*. Pittsburgh: University Center for Social and Urban Research.

Neiss, R. 1988. Reconceptualizing arousal: Psychobiological states in motor performance. *Psychlogical Bulletin* 103, no. 3: 345–66.

Nelson, C., M. Coovert, A. Kurtz, B. Fritzche, C. Crumley, and A. Powell. 1989. *Models of hurricane evacuation behavior*. Tampa: Department of Psychology, University of South Florida.

Neuman, R., and I. de Sola Pool. 1984. The flow of communications into the home. In *The future of the mass audience report*. Cambridge: MIT Press.

Nigg, J. M. 1987. Communication under conditions of uncertainty: Understanding earthquake forecasting. In *Sociology of disasters*, eds. R. R. Dynes, B. DeMarchi, and C. Pelanda, 103–17. Milan, Italy: Franco Angeli.

Nisbet, R. 1981. *Twilight of authority*. New York: Oxford University Press.

O'Neill, W. L. 1971. *Everyone was brave*. Chicago: Quadrangle Books.

Oberschall, A. 1973. *Social conflict and social movements*. Englewood Cliffs, NJ: Prentice-Hall.

Oberschall, A. 1978. The decline of the 1960s social movements. *Research in Social Movements, Conflicts and Change* 1: 257–89.

Oliver, P. E. 1989. Bringing the crowd back in: The nonorganizational elements of social movements. *Research in Social Movements, Conflicts and Change* 11: 1–30.

Oliver, P. E., and G. Marwell. 1988. The paradox of group size in collective action: A theory of critical mass II. *American Sociological Review* 53: 1–8.

Oliver-Smith, A. 1986. *The martyred city*. Albuquerque: University of New Mexico Press.

Oliver-Smith, A. 1990. Post-disaster housing reconstruction and social inequality: A challenge to policy and practice. Paper presented at the annual meetings of the Society for Applied Anthropology, York, England, March.

Palm, R. 1981. *Real estate agents and special study zone disclosure*. Boulder: Institute of Behavioral Science, University of Colorado.

Park, R. E., and E. W. Burgess. 1921. *Introduction to the science of sociology*. Chicago: University of Chicago Press.

Peacock, W., C. Killian, and F. Bates. 1987. The effects of disaster damage and housing aid on household recovery following the 1976 Guatemalan earthquake. *International Journal of Mass Emergencies and Disasters* 5, no. 1: 68–88.

Perrow, C. 1979. The sixties observed. In *The dynamics of social movements*, eds. M. N. Zald and J. D. McCarthy, 192–211. Cambridge, MA: Winthrop.

Perrow, C. 1984. *Normal accidents: Living with high-risk technologies*. New York: Basic.

Perrow, C. 1986. *Complex organizations: A critical perspective*. New York: Random House.

Perry, J., and M. D. Pugh. 1978. *Collective behavior: Response to social stress*. St. Paul, MN: West.

Perry, J., R. Hawkins, and D. M. Neal. 1983. Giving and receiving aid. *International Journal of Mass Emergencies and Disasters* 1: 171–88.

Perry, R. W. 1979a. Detecting psychopathological reactions to natural disaster—a methodological note. *Social Behavior and Personality* 2: 173–77.

Perry, R. W. 1979b. *The social psychology of civil defense*. Lexington, MA: Lexington Books.

Perry, R. W. 1979c. Evacuation decision-making in natural disasters. *Mass Emergencies* 4, no. 1: 25–38.

Perry, R. W. 1983. Population evacuation in volcanic eruptions, floods, and nuclear power plant accidents: Some elementary comparisons. *Journal of Community Psychology* 11 (January): 36–47.

Perry, R. W. 1985. *Comprehensive emergency management*. Greenwich, CT: JAI Press.

Perry, R. W. 1987. Disaster preparedness and response among minority citizen. In *Sociology of disasters*, eds. R. R. Dynes, B. DeMarchi, and C. Pelanda, 135–51. Milan, Italy: Franco Angeli.

Perry, R. W. 1989. Taxonomy, classification, and theories of disaster phenomena. In *Social structure and disaster*, ed. G. Kreps. Newark: University of Delaware Press.

Perry, R. W., and M. R. Greene. 1982. The role of ethnicity in the emergency decision-making process. *Sociological Inquiry* 52 (Fall): 309–34.

Perry, R. W., and M. R. Greene. 1983a. *Citizen response to volcanic eruptions: The case of Mt. St. Helens*. New York: Irvington.

Perry, R. W., and M. R. Greene. 1983b. *American minority citizens in disasters*. Seattle: Battelle Human Affairs Research Centers.

Perry, R. W., and M. K. Lindell. 1986. *Twentieth-century volcanicity at Mt. St. Helens: The routinization of life near an active volcano*. Tempe: School of Public Affairs, Arizona State University.

Perry, R. W., M. R. Greene, and M. K. Lindell. 1980. *Human response to volcanic eruptions*. Seattle: Battelle Human Affairs Research Centers.

Perry, R. W., M. R. Greene, and A. Mushkatel. 1983. American minority citizens in disasters. Final report to the National Science Foundation, Grant no. PFR-80-19297. Seattle: Battelle Human Affairs Research Centers.

Perry, R. W., M. K. Lindell, and M. R. Greene. 1981. *Evacuation planning in emergency management*. Lexington, MA: Lexington.

Perry, R. W., M. K. Lindell, and M. R. Greene. 1982. Threat perception and public response to volcano hazard. *Journal of Social Psychology* 116 (April): 199–204.

Peterson, W. A., and N. P. Gist. 1951. Rumor and public opinion. *American Journal of Sociology* 57: 159–67.

Pfuhl, E. H. 1980. *The deviance process.* New York: D. Van Nostrand.

Phinney, Col. N.d. MGI Vol. 2124, 215. Public Archives of Nova Scotia.

Pinard, M., and R. Hamilton. 1988. Intellectuals and the leadership of social movements: Some comparative perspectives. Montreal: McGill Working Papers on Social Behavior

Piven, F. F., and R. A. Cloward. 1977. *Poor people's movements.* New York: Pantheon.

The Plain Dealer. 1989. Soccer used as escape. *The Plain Dealer*, 23 April 1989, 9A. (*New York Times* Wire Service).

Prasad, J. 1935. The psychology of rumor: A study relating to the great Indian Earthquake of 1934. *British Journal of Sociology* 26: 1–15.

Price, J. L. 1972. *Handbook of organizational measurement.* Lexington, MA: D. C. Heath.

Prince, S. H. 1920. *Catastrophe and social change.* New York: Columbia University.

Qiu, Z. 1987. Social responses to and a preliminary analysis of an episode of earthquake misdissemination after the Dedu earthquakes of 1986. Trans. Q. Yang and W. Wang. *Recent Developments in World Seismology* 103: 19–22.

Quarantelli, E. L. 1954. The nature and conditions of panic. *American Journal of Sociology* 60 (November): 267–75.

Quarantelli, E. L. 1957. The behavior of panic participants. *Sociology and Social Research* 41: 187–94.

Quarantelli, E. L. 1960. A note on the protective function of the family in disasters. *Marriage and Family Living* 22: 263.

Quarantelli, E. L. 1966. Organizations under stress. In *Symposium on emergency operations*, ed. R. Brictson. Santa Monica, CA: System Development Corporation.

Quarantelli, E. L. 1970. Emergent accommodation groups: Beyond current collective behavior typologies. In *Human nature and collective behavior: Papers in honor of Herbert Blumer*, ed. T. Shibutani, 111–23. Englewood Cliffs, NJ: Prentice-Hall.

Quarantelli, E. L., ed. 1972. *Proceedings of organizational and community responses to disasters: Japan–United States disaster research seminar.* Newark: Disaster Research Center, University of Delaware.

Quarantelli, E. L. 1975. *The command post point of view in the local mass communication system.* Preliminary Paper no. 22. Newark: Disaster Research Center, University of Delaware.

Quarantelli, E. L. 1980a. Some research emphases for studies on mass communication systems and disasters. In *Disasters and Mass Media*, 293–99. Washington, DC: National Academy of Sciences.

Quarantelli, E. L. 1981. *Panic behavior in fire situations: Findings and a model from the English language literature.* Newark: Disaster Research Center, University of Delaware.

Quarantelli, E. L. 1982. *Sheltering and housing after major community disasters: Case studies and general conclusions.* Newark: Disaster Research Center, University of Delaware.

Quarantelli, E. L. 1983. *Delivery of emergency medical services in disasters: Assumptions and realities.* New York: Irvington.

Quarantelli, E. L. 1984a. Perceptions and reactions to emergency warnings of sudden hazards. *Ekistics* 309 (November–December).

Quarantelli, E. L. 1984b. *Organizational behavior in disasters and implications for disaster planning.* Emmitsburg, MD: National Emergency Training Center, Federal Emergency Management Agency.

Quarantelli, E. L. 1984c. *Evacuation behavior and problems: Findings and implications from the research literature.* Newark: Disaster Research Center, University of Delaware.

Quarantelli, E. L. 1985a. Social support systems: Some behavioral patterns in the context of mass evacuation activities. In *Disasters and mental health: Selected contemporary perspectives*, 122–36. Rockville, MD: DHHS Publication 85-1421.

Quarantelli, E. L. 1985b. Emergent citizen groups in disaster preparedness and recovery activities. Newark: Disaster Research Research Center, University of Delaware.

Quarantelli, E. L. 1985c. An assessment of conflicting views on mental health: The consequences of traumatic events. In *Trauma and its wake: The treatment of post-traumatic stress disorder*, ed. C. R. Figley. New York: Brunner/Mazel.

Quarantelli, E. L. 1987. Disaster studies: An analysis of the social historical factors affecting the development of research in the area. *International Journal of Mass Emergencies and Disasters* 5, no. 3: 285–310.

Quarantelli, E. L. 1988a. Community and organizational preparations for and response to acute chemical emergencies in the United States: Research findings and their wider applicability. In *Proceedings of European conference on emergency planning for industrial hazards*, 1–20. Ispara, Italy: Commission of the European Communities.

Quarantelli, E. L. 1988b. Characteristics of citizen groups which emerge with respect to hazardous waste sites. In *Psychological effects of hazardous toxic waste disposal on communities*, ed. D. L. Peck, 177–95. Springfield, IL: Charles C. Thomas.

Quarantelli, E. L., and R. R. Dynes. 1970. Editor's introduction: Issue on organizational and group behavior in disasters. *American Behavioral Scientist* 13: 325–30.

Quarentelli, E. L., and R. R. Dynes. 1972. When disaster strikes. *Psychology Today* 5: 66–70.

Quarantelli, E. L., and R. R. Dynes. 1976. Community conflict: Its absence and its presence in natural disasters. *Mass Emergencies* 1: 139–52.

Quarantelli, E. L., and R. R. Dynes. 1977. Response to social crisis and disaster. *Annual Review of Sociology* 3: 23–49.

Quarantelli, E. L., and R. R. Dynes. 1985. Community responses to disasters. In *Disasters and mental health: Selected contemporary perspectives*, ed. B. J. Sowder, 158–68. Rockville, MD: National Institute for Mental Health.

Quarantelli, E. L., and R. R. Dynes. 1989. Reconstruction in the context of recovery: Thoughts on the Alaskan earthquake. Paper presented at the Conference on Reconstruction after Urban Earthquarkes at the National Center for Earthquake Engineering Research, Buffalo, New York, 13–16 September 1989.

Quarantelli, E. L., and J. R. Hundley. 1965. A test of some propositions about crowd formation and behavior. Paper presented at the Annual American Sociological Association Meeting, Chicago.

Quarantelli, E. L., and J. R. Hundley. 1969. A test of some propositions about crowd formation and behavior. In *Readings in collective behavior*, ed. R. R. Evans, 538–55. Chicago: Rand McNally.

Quarantelli, E. L., and C. Pelanda, eds. 1989. *Proceedings of the Italy-United States seminar on preparations for, responses to and recovery from major community disasters*. Newark: Disaster Research Center, University of Delaware.

Quarantelli, E. L., and J. M. Weller. 1974. The structural problem of a sociological specialty: Collective behavior's lack of a critical mass. *American Sociologist* 9 (May): 59–68.

Quarantelli, E. L., B Baisden, and T. Bourdess. 1984. *Evacuation behavior and problems: Findings and implications from the research literature*. Newark: Disaster Research Center, University of Delaware.

Quarantelli, E. L., J. R. Ponting, and J. Fitzpatrick. 1974. Police department perceptions of the occurrence of civil disturbance. *Sociology and Social Research* 59: 30–39.

Raddall, T. H. 1948. *Halifax warden of the North*. Toronto: McClelland & Steward Limited.

Raphael, B., B. Singh, L. Bradbury, and F. Lambert. 1983. Who helps the helpers? The effects of disaster on rescue workers. *Omega* 14, no. 1: 9–22.

Report of services rendered by N.S. on the occasion of the disaster, 6 December 1917 the military forces at Halifax. N.d. RG 24 6349 HQ71-26-99-11, 15–23.

Riesman, D. 1950. *The lonely crowd*. New Haven: Yale University Press.

Roder, W. 1961. Attitudes and knowledge on the Topeka Flood Plain. In *Papers on flood problems*, ed. G. White. Chicago: Department of Geography, University of Chicago.

Rogers, D. L., D. A. Whetten, and Associates. 1982. *Interorganizational coordination: Theory, research, and implementation*. Ames: Iowa State University Press.

Rogers, G. O. 1985. *Human components of emergency warning*. Pittsburgh: University Center for Social and Urban Research, University of Pittsburgh.

Rogers, G. O., and J. Nehnevajsa. 1984. *Crisis conditions*. Pittsburgh: Center for Social and Urban Research, University of Pittsburgh.

Rogers, G. O., and J. Nehnrevajsa. 1987. Warning human populations of technological hazards. Proceedings of the ANS Topical Meeting on Radiological Accidents, CON-860932, 357–62.

Rosenthal, N., and M. Schwartz. 1989. Spontaneity and democracy in social movements. *International Social Movement Research* 2: 33–59.

Rosenthal, U. 1986. Collective behavior research in the Netherlands: From residual to partnership. Comparative methods and research on collective behavior and social movements, ed. G. Marx. *International Journal of Mass Emergencies and Disasters* 4, no. 2 (Special issue): 91–116.

Rosnow, R. L. 1976. *Rumor and gossip: The social psychology of hearsay*. New York: Elsevier.

Ross, G. A. 1976. The emergence and change of organization-sets: An interorganizational analysis of ecumenical disaster recovery organizations. Ph.D. diss. Columbus: Department of Sociology, Ohio State University.

Rossi, P. H., J. D. Wright, and S. R. Wright. 1981. Assessment of research on natural hazards reassessed in light of the SADRI Disaster Research Program. In *Social science and natural hazards*, eds. J. D. Wright and P. H. Rossi, 143–59. Cambridge, MA: Abt.

Rotter, J. B. 1966. Generalized expectancies for internal versus external control of reinforcements. *Psychological Monographs: General and Applied* 80: 1.

Rotter, J. B., and R. C. Mulry. 1965. Internal versus external control of reinforcement and decision time. *Journal of Personality and Social Psychology* 2: 598–604.

Rubin, C., and R. Palm. 1987. National origin and earthquake response: Lessons from the Whittier Narrows earthquake of 1987. *International Journal of Mass Emergencies and Disasters* 5, no. 3: 347–55.

Ruch, C., and L. Christensen. 1980. *Hurricane message enhancement*. College Station: Texas Sea Grant Program, Texas A&M University.

Rule, J. B. 1989. Rationality and nonrationality in militant collective action. *Sociological Theory* 7: 145–60.

Rule, J. B. 1988. *Theories of civil violence*. Berkeley: University of California Press.

Rule, J. B., D. McAdam, L. Stearns, and D. Uglow. 1980. *The politics of privacy*. New York: New American Library.

Rupp, L. J. 1980. Imagine my surprise: Women's relationships in historical perspective. *Frontiers* 5: 61–70.

Rupp, L. J., and V. Taylor. 1987. *Survival in the doldrums: The American women's rights movement, 1945 to the 1960s*. New York: Oxford University Press.

Rushton, J. P. 1980. *Altruism, socialization, and society*. Englewood Cliffs, NJ: Prentice-Hall.

Russell, S. T. 1989. Role enactment and disaster response: A methodological exploration. Master's thesis. Williamsburg, VA: College of William and Mary.

Ryan, B. P. 1989. No survivors—nothing to do? *Disasters* 13, no. 2: 175–76.

Saarinen, T. F., and J. L. Sell. 1985. *Warning and response to the Mount St. Helens eruption*. Albany: State University of New York Press.

Saltz, R. F. 1986. Searching for structure in beliefs. In *Rationality and collective belief*, ed. A. R. Harris, 81–130. Norwood, NJ: Ablex.

Scammell, R. L. N.d. History of the Victoria General Hospital. Unpublished Public Archives of Nova Scotia.

Scanlon, J. 1988. Disaster's little known pioneer: Canada's Samuel Henry Prince. *International Journal of Mass Emergencies and Disasters* (November): 213–32.

Scanlon, J. 1988. From the Titanic to the stars: Canada's pioneer in collective behavior: Samuel Henry Prince. Paper presented at the American Sociological Association, August, Atlanta.

Scanlon, T. J. 1989. Disaster planning. *Contact* 1, no. 2 (November): 4–8.

Schrank; R. 1977. *Snap, crackle, and popular taste*. New York: Dell.

Schuler, E. A., and V. J. Parenton. 1943. A recent epidemic of hysteria in a Louisiana high school. *Journal of Social Psychology* 17: 221–35.

Schuman, H., and J. Scott. 1989. Generations and collective memories. *American Sociological Review* 54 (June): 359–81.

Schur, E. M. 1971. *Labeling deviant behavior: Its sociological implications*. New York: Harper & Row.

Schur, E. M. 1984. *Labeling women deviant: Gender, stigma, and social control*. New York: Random House.

Schwartz, M. 1976. *Radical protest and social structure: The Southern Farmers' Alliance and Cotton Tenancy, 1880–1890*. New York: Academic.

Seeman, M. 1975. Alienation studies. *Annual Review of Sociology* 1: 91–123.

Selden, B. 5. 1979. Adolescent epidemic hysteria presenting as a mass casualty, toxic exposure incident. *Annals of Emergency Medicine* 18: 892–95.

Shepherd, M., and P. E. Hodgkinson. 1990. The hidden victims of disaster: Helper stress. *Stress Medicine* 6: 29–35.

Shibutani, T. 1966. *Improvised news: A sociological study of rumor.* Indianapolis, IN: Bobbs-Merrill.

Showalter, E. 1985. *The female malady: Women, madness, and English culture, 1830–1980.* New York: Pantheon Books.

Sime, J. D. 1983. Affiliative behavior during escape to building exits. *Journal of Environmental Psychology* 3: 21–41.

Simmel, G. 1950. *The sociology of Georg Simmel*, ed. K. Wolff. New York: Free Press.

Small, G. W., and J. F. Borus. 1983. Outbreak of illness in a school chorus: Toxic poisoning or mass hysteria? *New England Journal of Medicine* 308: 632–35.

Smelser, N. J. 1962. *Theory of collective behavior.* New York: Free Press.

Smelser, N. J. 1970. Two critics in search of a bias: A response to Currie and Skolnick. *Annals of the American Academy of Political and Social Science* 391: 46–55.

Smith, D. H. 1980. Determinants of individuals' discretionary use of time. In *Participation in social and political activities*, ed. D. H. Smith, J. Macaulay, and Associates. San Francisco: Jossey-Bass.

Smith, D. 1976. Primary group interaction in panic behavior: A test of theories. Paper read at the meetings of the Southern Sociological Society, Miami Beach.

Smith, S. K. N.d. *Heartthrobs of the Halifax horror.* Halifax: Gerald E. Weir.

Smith, T. W. 1982. General liberalism and social change in post-World War II America: A summary of trends. *Social Indicators Research* 10.

Smithson, M., P. R. Amato, and P. Pearce. 1983. *Dimensions of helping behavior.* Oxford: Pergamon.

Snook, J. W. 1990. Volunteer and private fire brigade resource in time of disaster. Technical paper presented at the South Pacific Fire Conference, Auckland, NZ.

Snow, D. A., L, A. Zurcher, and R. Peters. 1981. Victory celebrations as theater: A dramaturgical approach to crowd behavior. *Symbolic Interaction* 4: 21–42.

Snow, D. A., E. B. Rochford, S. K. Worden, and R. D. Benford. 1986. Frame alignment processes, micromobilization, and movement participation. *American Sociological Review* 51 (August): 464–81.

Song, S. 1986. The bad influence of earthquake rumors. *Jin Jin Daily*, July 21.

Song, S. 1987. Responses to earthquake rumors. Trans. Q. Yang and W. Wang. Paper presented at the Second International Earthquake Conference, Los Angeles.

Sorensen, J. H. 1982. *Evacuation of the emergency warning system at the Fort St. Vrain nuclear power plant.* ORNL/TM, Oak Ridge, TN: Oak Ridge National Laboratory.

Sorensen, J. H. 1985. Knowing how to behave under the threat of disaster: Can it be explained? *Environment and Behavior* 15 (July): 438–57.

Sorensen, J. H., and B. Richardson. 1983. Risk and uncertainty as determinants of human response in emergencies: Evacuation at TMI reexamined. Paper presented at the Society for Risk Analysis Annual Meeting, Knoxville, TN.

Spector, M., and J. I. Kitsuse. 1973. Social problems: A re-formulation. *Social Problems* 21: 145–59.

Spector, M., and J. I. Kitsuse. 1977. *Constructing social problems*. Menlo Park, CA: Cummings.

Spilerman, S. 1970. The causes of racial disturbances: A comparison of alternative explanations. *American Sociological Review* 35: 627–49.

Staggenborg, S. 1988. Consequences of professionalization and formalization in the pro-choice movement. *American Sociological Review* 53: 585–606.

Staggenborg, S. 1989a. Stability and innovation in the women's movement: A comparison of two movement organizations. *Social Problems* 36: 75–92.

Staggenborg, S. 1989b. Organizational and environmental influences on the development of the pro-choice movement. *Social Forces* 68: 204–40.

Stahl, S. M., and M. Lebedun. 1974. Mystery gas: An analysis of mass hysteria. *Journal of Health and Social Behavior* 15: 44–50.

Stallings, R. A. 1978. The structural patterns of four types of organizations in disaster. In *Disasters*, ed. E. L. Quarantelli, 87–103. Beverly Hills, CA: Sage.

Stallings, R. A. 1987. Organizational change and the sociology of disaster. In *Sociology of disasters: Contribution of sociology to disaster research*, eds. R. R. Dynes, B. DeMarchi, and C. Pelanda, 239–57. Milan, Italy: Franco Angeli.

Stallings, R. A. 1989. Volunteerism inside complex organizations: Off-duty hospital personnel in a disaster. *Nonprofit and Voluntary Sector Quarterly* 18: 133–45.

Stallings, R. A., and E. L. Quarantelli. 1985. Emergent citizen groups and emergency management. *Public Administration Review* 45: 93–100.

State Seismological Bureau. 1985. *Earthquake countermeasures*. Trans. Q. Yang and W. Wang. Beijing, China: State Seismological Bureau.

Staub, E., and H. K. Feinberg. 1980. Socialization and prosocial behavior in children. In *Participation in social and political activities*, eds. D. H. Smith, J. Macaulay, and Associates, 280–305. San Francisco: Jossey-Bass.

Steele, C. 1989. In defence of black humour (Letter 1). *The Psychologist* (November): 488.

Steinberg, S. N.d. *The ethnic myth*. New York: Athenum.

Strauss, A. 1947. Research in collective behavior: Neglect and need. *American Sociological Review* 12: 352–54.

Strickland, B. R. 1965. The prediction of social action from a dimension of internal-external control. *The Journal of Social Psychology* 66: 353–58.

Su, G., J. Li, J. Zhang, and Z. Wang. 1986. A general account of the rumor occurring in Shanghai during May–July of 1986 and the countermeasures against it. Beijing, China: Institute of Geophysics, State Seismological Bureau.

Sundeen, R. A., and G. B. Seigel. 1986. The uses of volunteers by police. *Journal of Police Science and Administration* 14: 49–61.

Swidler, A. 1986. Culture in action: Symbols and strategies. *American Sociological Review* 51: 273–86.

Symons, I., and D. Glowaski. 1988. Stress issues relating to the country fire authority. Paper presented at the conference, Dealing with Stress and Trauma in Emergency Services, Melbourne, Australia.

Synopsis. N.d. RG 24 4548 86-1-3-1. National Archives of Canada.

Tarrow, S. 1983. *Struggling to reform: Social movements and policy change during cycles of protest.* Western Societies Paper, no. 15, Ithaca: Cornell University.

Taylor, A. J. W., and A. G. Frazer. 1982. The stress of postdisaster body handling and victim identification work. *Journal of Human Stress* 8: 4–12.

Taylor, P. 1989. The Hillsborough Stadium disaster. Interim Report, August. London: Her Majesty's Stationery Office.

Taylor, P. 1990. The Hillsborough Stadium disaster. Final Report, January. London: Her Majesty's Stationery Office.

Taylor, V. 1989. Social movement continuity: The Women's movement in abeyance. *American Socioloġlcal Review* 54: 761–75.

Taylor, V., and N. Whittier. 1992. The new feminist movement. *Feminist Frontiers III*, ed. L. Richardson and V. Taylor. New York: McGraw-Hill.

Thompson, J. D. 1967. *Organizations in action.* New York: McGraw-Hill.

Tierney, K. J. 1982. The battered women movement and the creation of the wife beating problem. *Social Problems* 29: 207–20.

Tierney, K. J. 1987. Chemical emergencies, offsite exposures and organizational response. Research report to the Natural Hazards Research Applications and Information Center, Boulder: Institute of Behavloral Science, University of Colorado.

Tierney, K. J. 1980. Emergent norm theory as theory: An analysis and critique of Turner's formulation. In *Collective behavior: A Source Book*, ed. M. D. Pugh. St. Paul, MN: West.

Tilly, C. 1978. *From mobilization to revolution.* Reading, MA: Addison-Wesley.

Tilly, C. 1979. Social movements and national politics. Working paper, no. 197. Ann Arbor: Center for Research on Social Organization, University of Michigan.

Tilly, C., L. Tilly, and R. Tilly. 1975. *The rebellious century, 1830–1930.* Cambridge: Harvard University Press.

The Times. 1989. Catalogue of terrace tradgedy. *The Times*, 17 April, 4.

The Times. 1990. Inquest told of mass hooliganism. *The Times*, 19 December, 6.

Titchener, J. L., and F. T. Kapp. 1976. Family and character change at Buffalo Creek. *American Journal of Psychiatry* 133 (March): 295–99.

Tooke, F. T. N.d. An experience through the Halifax disaster. *The Canadian Medical Association Journal* 8: 308–20.

Touraine, A. 1985. An introduction to the study of social movements. *Social Research* 52, no. 4: 749–87.

Tubbesing, S. 1989. Earthquake reconstruction: Emerging lessons for a safer future. Paper presented at the Conference on Reconstruction after Urban Earthquakes at the National Center for Earthquake Engineering Research, Buffalo, New York, 13–16 September.

Tumin, M. M., and A. S. Feldman. 1955. The miracle at Sabana Grande. *Public Opinion Quarterly* 19: 124–39.

Turner, R. H. 1964. New theoretical frameworks. *The Sociological Quarterly* 5: 122–32.

Turner, R. H. 1964. Collective behavior. In *Handbook of modern sociology*, ed. R. E. L. Faris. Chicago: Rand McNally.

Turner, R. H. 1969. The public perception of protest. *American Sociological Review* 34: 815–31.

Turner, R. H. 1970. Determinants of social movement strategies. In *Human nature and collective behavior*, ed. T. Shibutani. Englewood Cliffs, NJ: Prentice-Hall.

Turner, R. H. 1978. Earthquake prediction volunteers: What can the United States learn from China? *Mass Emergencies* 3: 142–60.

Turner, R. H. 1981. Collective behavior and resource mobilization as approaches to social movements: Issues and continuities. *Research in Social Movements, Conflict, and Change* 4: 1–24.

Turner, R. H. 1983. Waiting for disaster: Changing reactions to earthquake forecasts in Southern California. *International Journal of Mass Emergencies and Disasters* 1, no. 2: 307–34.

Turner, R. H. 1990. The use and misuse of rational models in collective behavior and social psychology. Unpublished manuscript.

Turner, R. H., and L. M. Killian. 1957. *Collective behavior.* Englewood Cliffs, NJ: Prentice-Hall.

Turner, R. H., and L. M. Killian. 1972. *Collective behavior.* 2d ed. Englewood Cliffs, NJ: Prentice-Hall.

Turner, R. H., and L. M. Killian. 1987. *Collective behavior.* 3d ed. Englewood Cliffs, NJ: Prentice-Hall.

Turner, R. H., J. M. Nigg, and D. H. Paz. 1986. *Waiting for disaster: Earthquake watch in California.* Berkeley: University of California Press.

Turner, R. H., J. M. Nigg, D. H. Paz, and B. S. Young. 1979. *Earthquake threat: The human response in Southern California.* Los Angeles: Institute for Social Science Research, University of California.

Turner, R. H., J. M. Nigg, D. H. Paz, and B. S. Young. 1981. *Community response to earthquake threats in Southern California.* Los Angeles: University of California Press.

U.S. Department of Commerce. 1980. *Social indicators III.* Washington, DC: G.P.O.

U.S. Department of Commerce. 1982 and 1989. *Statistical abstract of the United States.* Washington, DC: G.P.O.

Useem, B., and P. Kimball. 1989. *States of Seige: U.S. prison riots, 1971–1986.* New York: Oxford University Press.

Useem, B., and M. N. Zald. 1982. From pressure group to social movement: Organizational dilemmas of the effort to promote nuclear power. *Social Problems* 30: 144–56.

Vaillant, G. E. 1977. *Adaptation to life.* Boston: Little Brown.

van der Kolk, B. A. 1988. The trauma spectrum: The interaction of biological and social events in the genesis of the trauma response. *Journal of Traumatic Stress* 1, no. 3: 273–90.

Vitaliano, P. P., R. D. Maiuro, P. A. Bolton, and G. C. Armsden. 1987. A psychoepidemiologic approach to the study of disaster. *Journal of Community Psychology* 15: 99–121.

Wallace, A. F. C. 1956. *Tornado in Worcester.* National Research Council Disaster Study, no. 3. Washington, D.C.: National Academy of Sciences.

Walsh, E. J. 1978. Mobilization theory vis-a-vis a mobilization process: The case of the United Farm Workers' movement. *Research in Social Movements, Conflicts and Change* 1: 155–77.

Walsh, E. J. 1981. Resource mobilization and citizen protest in communities around Three Mile Island. *Social Problems* 29 (October): 1–21.

Wang, P., X. Zheng, Y. Chen, X. Xie, and M. Yu. 1987. Feature analysis of and problem ponderation from the episode of fleeting earthquake danger caused by an earthquake rumor which went around in the University for Overseas Chinese. Trans. Q. Yang and W. Wang. *Recent Developments in World Seismology* 103: 16–19.

Warren, C. A. B. 1987. *Madwives: Schizophrenic women in the 1950's.* New Brunswick, NJ: Rutgers University Press.

Waterstone, M. 1978. Hazard mitigation behavior of flood plain residents. Natural Hazards Working Paper, no. 35. Boulder: Institute of Behavioral Science, University of Colorado.

Weatherbe, P. N.d. Formation of the Medical Relief Committee. In *A history of the medical aspects of the disaster and Halifax and Dartmouth, Nova Scotia,* ed. D. F. Harris. Unpublished, Public Archives of Nova Scotia.

Webber, M. 1968. The post-city age. *Daedalus* 97.

Weber, F. 1987. Whittier Narrows earthquakes. *California Geology* (December): 275–81.

Weller, J. M., and E. L. Quarantelli. 1973. Neglected characteristics of collective behavior. *American Journal of Sociology* 79: 665–85.

Wenger, D. E. 1987. Collective behavior and disaster research. In *Sociology of disasters: Contribution of sociology to disaster research,* eds. R. R. Dynes, B. DeMarchi, and C. Pelanda, 213–38. Milan, Italy: Franco Angeli.

Wenger, D. E. 1978. Community response to disasters: Functional and structural alterations. In *Disasters, theory and research,* ed. E. L. Quarantelli, 17–47. Beverly Hills: Sage.

Wenger, D. E., and E. L. Quarantelli. 1973. A voice from the 13th century: The characteristics and conditions of the emergence of a Ouija board cult. *Urban Life and Culture* 1: 379–400.

Wenger, D. E., T. F. James, and C. F. Faupel. 1980. *Disaster beliefs and emergency planning.* Newark: Disaster Research Center, University of Delaware.

White, R. 1988. Commitment, efficacy, and personal sacrifice among Irish republicans. *Journal of Political and Military Sociology* 16: 77–90.

Whittier, N. 1991. Feminists in the post-feminist age: Collective identity and the persistence of the Women's movement. Ph. D. diss. Ohio State University.

Whyte, W. H. 1956. *The organization man.* New York: Simon & Schuster.

Wijkman, A., and L. Timberlake. 1988. *Natural disasters: Acts of God or acts of man?* Santa Cruz, CA: New Society.

Williams, H. B. 1957. Some functions of communication in crisis behavior. *Human Organization* 16: 15–19.

Wilson, J. 1973. *Introduction to social movements.* New York: Basic.

Wimberly, R. C., T. C. Hood, C. M. Lipsey, M. Clelland, and M. Hay. 1975. Conversion in a Billy Graham crusade: Spontaneous event or ritual performance? *The Sociological Quarterly* 16: 162–70.

Windam, G., E. Posey, P. Ross, and B. Spencer. 1977. *Reactions to storm threat during Hurricane Eloise.* State College: Mississippi State University.

Wolensky, R. P. 1979. Toward a broader conceptualization of volunteerism in disaster. *Journal of Voluntary Action Research* 8: 3–4, 33–43.

Wood, J. L., and M. Jackson. 1982. *Social movements: Development, participation, and dynamics.* Belmont, CA: Wadsworth.

Worth, M. F., and B. F. McLuckie. 1977. *Get to high ground! The warning process in the Colorado floods, June 1965*. Historical and Comparative Disaster Series. Newark: Disaster Research Center, University of Delaware.

Wright, S. 1978. *Crowds and riots: A study in social organization*. Beverly Hills: Sage.

Wrong, D. 1961. The oversocialized conception of man. *American Sociological Review* 26 (April): 183–93.

Wuthnow, R., J. D. Hunter, A. Bergesen, and E. Kurzweil. 1984. *Cultural analysis: The work of Peter L. Berger, Mary Douglas, Michel Foucault, and Jurgen Habermas*. London: Routledge & Kegan Paul.

Xu, Y., A. Xie, and P. Zhao. 1989. The earthquake rumor spreading in Jiujiang and measures for rumor spiking and earthquake preparedness. Trans. Q. Yang and W. Wang. *Recent Developments in World Seismology* 128: 15–18.

Yamamoto, Y., and E. L. Quarantelli. 1982. *Inventory of the Japanese disaster literature in the social and behavioral sciences*. Newark: Disaster Research Center, University of Delaware.

Yankelovich, D. 1981. *New rules*. New York: Bantam.

Ye, M. 1989. Societal responses to earthquake rumors, misdisseminations of information about earthquakes and earthquake-phobia psychology and countermeasures against them. Trans. Q. Yang and W. Wang. *Recent Developments in World Seismology* 122: 20–23.

Yin, R. K. 1984. *Case study research: Design and methods*. Beverly Hills: Sage.

Yin, R. K., and G. B. Moore. 1985. *The utilization of research: Lessons from the natural hazards field*. Washington, DC: Cosmos Corporation.

Zablocki, B., and R. Kanter. 1976. The differentiation of life-styles. *Annual Review of Sociology* 2: 269–298.

Zald, M. N. 1988. Looking backward to look forward: Reflections on the past and future of the Resource Mobilization Research Program. Plenary address, Workshop on Frontiers of Social Movement Theory, Ann Arbor, Michigan, June 8–10.

Zald, M. N., and R. Ash. 1966. Social movement organizations: Growth, decay, and change. *Social Forces* 44: 327–41.

Zald, M. N., and J. D. McCarthy. 1979. *The dynamics of social movements*. Cambridge, MA: Winthrop.

Zech, C. E. 1982. Citizen willingness to assist as volunteers in the provision of local public goods. *American Journal of Economics and Sociology* 41, no. 3: 303–14.

Zelman, P. G. 1982. *Women, work, and national policy: The Kennedy-Johnson years*. Ann Arbor: UMI Research.

Zheng, X., and C. Xiande. 1987. General account and countermeasures against the episode of fleeing the earthquake danger caused by an earthquake rumor which went around in the University of Overseas Chinese. Trans. Q. Yang and W. Wang. *Recent Developments in World Seismology* 103: 12–16.

Zurcher, L. A. 1968. Socio-psychological functions of ephemeral roles: A disaster work crew. *Human Organizations* 27: 281–97.

Zymunt, J. F. 1986. Collective behavior as a phase of societal life: Blumers' emergent views and their implications. *Research in Social Movements, Conflicts, and Change* 9: 25–46.

List of Contributors

BENIGNO E. AGUIRRE is Professor of Sociology at Texas A&M University. He is currently writing a book on the Marielito migration from Cuba to the United States in the spring of 1980. He is also conducting research on the relationship between ecological characteristics of counties and the effects of tornadoes on populations and the built environment.

FREDERICK L. BATES has been Professor of Sociology at the University of Georgia since 1963 and also serves as director of the International Laboratory for Socio-Political Ecology. His work on disasters includes articles and monographs on hurricanes Audrey and Carla and on the Guatemalan earthquake of 1976. His most recent disaster research has been concerned with designing and testing a household level disaster impact measure that can be applied cross-culturally. He has also published extensively on the topic of social structure and is now completing a book that will develop an ecological approach to social structure and social organization.

ROBERT BOLIN is a Professor of Sociology in the Department of Sociology and Anthropology, New Mexico State University. He recently completed a monograph on long-term household and community recovery from California earthquakes. His other research concerns risk perceptions and the psychosocial impacts of technological hazards, with a focus on oil spills and nuclear waste.

NEIL R. BRITTON was Associate Professor of Disaster Management and director of the Centre for Disaster Management at the University of New England, Armidale, Australia. His current research projects include an examination of organizational and community response to natural and technological hazards; emergency management policy development and implementation; the nature of volunteerism in emergency services; hazard insurance policies and practices; and hazard risk assessment and mitigation options. He is currently the Chief Advisor to the Emergency Management Office, Wellington City Council, New Zealand.

BEN CORREY was employed as a research assistant to Drs. Britton and Moran during the time of writing the article that appears in this publication.

Trained as a hazards geographer at the University of Sydney, he focused his interest on wildfire planning and management, with specific reference to the organization of Australian volunteer bushfire brigades.

THOMAS E. DRABEK is Professor of Sociology at the University of Denver and president of the Research Committee on Disaster, International Sociological Association. His long-term research focus has been on organizational and group responses to disasters. His most recent book is on evacuation decision making by executives responsible for tourist-oriented businesses.

RUSSELL R. DYNES is Research Professor, Disaster Research Center, University of Delaware. He is cofounder and former codirector of the Disaster Research Center. He recently coauthored a monograph on the Mexico City earthquake and currently is working on a project, related to the International Decade of Natural Disaster Reduction, on the development of training materials for UNESCO.

WILLIAM E. FEINBERG, Professor of Sociology at the University of Cincinnati, conducts collaborative research with Norris Johnson on computer simulation models of the development of consensus in crowds. He is also pursuing a simulation investigation of various structural and cultural interrelationships incorporated in Blau's structural theory.

COLLEEN FITZPATRICK is a research scientist with the Hazards Assessment Laboratory at Colorado State University. She recently completed a dissertation in the Department of Sociology on emergent cognitions theory and the risk communication process with data from a study of public response to the Parkfield earthquake prediction experiment. She is also coauthoring a book on risk communication and public action with Dennis Mileti.

CHARLES E. FRITZ is a disaster research consultant residing in Bethesda, Maryland. His experience in disaster studies started in 1945. His past positions include associate director of the Disaster Research Project at the University of Chicago, staff member of the National Academy of Sciences Committee on Disaster Studies, assistant director of the Disaster Research Group, Associate Professor of Psychiatry at the University of Florida, and staff member at the Institute of Defense Analysis. He was executive secretary of the Advisory Committee in Emergency Planning, NAS, from 1982–87. He is author of over a hundred books, monographs and articles on human and organizational behavior.

THOMAS F. JAMES is currently studying a successful grassroots environmental social movement that developed in opposition to a solid waste disposal plan.

He is examining the movement's organization, leadership, and relationship to the media over its life history. He is Associate Professor of Sociology and department chair at Saint Lawrence University in Canton, New York.

NORRIS R. JOHNSON is Professor of Sociology at the University of Cincinnati. His continuing interest in theories of crowd processes is reflected in collaborative work with William E. Feinberg on theoretical and computer simulation models of crowd interaction. Development of more complex simulation models is in process and planned extensions will consider simulation of the form of collective flight examined in the chapter included in this volume.

DRUE M. JOHNSTON is an administrator in the College of Evening and Continuing Education at the University of Cincinnati, where she is also a doctoral student in the Department of Sociology. She has published research on the functioning of employee roles in the Beverly Hills Supper Club fire; her current research interest is in the social psychology of obesity.

MICHAEL L. KELSEY is Professor and Chair, Department of Geography and Economics, Amis Community College, Greeley, Co.

LEWIS M. KILLIAN is Professor Emeritus of Sociology, University of Massachusetts, Amherst, and since 1985 has been Faculty Associate in Sociology at the University of West Florida, Pensacola. He is chair of the Florida Human Rights Advocacy Committee serving four counties in the western part of the state. He recently completed an autobiography titled *Black and White: Reflections of a White Southern Sociologist on Race Relations.*

GARY KREPS is Professor of Sociology, College of William and Mary. He has collaborated with Susan L. Bosworth and several graduate student colleagues over the past ten years to produce structural data from the archives of the Disaster Research Center. The goal of this research program is to develop an empirically grounded formal theory of the process of organizing and role enactment during the emergency period of disaster.

JERRY M. LEWIS is Professor of Sociology at the Kent State University, where he joined the faculty in 1966. In recent years, his work on collective behavior topics has focused on research on soccer crowds. Previous to that interest, he wrote extensively on the 4 May shootings at Kent State University.

GARY T. MARX is Professor of Sociology at the University of Colorado and was previously at the Massachusetts Institute of Technology. He is author of *Protest and Prejudice* and *Undercover*. In 1986, he edited a special issue of the

International Journal of Mass Emergencies and Disasters on "Collective Behavior and Social Movements." He is currently engaged in research on technologies for the collection of personal information.

DENNIS S. MILETI is Professor of Sociology and Director of the Natural Hazards Research and Applications Information Center, University of Colorado, Boulder. He has researched public response to hazard and disaster warnings in reference to a variety of climatological, geological, and technological emergencies. He is currently organizing an assessment of research and applications on natural hazards.

CARMEN MORAN is a senior lecturer in Psychology at Cumberland College of Health Sciences, The University of Syndey, Australia, and honorary research associate with the Centre for Disaster Management, The University of New England. Her current research interests include stress and coping in extreme circumstances, and the use of humor as a coping strategy.

CARLO PELANDA is Professor of "Strategic Scenarios" at LUISS University, Rome, Italy, and Adjunct Professor of Sociology at the University of Georgia. He serves as associate director of the "International Laboratory of Socio-Political Ecology" (INTERLAB-UGA).

RONALD W. PERRY is Professor of Public Affairs at Arizona State Universlty. He serves, with Neil Britton, as coeditor of the *International Journal of Mass Emergencies and Disasters*. His past research has focused on models for disaster warning systems, and his current research interests focus on the development of models of the process whereby citizens adopt protective measures in response to disasters.

T. JOSEPH SCANLON is a member of the School of Journalism and head of the Emergency Communications Research Unit at Carelton University in Ottawa, Ontario. Among his research interests are exploring the documentary sources of Canada's worst disaster, the Halifax explosion.

ROBERT A. STALLINGS is Associate Professor of Public Policy and Sociology at the University of Southern California. He is completing a quantitative analysis of local governments' efforts to reduce the risks of unreinforced masonry buildings in earthquakes in 336 California cities. He looks forward to resuming an examination of power, risk, and public policy-making related to the earthquake hazard.

VERTA TAYLOR is Associate Professor of Sociology at Ohio State University. She is coauthor, with Leila J. Rupp, of *Survival in the Doldrums: the Ameri-*

can Women's Rights Movement 1945 to the 1960s (New York: Oxford University Press) and coeditor with Laurel Richardson of *Feminist Frontiers II: Rethinking Sex, Gender, and Society* (New York: Random House, 1989). In addition to her continuing interest in historical and contemporary aspects of the American women's movement, gender, and collective behavior and social movements, she is currently working on a book that analyzes the experience of postpartum depression and the self-help movement, which has arisen to bring the condition to prominence.

KATHLEEN J. TIERNEY is Associate Professor of Sociology and Research Director, Disaster Research Center, University of Delaware. Among her current projects are studies on community earthquake hazard reduction, intergovernmental and interjurisdictional coordination in major disasters, community disaster recovery, and organizational planning and response in major oil spills. She recently coedited a special issue of the journal *Industrial Crisis Quarterly* that focused on the Exxon Valdez oil spill.

RALPH H. TURNER is Professor Emeritus of Sociology at the University of California (Los Angeles). He is currently working on a fourth edition of *Collective Behavior*, with Lewis Killian and David Snow. He is also conducting investigations on the genesis of emergent norms, the situational contexts for self-feeling, and the theory of social roles.

DENNIS E. WENGER is the Director of the Hazard Reduction & Recovery Center at Texas A&M University. He holds faculty positions in both the Department of Urban and Regional Planning and the Department of Sociology. Professor Wenger's current research focuses on search and rescue and the epidemiology of death and injury during disasters, volunteer and collective action during crises, and cross-cultural studies of disaster response and recovery. From 1985–1989 he was Co-Director of the Disaster Research Center at the University of Delaware with Russell R. Dynes and E. L. Quarantelli.

Index

Adams, D., 233
Adorno, T., 274
Aguirre, B., 229, 257–272; and E. L. Quarantelli, 17, 260, 261; E. L. Quarantelli and J. L. Mendoza, 15, 227, 259, 269
Aldrich, H., 33, 103
Alexander, D., 32
Alexander, J. C., 63
Allport, G. W., and L. Postman, 244, 245, 250
Altheide, D., 207, 213, 217
Alston, J., 257
Amato, P. R., R. Ho, and S. Partridge, 130
Ambrosini, P. J., 246
American Psychiatric Association, 208
American Sociological Association: Section on Collective Behavior and Social Movements, 16–17
Anderson, W. A., 73, 229; R. R. Dynes and E. L. Quarantelli, 15
Andrews, G., C. Pollack, and G. Stewart, 134
Antonovsky, A., 139
Aveni, A., 169, 289

Baker, E. S., 74, 78, 81, 103
Barkan, S. E., 293
Barton, A., 142, 230, 232
Bates, F. L., 116, 118, 119, 150; and C. C. Harvey, 59, 61, 151; and C. Pelanda, 17, 25, 145–159; and W. G. Peacock 117, 120, 272
Bauer, R. A., and D. B. Gleicher, 245
Bauman, D., 72
Becker, H. S., 222, 227–228
Bell, F. M., 100, 108, 109, 111
Bellamy, L. J., 73, 80
Benson, J. K., 34
Berah, E. F., H. J. Jones, and P. Valent, 137
Berk, R., 86
Best, J., 207
Beverly Hills Supper Club fire, 168–189; social bonds in, 174–185; social bonds and patrons' responses, 179–185
Blalock, H., 86, 89, 92
Blanshan, S., and E. L. Quarantelli, 15
Blau, P., 34
Blumer, H., 10, 59, 213, 218, 221, 225, 264, 265, 282
Bolin, R., 24, 115–127; and P. Bolton, 116, 120, 121; and L. Stanford, 116, 117, 121, 126, 127
Bolton, P., 117, 122
Bosworth, S., 47; and G. Kreps, 59, 227
Bramson, L., 225
Brauer, C. M., 292, 300
Breslau, N., and G. C. Davis, 136
Britton, N., 129, 140, 141, 142; C. C. Moran, and B. J. Correy, 25, 128–44
Brodsky, C. M., 211, 212, 218, 226
Bromley, D. G., and A. D. Shupe, 283
Brown, M., and A. Goldin, 264
Browning, G., 257
Bucher, R., C. E. Fritz, and E. L. Quarantelli, 11
Buckner, H. T., 246

Cable, S., 293
Calhoun, C., 260
Campbell, D. T., 33; and J. S. Stanley, 34
Caplan, N. S., and J. M. Paige, 230
Caplow, T., 245; H. M. Bahr, and B. A. Chadwick, 31
Caporale, R., 116, 118
Carden, M., 284
Carr, A. T., 132
Carroll, G. R., 32
Carter, T. M., 78
Charles, M. T., and J. C. K. Kim, 33

"Circular reaction," 218, 225, 264
Cisin, I. H., and W. B. Clark, 33
Clifford, R. A., 73, 75, 80
Cohen, J., 267
Coleman, James, 262
Collective behavior: convergence theories of, 220, 229–230; Chicago School, 10, 263–266, 273–276, 278; emergent norm theory of, 231, 278–280; ideology and, 279; incorporation of disaster research in, 69; misinterpretations of scholarship on, 259–266, 274–278; rationality and, 276–278; resource mobilization theory and, 257–263, 267–270; and the study of social movements, 257–272, 273–280, 282; value added theory of, 191–193, 219–220
Collins, R., 223, 277
Condie, B., 197, 203
Connelly, W. J., 101
Conrad, P., and J. W. Schneider, 213, 224
Convergence theories: and volunteer behavior following disasters, 242; of collective behavior, 220, 229–230
Cook, B. W., 291
Copp, S., 257
Coser, Lewis, 329
Costner, H., 86
Couch, C., 262, 274
Coughlan, R., 281
Coutu, W., 280
Craze, 190; normative regulation and, 193; value added theory and, 191–193
Crowds: panic behavior, 168–188; riots, 221, 230–231, 242; soccer violence, 190–206
Currie, E., and J. H. Skolnick, 274
Cutter, S., and K. Barnes, 80

Dahrendorf, R., 277
Danzger, M. H., 221
Danzig, E. R., P. W. Thayer, and L. R. Galater, 74
Davis, I., 118
Davis, W., and E. J. Phares, 81
Dawson, D., 131
DeFleur, M. L., 246
Demerath, N. J., 73
Diagnostic and Statistical Manual (DSM III), 208
Diggory, J. C., 79, 80
Disaster: ambiguity of term, 23; convergence in, 232; distinction between consensus and conflict, 14, 231–232; initial definition by DRC, 12; mass assault, 242–243; role enactment in, 59–68; stages, 24, 115
Disaster in Aisle 13, 23, 26, 44; conclusions, 27–30; implications for organizational theory, 30–33; influence on application, 36; methodological implications, 33–36
Disaster research: archival studies, 46; ecological approach to, 147–158; influence on policy, 44; as multidisciplinary field, 145
Disaster Research Center, 9, 12, 26, 27, 35, 37, 45, 145; archives of, 46–70; initial focus of, 16; organization of, 12, 16; typology, 13, 31
Disaster study sites: Abilene, TX, 91; Alaska, 13; Beverly Hills, KY, 168–89; Coalinga, CA, 120; Halifax, NS, 99–114; Indianapolis, IN, 13, 26–44; Mexico City, 229–43; Mt. Vernon, WA, 92; New South Wales, Australia, 130; Niigata, Japan, 13; Santa Cruz, CA, 120; Sheffield, England, 190; Toutle, WA, 92; Watsonville, CA, 126; White County, AR, 11; Whittier, CA, 126
Disaster volunteers: coping strategies of, 133–136; factors affecting behavior of, 129–131; in Mexico City Earthquake, 234–243; task performance of, 136; as stress management, 131; within emergency organizations, 140–143; research on, 232–233
Dollard, J. L., L. W. Doob, N. E. Miller, O. H. Mowrer, and R. R. Sears, 229
Drabek, R. A., 26
Drabek, T. E., 13, 23, 26–44, 45, 51, 73–76, 80, 81, 89, 90, 116, 118, 130, 130, 230, 269; and E. L. Quarantelli, 15; G. J. Hoetmer, 42; H. L. Tamminga, T. S. Kilijanek and C. R. Adams, 52, 232, 233, 241; and J. S. Stephenson, 74–76, 80, 90; R. Braito, C. C. Cook, J. R. Powell, and D. Rogers, 34

Dubin, R., 50, 57
Duncan, O. D., 86
Dunning, C. M., 133, 136
Durham, T. W., S. L. McCammon, and E. J. Allison, 135, 136
Dynes, R. R., 9, 12, 25, 31, 41, 45, 46, 142, 145; A. H. Purcell, D. E. Wenger, P. E. Stern, R. A. Stallings, and Q. T. Johnson, 78, 81; B. DeMarchi and C. Peland, 145; B. E. Aguirre, 35; and E. L. Quarantelli, 13, 15, 80, 137, 227, 232, 242; E. L. Quarantelli and D. E. Wenger, 239; E. L. Quarantelli and G. A. Kreps, 39, 234; K. Tierney and C. Fritz, 9–17

Earthquake prediction rumors; in California, 247–251; in China, 251–256
Edwards, L. P., 282
Eckstein, S., 306
Eisenstat, R. A., and R. D. Felmen, 131
"Elite Sustained" movement, 281, 283–284; women's movement as, 285–298, 302, 303
Emergent groups in disaster situations, 233
Emergent norm theory: as applied to evacuation behavior, 89; as applied to looting, 14; and mass hysteria, 219–220; and volunteer behavior in disasters, 231
Enoch, J. K., 46
Erikson, K. T., 34, 117
Estepp, M. H., 140
Ettedgui, E., and M. Bridges, 136
Evacuation behavior: causal model of compliance, 85–91
Evans, R. R., 218, 298

Farmer, E. Z., 56, 58
Ferree, M. M., and B. B. Hess, 298
Ferris, J. M., 129
Festinger, L., D. Cartwright, K. Barker, J. Fleischl, J. Cottsdanker, A. Keysen, and G. Leavitt, 246
Feuer, L., 274
Fireman, B., W. A. Gamson, S. Rytina, and B. Taylor, 271
Firth, R., 245
Fischer, C., 306
Fitzpatrick, C., 24
Flexner, E., 282
Flynn, C. B., 78, 80, 81
Flynn, C. B., and J. A. Chalmers, 74, 80
Form, W. H., and S. Nosow, 28, 232
Forrest, T., 227
Fossett, M., 257
Foster, T. A., 73, 76, 101
Frame alignment, 267–268
Freeman, J., 284, 289, 298–300
Francis, P., and G. A. Kreps, 52
French, S., C. Ewing, and M. Isaacson, 121
Friedan, B., 281
Friedrichs, R., 277
Friedsam, J. J., 81
Friedson, E., 224, 228
Fritz, C. E., 9, 11, 12, 15, 73; and E. S. Marks, 78, 89; and J. H. Mathewson, 232
Fritz, N. J., and D. J. Altheide, 213
Fromm, E., 327

Gallino, L., 151
Gamson, W. A., 261, 271, 274, 283, 284, 294; and A. Modigliani, 215
Garfinkel, H., 222
Garnica Andradi, A., 229
Garner, R. A., and M. N. Zald, 298
Gehlen, F. L., 228
Gelb, J., and M. L. Palley, 284
Gergen, K., M. Gergen, and K. Meter, 131
Gibbs, M. S., 144
Gillespie, D. F., M. W. Sherraden, C. L. Streeter, and M. J. Zakour, 31, 35
Gleser, G. C., B. L. Green, and C. Winget, 136
Golec, J., 117, 125
Goffman, E., 207
Gormley, F. A., 102
Grey, J., and E. L. Quarantelli, 15
Gruntfest, E. C., 78, 80
Guardian (London), 196
Gusfield, J., 209, 213, 270, 275, 304

Haas, J. E., 12; and P. Trainer, 71; R. Kates, and M. Bowden, 120
Hage, J., and M. Aiken, 34
Haicheng Earthquake Study Delegation, 254–255
Haines, H. H., 284

Halifax Explosion, 99; emergence of E.M.S. in, 106–113; impact on medical services, 100–102; initial medical situation, 102–104
Hall, R. H., 34
Hannigan, J., 267
Hannan, M. T., and J. Freeman, 33
Harris, D. F., 101, 102, 104, 107, 109, 110, 112
Harrison, C. E., 292
Hartman, E., 86
Hayes, M. R., 102
Heise, D., 86
Helfgot, J., 284
Henig, J. R., 260
Hillsborough Soccer Crowd Crush, 190–206; generalized belief and, 202–203; mobilization and, 203–204; social control and, 204–205; spectator types and, 200–203, 205–206; structural conduciveness and, 196–200; structural strain and, 200–202; Taylor Report on, 194–195, 198, 201–203
Hiroi, E., S. Mikami, and K. Miyata, 79
Hirsch, E. L., 261–263
Hoffer, E., 274
Hoffman, L. M., 284
Hogg, S., 120
Holland, C. J., 233
Hood, C., 140; and M. Jackson, 140
Hopper, R. D., 282
Hull, R. B., and D. E. Wenger, 232, 241

Incident Command System (ICS), 37
Inglis, S., 193
International Decade of Natural Disaster Reduction, 17
International Journal of Mass Emergencies and Disasters, 16
International Sociological Association, Research Committee on Disasters, 16

Jacobs, N., 218, 221
Jackman, N. R., 270
Jara Elias, Jose Ruben, 229
Jenkins, J. C., 271, 282, 289, 298; and C. M. Eckert, 282, 283, 303
Johnson, D. M., 221
Johnson, N., 32, 218, 269; W. E. Feinberg, and D. M. Johnston, 168–189
Johnston, D., and N. Johnson, 169, 170
"June Bug" epidemic, 218–220, 221

Kapferer, J., 246
Katz, E., 249
Keating, J. P., 168, 170
Kelley, H. H., J. Condry, A. Dahlke, and A. Hill, 169
Kemper, T. D., 129
Kerckhoff, A. C., 218; and K . W . Back, 218, 220, 221, 222; K. W. Back, and N. W. Miller, 218, 226
Killian, L., 11, 34, 89, 273–280
Kingswood, J. E., 134
Kitz, J. F., 101
Klandermans, B., 271, 298
Klapp, O. E., 265, 266
Knopf, T. A., 245
Koenig, F., 244
Kornhauser, W., 274
Kreps, G. A., 13, 23, 31, 33, 35, 41, 45–70, 97, 98, 227; and S . Bosworth, 52; J. G. Crooks, and J. R. Linn, 56
Kunreuther, H., 71; and E. Ley, 35

Lang, K., and G. E. Lang, 274, 282
LaPiere, R. T., 273
Landry, T., and G. Rogers, 73, 81
Lazarsfeld, P., B. Berelson, and H. Gaudet, 249
Lazarus, R., and S. Folkman, 132, 134
LeBon, G., 263, 264, 273
Lechat, M. F., 232, 241
Lefcourt, H. M., 81
Lehto, M. R., J. M. Miller, 73
Leik, R. K., et. al., 35, 77
Lemert, E. M., 228
Lenin, V. I., 284
Leonard, V. A., 89
Lewis, J. J., 198, 269; and M. Kelsey, 190–206
Lieberson, S., 86, 87; and A. R. Silverman, 221
Lifton, R. J., and E. Olson, 34
Lind, A., 244
Linder, S. B., 258
Lindsey, R. K., 199
Linn, J., and G. Kreps, 56
Lipset, S. M., 329; and W. Schneider, 315
Lipsky, M., 284
Livermore, D., and J. P. Wilson, 81

Lofland, J., 269, 278, 283; and V. Johnson, 272
Lofland, L., 259
Loizos, P., 118
Los Angeles Times, 213–218, 222, 223, 248, 249

Mack, R. W., and G. W. Baker, 80
MacLennan, H., 101
Mann, L., T. Nagel, and P. Dowling, 170
Manning, R., 274
Marger, M., 284
Marwell, G., and P. Oliver, 272
Marx, G., 268, 306–329
Mass Emergencies, 16
Mass hysteria, 207–228; mass media framing of, 213–218; psychiatric conceptions of, 208–213; as a social construction, 228; sociological approaches to, 218–220
May, P., 32
Mayer, M., 258, 262
Mayhew, B., 51
McAdam, D., 282–284, 289, 293, 297
McCammon, S., T. W. Durham, E. J. Allison, and J. E. Williamson, 135
McCarthy, J. D., 257, 261; and M. N. Zald, 283, 284, 294, 303
McFarlane, A. C., 135
McIntosh, C., 102
McKeen, J. C., 104
McLuckie, B., 74, 76
McPhail, C., 230, 265, 269, 272; and D. Miller, 269
Medalia, N. Z., and O. N. Larsen, 218, 221
Mexico City Earthquake, 229–243; social characteristics of volunteers in, 234–243; types of volunteer activity, 237–240; volunteer behavior in, 234–243
Melucci, A., 267
Merton, R., 59, 151
Metson, G., 100, 101
Mikami, S., and K. Ikeda, 78
Mileti, D., 24, 33, 73, 85, 90; and C. Fitzpatrick, 71–84; and E. M. Beck, 78; and J. H. Sorenson, 73, 74; and P. O'Brien, 232, 233, 241; S. Hutton and J. H. Sorenson, 35, 71, 75, 77, 80, 81; T. E. Drabek and J. E. Haas, 45, 73, 74, 76, 80, 232
Miller, D. L., 229, 269, 283, 284
Miller, G. A., D. L. Anderson, and J. Conaty, 34
Mills, C. W., 306
Mintz, A., 230
Mintzberg, H., 141
Mitchell, S. J., 133, 135
Mogil, M., and H. Groper, 90
Moore, H. E., F. L. Bates, M. V. Layman, and V. J. Parenton, 73, 80
Mooney, J. A., 59
Moran, C., 130, 131, 133, 134; N. R. Britton, and B. J. Correy, 34
Morris, A., 262, 282, 283, 289, 303; and C. Herring, 257, 264
Morris, D., 206
Morrissey, J. P., R. H. Hall, and M. L. Lindsey, 34
Mueller, C. M., 284
Myambo, K., 131
Myers, K. A., 59

National Academy of Sciences/National Research Council, 12, 13, 72; Board of Natural Disasters, 17; Committee on Disaster Studies, 12
National Opinion Research Center, 10, 13, 45, 309
National Organization for Women (NOW), 298–301
National Science Foundation (NSF), 26, 44, 115
National Women's Party, 285–288, 291–292, 296–300
Neal, D., 32
Nehnevajsa, J., 233, 243
Neiss, R., 136
Nelson, C., M. Coovert, A. Kurtz, B. Fritzche, C. Crumley, and A. Powell, 85
Neuman, R., and I. deSala Pool, 306, 319
Nigg, J., 74, 78, 79, 81

O'Neill, W., 282
Oberschall, A., 263, 275, 276, 282–284, 289
Ogburn, W., 273
Oliver, P. E., 271; and G. Marwell, 294
Oliver-Smith, A., 115–117, 120

Palm, R., 72
Panic, 11–12, 168–189; in Beverly Hills Supper Club fire, 185–188; models of, 168–170
Park, R., 10, 263, 264, 267, 273; and E. W. Burgess, 218, 225
Peacock, W., C. Killian, and F. Bates, 119, 120
Pelanda, C., 150
Perrow, C., 35, 275, 292
Perry, J., and M. D. Pugh, 229, 230
Perry, R., 24, 35, 74, 80, 81, 85–98; and M. Greene, 76–78, 89, 91; and M. K. Lindell, 73, 80; M. K. Lindell and M. R. Greene, 73, 77, 81, 90, 233; M. R. Greene, and A. Mushkatel, 74, 78
Peterson, W. A., and N. P. Gist, 246
Pfuhl, E. H., 213
Pinard, M., and R. Hamilton, 284
Piven, F. F., and R. A. Cloward, 284
Plain Dealer, 196
Prasad, J., 245
Prince, S. H., 24, 34, 102, 112, 113

Quarantellil E. L., 9–17, 24, 31, 32, 41, 45, 73, 74, 76, 78, 80, 85, 88, 89, 102, 113, 115–20, 123, 127, 142, 145, 146, 168, 179, 183, 185, 186, 214, 227, 262, 278, 306, 323; comparative interests, 16; interest in policy research, 17; interests emerging from NORC study, 11; and J. M. Weller, 262, 273; and J. R. Hundley, 15, 261, 264; J. R. Ponting, and J. Fitzpatrick, 15; orientation to sociological theory, 9–10; and R. R. Dynes, 14, 15, 45, 116, 139, 170, 231, 243; views on personality consequences of disaster, 14

Raddall, T. H., 100
Raphael, B., B. Singh, L. Bradbury, and F. Lambert, 118, 135
Resource Mobilization Theory: applied to the women's movement, 1940's to 1960's, 289–304; and social movements, 257–263, 267–269, 274–277, 282–285; as a type of fad, 259–260
Riesman, D., 306
Risk Communication as hazard information, 71; as warning, 71
Roder, W., 71
Rogers, G. O., 76, 78, 80, 81; and J. Nehnevajsa, 80
Rosenthal, N., and M. Schwartz, 231
Rosnow, R. L., and G. Fine, 246
Ross, G. A., 227
Rossi, P. H., J. D. Wright, and S. R. Wright, 33
Rotter, J. B., 81; and R. C. Mulry, 81
Rubin, C., and R. Palm, 122
Ruch, C., and L. Christensen, 72
Rule, J., 261–263, 271, 275
Rumors: 244–256, 270; Allport and Postman's formulation on, 244–246, 250–251; conditions facilitating, 244–245; involving earthquake predictions in California, 247–251; involving earthquake predictions in China, 251–256; theories of, 244–247
Rupp, L., 290; and V. Taylor, 300
Rushton, J. P., 129
Russell, S. T., 59

Saarinen, T. F., and J. L. Sell, 80
Saenz, R., 257
Saunders, S. L., and G. Kreps, 70
Scanlon, J., 24, 99–114
Schuler, E. A., and V. J. Parenton, 218
Schuman, H., and J. Scott, 267
Schur, E. M., 228
Selden, B., 208–211
Sheperd, M., and P. E. Hodgkinson, 135
Shibutani, T., 245, 247, 270
Showalter, E., 212
Sime, J. D., 168, 169
Simmel, G., 329
Small, G. W., and J. F. Borus, 209, 211, 218
Smelser, N., 169, 190–193, 202, 206, 220, 230, 264, 274
Smith, D., 132, 170, 306, 316
Smithson, M., P. R. Amato, and P. Pearce, 129
Snook, J. W., 140
Snow, D. A., 268; E. B. Rochford, S. K. Worden, and R. D. Benford, 267
Social attachment, 308–322; attitudes indicative of, 315–316; mass media and, 311–313; prosocial behavior and, 314–315; social participation and, 308–322

Social diversity, 316–320; and social attachment, 316–322, 325–326; societal consequences of, 324–329
Social movements, 257, 272, 273–280, 281–304; organization and, 282–285; rationality and, 276–278; resource mobilization approaches to, 257–263, 267–269, 274–277
Social organization: means of, 54–55; origins of, 47; restructuring of, 56–58 structural code, 47–48
Song, S., 252
Sorenson, J. H., 76
Spector, M., and J. I. Kitsuse, 210
Spilerman, S., 220
Staggenborg, S., 283
Stahl, S. M., and M. Lebedun, 218
Stallings, R., 31, 207–228; and E. L. Quarantelli, 15
Steele, C., 134
State Seismological Bureau (China), 251–252, 254
Strauss, A., 273
Strickland, B. R., 81
Stryker, S., 59
Sundeen, R. A., and G. B. Seigel, 140
Swidler, A., 207, 228
Symons, I., and D. Glowaski, 132

Tarrow, S., 298
Taylor, P., 194, 198, 201, 203, 205
Taylor, V., 281–305; and N. Whittier, 303
Thompson, J. D., 33, 55
Tierney, K., 9, 80, 190, 257, 264, 283
Tilly, C., 288, 289, 294; L. Tilly, and R. Tilly, 283
Times (London), 199–206
Titchener, J. L., and F. T. Kapp, 34
Tooke, F. T., 101, 105
Touraine, A., 267
Tubbesing, S., 122, 127
Tumin, M. M., and A. S. Feldman, 218, 221
Turner, R., 11, 59, 61, 62, 78, 83, 88, 231, 244–56, 258, 262–64, 271, 293, 294; and B. S. Young, 71, 81; J. M. Nigg and D. H. Paz, 73, 77, 247, 253; and L. M. Killian, 169, 228, 231, 274, 276, 293; and S. Surace, 270

Useem, B., and M. Zald, 293, 294, 297

Value-added theory, 190–193, 219; applied to Hillsborough crowd crush, 196–206; and mass hysteria, 219
Vitaliano, P. P., R. D. Maiuro, P. A. Bolton, and G. C. Armsden, 133
Volunteers: in disaster situations, 128–144, 229–243; in disaster situations compared with riots, 242; in Mexico City Earthquake, 234–243; permanent, 25, 128–144; social characteristics of, 232–233, 236–237; types of, 232–233

Wallace, A. F. C., 73
Walsh, E., 32, 261
Warning: causal model of, 82; determinants of risk perception and behavior, 75–81; as a social process, 72; stages of, 73–75
Warren, C. A. B., 207, 222
Waterstone, M., 72
Weber, F., 122
Weller, J. M., and E. L. Quarantelli, 17, 227, 264
Wenger, D. E., 46, 232, 242, 257; and E. L. Quarantelli, 15, 227; and T. F. James, 25, 229–243; T. F. James and C. F. Faupel, 39
White, R., 289
Whittier, N., 283
Whyte, W. H., 306
Williams, H. B., 73, 76, 90
Wilson, J., 275
Wimberly, R. C., T. C. Hood, C. M. Lipsey, M. Clelland, and M. Hay, 270
Windam, G., E. Posey, R. Ross, and B. Spencer, 89
Wolensky, R. P., 130
Women's Movement, 1940 to 1960: 281–304; constraints on, 292–293; key organizations in, 285–288; organizational structure of, 285–291; strategies of, 292–298
Wood, J. L., and M. Jackson, 264, 276, 306
Worth, M. F., and B. F. McLuckie, 80
Wright, S., 264
Wuthnow, R. J., D. Hunter, A. Bergesen, and E. Kurzweil, 228

Yamamoto, Y., and E. L. Quarantelli, 73, 81

Yin, R. K., 33; and G. B. Moore, 42

Zald, M. N., 262, 268; and R. Ash, 166, 283; and J. D. McCarthy, 265, 282

Zech, C. E., 140
Zelman, P. G., 300
Zurcher, L. A., 233
Zymunt, J. F., 257, 264